Orlando D. Miller

Har-Moad

The mountain of the assembly - a series of archaeological studies - chiefly from the

stand-point of the cuneiform inscriptions

Orlando D. Miller

Har-Moad
The mountain of the assembly - a series of archaeological studies - chiefly from the stand-point of the cuneiform inscriptions

ISBN/EAN: 9783337286866

Printed in Europe, USA, Canada, Australia, Japan

Cover: Foto ©Andreas Hilbeck / pixelio.de

More available books at **www.hansebooks.com**

O. D. Miller

הַר־מוֹעֵד

HAR-MOAD

OR

THE MOUNTAIN OF THE ASSEMBLY

A SERIES OF ARCHÆOLOGICAL STUDIES, CHIEFLY
FROM THE STAND-POINT OF THE
CUNEIFORM INSCRIPTIONS

BY

REV. O. D. MILLER, D.D.

MEMBER OF THE AMERICAN ORIENTAL SOCIETY; OF THE ARCHÆOLOGICAL INSTITUTE
OF AMERICA; OF THE VICTORIA INSTITUTE, OR THE PHILOSOPHICAL
SOCIETY OF GREAT BRITAIN, ETC.

*WITH PORTRAIT OF THE AUTHOR AND
PLATE ILLUSTRATIONS*

NORTH ADAMS, MASS.
PUBLISHED BY STEPHEN M. WHIPPLE
110 MAIN STREET
1892

PREFACE.

I SUBMIT to-day to the judgment of the public, and especially of learned critics, the results, in part, of protracted and laborious researches in the department of antiquities. The particular field of antiquarian studies to which in the main my inquiries have been directed comprehends the entire prehistoric period, especially in Asia, including the primitive traditions of mankind and the origin of the ancient civilizations. One of the principal objects which I have had in view has been to ascertain the real character, and to trace the actual origin, of those ideas that formed the theoretical basis of the religious, political, and social institutions of the ancient world. But a still more definite aim in this direction has been to discover that primitive stratum of conceptions and doctrines which may be regarded as fundamental to the two religions of the Bible, constituting historically the germ of their development. Another prominent object has been to determine the locality, geographically, from which these traditionary ideas, inherited alike by nations widely separated, had been at first derived; the locality, in fact, from which the different races had departed toward the countries occupied by them since the opening of the historical period. In connection with these matters I have made the attempt, however hazardous it might at first seem, to fix chronologically the epoch by means of certain astronomical data, to which the primeval traditions definitely appertained. Aside from these more general topics other questions have been treated, sometimes to the extent of entire chapters, for the reason that they were important in themselves, and helped to complete the view of antiquity embodied in the present work.

As regards the spirit in which these investigations have been conducted, the reader will be of course the better judge; but it has been my desire, so far as the nature of the subjects treated would permit and my own qualifications allow, to regard everything from the scientific point of view. In verifying the facts, the rule has been to depend only upon the latest and best authorities at my command, including, however, such writers of former periods as were generally held to be reliable. To a considerable extent what are termed the original sources have been consulted, especially the cuneiform inscriptions; but where this was impossible, in covering so wide a field of investigation, I have usually relied upon those authors only who are familiar with such sources. In all cases I have labored to ascertain precisely the facts, and to place before the reader the most available means of substantiating them. But the same facts in the hands of various writers will often receive quite different constructions; and it is here principally, if anywhere, that the critical ability and scientific spirit of an author exhibit themselves. I believe that my readers will freely accord to me a fair degree of merit in this respect. It is my misfortune, perhaps, in conducting these researches to have been so little influenced by the opinions and theories entertained by many very eminent authorities, and thus to have arrived at a general view of antiquity equally opposed in some important respects to those received by the different and conflicting schools of modern investigators.

The materials that have been actually of the most service to me in attempting to solve the more difficult problems, pertaining especially to high antiquity, affording oftentimes the only key to their solution, are those derived from the valleys of the Euphrates and Tigris, to a great extent embodied in the cuneiform inscriptions. Aside from the books of Moses, I think these texts contain the most ample and reliable notices pertaining to the primitive ages of humanity of any known sources. Although they are neither so voluminous nor ancient as the hieroglyphic inscriptions of Egypt, for instance, they certainly offer a better reflection of the original Hamite development of Asia; and this, as must be admitted, preceded both the Aryan and Semitic. In consulting these texts I have availed myself of the previous labors of cuneiform scholars in

Europe so far as possible, and my indebtedness to them has been continually very great. But in the present state of cuneiform science every student has to rely chiefly upon himself; and it has been from the astronomical and mythological texts themselves that I have derived the most valuable hints relative to the prehistoric ages.

In writing the present treatise the attempt has been made to adapt it as well to the ordinary reader as to those who have made antiquities a special study. For the one class the extracts from authors in foreign languages have been put into English, while for the other class the references simply to the works cited would have been sufficient, and even preferable. For similar reasons an English transliteration of foreign terms has been given, without regard to any uniform system, however, as the original text has been also given in nearly all cases. For the convenience of those who may wish to consult the authorities cited, a list of them has been prepared in which the titles of their works are given in full.

I have to regret exceedingly the state of circumstances which has deprived me of the advice and assistance of competent critics in working out those more difficult problems which I have deemed it important to investigate, even without such aid. Nevertheless, I believe that, by a singularly good fortune, certain important facts have been established pertaining to the earliest periods of history, the result of which must be to modify essentially the prevailing theories respecting antiquity. Thus I shall indulge the hope that my labors will contribute in some degree to advance our knowledge of the remote past.

THE AUTHOR.

WELLESLEY, MASS., *November*, 1876.

HOW TO READ THIS BOOK.

HAVING read the preface, turn to page 413, and read " A Summary of Results," and take special note of " the Cabiri " in section 172, amplified in chapter three. Then look at section 174, amplified in chapter seven, and note the "*fourth*" region, or under world, and its probable nonentity, page 201. Then, too, in the closing paragraph of section 177, note the relation of Christ to the " traditionary ideas of antiquity," that " all is fulfilled and realized in Him," page 435. Turn, then, to page 16, and read the Scripture there quoted, and note that the great Hamite-Cushite race from the East was the *first* to settle Egypt and Babylon ; that this Bible race can be traced to Mount Meru, on the high table-lands of Central Asia ; that the races emigrating from that region can be traced from thence to the four quarters of the earth, and that however widely separated, they can be *re*-traced to their common home, — the Eden of Genesis.

For " the fundamental law of mind and nature, as well as of all historical development," read the first paragraph of section 57, page 148, and note " *identity in the different,* assimilation of things divided, separated." Then pass to section 100, for another law, and note that " *harmony* is born of the *reaction of contraries ;* " yea, " the *identity of contraries,*" page 266.

Make Plate IV., page 372, a study. Note on page 377 " the principle of *union in opposition,*" and its application to the signs and constellations of the zodiac, with comments in section 152, page 376.

The author's philosophy crops out repeatedly in his work, and may be specially noted in the last two sections.

Finally, reader, make this book a study, comprehend its thought, *live in that thought,* and you will come to know that man once lived in *conscious* unity with God and all created things.

S. M. WHIPPLE.

NORTH ADAMS, MASS., *August,* 1891.

WHAT THIS BOOK TEACHES.

FIRST and foremost, it teaches the personality of God; that He *dwells* somewhere, the same as man.

Symbolism is at the bottom of everything ancient. "Symbolic writing was the most ancient among the cultured nations of antiquity" (page 32).

The *shovel* was the symbol of the primitive worship of mankind. "The hearth, and the divinity of the hearth, constituted the focus of all the ancient civilizations" (page 33).

"The God of the hearth was really the paternal head of the household, and its members were his family" (page 34).

"The national God of the Jews was originally one with the ancient Accadian or Cushite divinity of the hearth" (page 38).

"It was there that the institutions and civilizations of the ancient world were cradled." "The consecrated hearth was His focus, His altar, His house, His table, His fireside" (page 41).

"Such was one of the original conceptions of the Jewish theocracy." "This grand idea had been taught the world, earlier than the time of Abraham, earlier than the Tower of Babel, and while the Hamite and Semite, the Turanian and Aryan, were yet as one family" (page 42).

"The notion that God *dwells, inhabits*, the same as man, was everywhere fundamental; and it was for this reason that the national temple was considered *God's house*, and its altar the *National Hearth*" (page 43).

"With the men of high antiquity God was not conceived as wholly distinct from created nature. On the contrary, nature was considered as the Face, the Name, the external manifestation of divinity. As something purely universal and abstract, far removed from the work of his own wisdom and power, the Deity was almost wholly unknown in the first ages of humanity. . . . The Divine Mind was everywhere present in the outward world, and everything that had life was a symbol of God. . . . **God** is universal, infinite, but not as an abstraction. He alone is really universal who is present in his fullness in each and every *particular*. If the Infinite dwells not in this tree and in this stone, He is nowhere. If it is only beneath temple domes that the Deity takes up his abode, primitive humanity was without a God, for then there were no temples."

"It is mind alone that dwells, and it is matter alone that constitutes the dwelling. . . . God *must dwell somewhere*, the same as man. This eternal law of all mind, which philosophy, science, and speculative theology have now well-nigh forgotten, is that alone which gives meaning to the phrase, 'House of God'" (pages 61, 62).

"The assumption of the entire expanse of the sky as the abode of the Heavenly Father was just that pantheistical conception expressly discountenanced by the sacred writers, and the assumption of a purely ideal region beyond the material heavens as such abode found no justification in the sacred tradition; this ideal region was an *additional* world, superinduced upon the real cosmos, when philosophy and science had consummated the divorce between mind and matter, *and it is just this divorce* which is to-day undermining the faith of mankind" (page 440).

Next to the personality of God is the chronology of creation, a *zodiacal* chronology, the chronology of the stars; whereon God in his providence inscribed man's early history: his temptation, "the fall," and a promised Redeemer in the seed of the woman; — all this, before there was an Abraham, a Moses or a Jew, or even a written language. "The first prophecy ever uttered to man . . . may be read to-day, as plainly inscribed on the celestial sphere as in the third chapter of Genesis" (page 417).

It is my purpose now to interpret that early history and subsequent revelations, in harmony with this book, the Old and New Testament scriptures, and my own individual experience.

I hazard nothing in saying that the spirit and intent of this book is to teach *a primitive revelation written in the heavens:* first, inscribed astronomically and zodiacally on the eastern sky; then, historically on the northern heavens which overlook "the Mountain of the Assembly," the "Har-Moad" of Isaiah, the Olympus of all Asia. Out from the Gan-Eden of Genesis — to the east, to the west, to the south — the author has traced the earliest traditions of mankind; then, from the widely separated countries to which the cultured races of antiquity had migrated, he has retraced the same traditions to their common origin on the high table-lands of Central Asia, "the home of all the traditions, the birthplace of all the mysteries."

The traditions themselves fix the epoch of their birth, by revealing the state of the heavens to which they all appertain, and thus determine the chronology of creation.

That chronology is a three-strand cord, scientifically conceived and inseparably bound together: one strand broken, all broken; one verified, all verified!

The first strand may be termed the creative epoch; the second, the celestial Eden; the third, the primitive pole. Admit either of these prime factors, and the other two have to be admitted. These factors had their fulfillment more than ten thousand years before our era (page 411).

Then, at the winter solstice, the sign Capricorn was in the constellation Gemini, marking the creative epoch, the birth of the *organized* world and the primitive man.

Then, at the vernal equinox, the sign Taurus was in the constellation Libra, or the Pincers of Scorpio, marking the celestial Eden of the East, — the sacred mountain, the source of the four rivers, the trees of life and knowledge, the first human pair, the serpent, the great transgression, the fall, the expulsion, the cherubim, the flaming sword, the seed of the woman, the promised Redeemer, the hope of the world in all ages.

Then, at the creative epoch, the star Vega, in the constellation Lyra, was marking the celestial Eden of the north, — the man, the woman, the serpent, the temptation, the great calamity, the Hercules, the expected Redeemer, the hope of the heathen world.

These factors were in their respective places by virtue of the immutable law, *precession* — the great "celestial clock" that keeps time by centuries, and will keep it while the earth has an orbit round the sun or the sun an orbit round a "vaster sun" — God! whose presence fills immensity; whose substance permeates everything, animate or inanimate; whose power energizes the universe of mind and matter in everlasting unity; whose attributes are limited by his own creation, in which He lives and moves and has his being, as man lives and moves and has his being in Him, the supreme personality; and yet, *in* everything, *for* everything, ruling everything in law, order, and special providence; the creator and destroyer of dynasties, of kingdoms and empires according to His own will; the foster-father of every civilization, prescribing its boundaries; the patron of every religion, doing with His own whatsoever He will, in *justice-love* [1] to all; the hearth god of every dwelling, providing, directing, reforming, blessing; rejoicing with those who rejoice, sorrowing with those who sorrow; a present help in time of need; the *divinity* in every soul created in His image; *disciplining* that soul-image, (divine and human,) into the *likeness* of Himself, Christ the supreme example, — *this* is my conception of God; [2] it is the underlying conception of Dr. Miller's life-work; the restoration to the modern world of the ancient order of things; a vindication of the Old and New Testament scriptures; a revelation of God's immanent presence in the affairs of men, in the church, the state, the home, the heart; subduing, correcting, purifying, elevating, restoring man to his first estate, through a crucified and risen Redeemer, the Christ, the Hercules, the seed of the woman that shall bruise the serpent's head and redeem the human race.

THE PUBLISHER.

November, 1891.

---

[1] The justice of God is the love of God; hence, *justice*-love : the ground thought of reform; the underlying law of moral and spiritual growth ; the unseen hand that smites to quicken, probes to cure, lacerates to heal ; *kills*, even, to make alive. Such has been my experience, to know *God's justice*-love ! Not man's love, nor woman's love, both partial and unjust, seeking its own and not another's good. Divine love is not human love, nor is human love divine love, till *both blend in justice-love to all*, incarnated in the primitive man, and in all men, as witnesses of His justice and judgment, here and hereafter, in the body or out, in Heaven or in Hades.

[2] As defined in this sentence of many members, and in the above note.

A BRIEF SKETCH OF THE AUTHOR'S LIFE AND WORK, AND THE WRITER'S CONNECTION WITH HIM IN THAT WORK.

ORLANDO DANA MILLER was born at Woodstock, County of Windsor, State of Vermont, October 18, 1821.[1]

In boyhood, he joined the Methodist Episcopal Church, of which his father was a member to the close of his life in 1878.

In 1845, he was graduated at Norwich University, Vermont; was accorded the degrees of Bachelor of Arts, Master of Military Science and Civil Engineering, and subsequently the degree of Master of Arts. After graduation, he entered the field of civil engineering; finding it unsuited to his tastes and mode of thought, he left it for the law; and finding that profession not adapted to his aspirations, he entered the Christian ministry.

In 1848, he married Miss Cornelia M. Burton, Norwich, Vermont. To them were born three daughters. The youngest died in infancy: the remaining two reside with their mother at South Merrimack, New Hampshire.

For the first fifteen years after graduation, aside from the practice of civil engineering and the study of law, he devoted his time to Biblical studies and the modern languages. For twelve years of that time, he had charge of parishes in the States of Vermont, Ohio, New York, Massachusetts, and New Hampshire. He was a member of the Masonic order, and a Royal Arch Mason.

But it became evident to him that he was adapted to some other work than the ministry, whereupon he resigned the pastorate of the First Universalist Society of Nashua, N. H., and after an interval of about five years, settled down to the study of the ancient languages, especially the cuneiform and kindred tongues, that he might become his own interpreter of Oriental thought.

---

[1] The writer, Stephen Munson Whipple, was born at Whipple's Corners, town of Pownal, County of Bennington, State of Vermont, May 6, 1821.

After fifteen years' labor, which knew little abatement, his scholarship, began to be recognized.

In 1875, he was elected a member of the American Oriental Society, and subsequently a member of the Archæological Institute of America.

In 1880, the secretary of the Victoria Institute, or the Philosophical Society of Great Britain, thus wrote: —

"Sir, — The council presents its compliments, and sends a copy of a paper recently read before this society, which it invites you to join. . . . Should you kindly consent to prepare a paper for it, the council will be much gratified."

In June, 1881, the secretary again wrote: —

"Sir, — I have the honor to convey to you the president and council's invitation for you to join this society as a member or as an associate."

To be elected, under the rules of that society, required the payment of twenty guineas. This sum Dr. Miller did not feel able to spare from his limited means, whereupon the Institute elected him an honorary life member. In return for the honor thus conferred upon him, he prepared a paper, which was read before the council and printed in London. Thereafter, he was invited, through Professor Schrader, of Berlin University, to attend the world's congress of Orientalists at Berlin; and subsequently to attend the same congress at St. Petersburg.

In 1882, he received notice from the Faculty of Tufts College that, "with the unanimous approval of the Faculty and Board of Trustees, our President honors our college by conferring upon you the honorary degree of Doctor of Divinity. . . . Permit me," writes the secretary of the Faculty, "to express my regret that we have been so tardy in recognizing your merits. But the fact is, we have been so attentive to our special work, and you have done your work so quietly, that we were not aware, until quite recently, what you were doing and what recognition your labors have received by the scholars and learned bodies of America and Europe. The honors you have received are indeed a very complimentary testimony to your scholarship, and, I have no doubt, most worthily bestowed."

It may not be without interest to the reader to know what papers have been given to the public by Dr. Miller, and the circumstances under which they were called forth, and where first published.

In the **April** number of the "Bibliotheca Sacra" for 1875, there appeared, in the editorial correspondence, two letters on "The Raw-linson Theory respecting the Site of Ur of the Chaldees," — the first from the pen of Rev. Lucien H. Adams, and the other, in reply, embracing sundry criticisms, from the pen of Rev. Selah Merrill. When these letters came under the observation of Rev. O. D. Miller, he wrote a letter to the publisher, in which he says, " Rev. Mr. Merrill's remarks, touching Ur of the Chaldees, are well put, and may be regarded as embodying the results of investigations to the present time." This letter the publisher handed to Mr. Merrill, who in a letter to Mr. Miller, under date of April 16, 1875, among other things, said : " Our introduction has certainly been a novel one, but none the less pleasant. From the tone of your letter — which, by the way, there are not six men, and probably not over four, in America who could have written the criticisms it con-tained," etc. And further, "By the way," Mr. Merrill inquired, " to what denomination do you belong?" To which Mr. Miller replied, " I am with the Universalists on the final destiny of man ; in all else, I differ."

Thereafter, Professor Merrill took a friendly interest in Dr. Mil-ler, and commended him to the favorable notice of cuneiform scholars and others of kindred thought.

When Rev. Mr. Peet proposed to publish an Oriental and Bibli-cal Journal, at Chicago, Illinois, Professor, now Dr. Merrill favored the enterprise ; and when it was brought to the attention of Rev. Mr. Miller, he consented to prepare articles on Oriental subjects to be published in that journal, five of which appeared in the first vol-ume for 1880, as follows : —

" The Assyrio-Babylonian Doctrine of the Future Life, following the Cuneiform Inscriptions ; " " The Antiquity of Sacred Writings in the Valley of the Euphrates ; " " Accadian or Sumerian ; " " The Gan-Eden of Genesis ; " " The Pyramidal Temple." For 1881, three articles : " Solar Symbolism in the Ancient Religions ; " " Symbolic Geography of the Ancients ; " " Dr. Brugsch-Bey, on the Origin of the Egyptians, and the Egyptian Civilization." For 1882, " The Divinity of the Hearth ; " and to it may be added, " Testimony of the Cuneiform Texts to the Antediluvian Period of the Mosaic History," which, as already noted, was read before the council of the Victoria Institute, and printed in London.

These papers have been translated into other languages, and have
given to Dr. Miller whatever reputation he may have as an Assyriologist and cuneiform scholar.

Honors multiply. Among them, one may be noted here, from
"The Cumberland Presbyterian Quarterly Review," edited by the
Theological Faculty of Cumberland University, Tenn.

In that review for April, 1883, in an article on "Eden: its
Location and Geography in the Bible and out of it," Rev. Dr. Bu-
chanan says: " For translations of passages from the French found
below, I am under obligations to Rev. O. D. Miller, of Nashua,
N. H., whom Rev. S. D. Peet, editor of ' The American Antiquarian
and Oriental Journal,' says is the finest Assyriologist in America.
For several years, I have watched his papers in ' The American Antiquarian and Oriental Journal,' with great profit and interest, and
can commend him as to candor and learning in antiquarian
studies."

It may be said that these honors, from first to last, were wholly
unsought by the recipient; he was too modest and retiring to think
of preferment, always shrinking from contact with the world, and,
in his latter days, from men, except in his study; *there*, scholars
were always welcome. One, in a letter addressed to the writer
dated Andover, Mass.. September 1, 1890, says of him : —

" Dr. Miller was conspicuous for his modesty in his judgment of
himself and his work. He was sincere in his search after truth and
patient in all his investigations. He was brave and hopeful under
very trying pecuniary difficulties. He was appreciative of the services and work of other scholars to a degree far beyond what is
usually found among close students. In his faithfulness as a friend,
there was something manly and inspiring. His words were an encouragement to perseverance and renewed effort in study, and from
his quiet home and life there went forth a perpetual blessing.

Yours sincerely,

SELAH MERRILL."

It was in his study, in our walks. in the forest, that I heard over
and over again, a hundred times repeated, the story of his life-work,
until I became familiar with its leading thought ; and being myself
an expert, I scanned his facts, probed his theory. and believed in
the outcome of his more than human effort to recover from oblivion
the " primitive revelation written in the heavens."

The reader may naturally ask, How did Dr. Miller maintain himself and those dependent upon him after withdrawing from the ministry? How did he obtain books and periodicals to carry on his work? These questions are answered, in part, in a "Memorial of a Scholar," published soon after his death, by Rev. Dr. Flanders : —

"I am glad," says the doctor, "to record that in his labor of love he was assisted by Mr. S. M. Whipple, an old and devoted friend, who not only sympathized with his desire for knowledge and his lofty aims, but for many years took upon himself the maintenance of his family, that he might pursue his cherished studies free from care."

It may be further said that rare and costly books, catalogued in the libraries of Europe, were sought out and added to his library, for study and reference, to fortify and strengthen his method of interpreting Oriental thought.

My acquaintance with Dr. Miller commenced in January, 1852, and continued with slight interruptions to the close of his life in 1888.

In the last days of 1863, — after the termination of a business venture which proved disastrous in the extreme, without which, perchance, this book had not been written, — I went to see him, to persuade him, if possible, to enter upon the work foreshadowed to him at Albany, N. Y., in the spring of 1851: "There is a primitive revelation written in the heavens. Obey God, and thou shalt read it, and leave it a legacy to the world."

It was here that my work commenced which enabled him to realize the promise made to him, thirteen years before. For a term of years, he pursued his life-work in the way of preparation.

In October, 1869, he moved from New Hampshire, to Wellesley, Mass., there to resume his work and carry it to completion. His surroundings were all that he could wish or desire, — free from care or reasonable anxiety; free to follow the lead of the Divine Power which he claimed was ever present to aid and direct his work. That work was finished, and the preface written, in November, 1876, and is now before the reader.

In the spring of 1877, the author conceived the idea that it was not well for him to be longer dependent upon another, and proposed a separation, he naming the terms, all of which were complied with; and thus he was enabled to purchase a farm in the vicinity of Nashua, N. H., and move on to it in June of that year, with every prospect of enjoying that independence which his being ever craved, but never realized.

Here it is due to both to say that in 1865 the twain entered into a verbal covenant, that if one would devote his time and talent to recover from oblivion "the primitive revelation written in the heavens," the other would share equally with him the proceeds of his labors, as God should prosper him. And let me add, by way of explanation, that, for twelve years, something more than one half the proceeds of my labors passed directly or indirectly through his hands, as the accounts kept by me clearly show. I take no credit to myself for what I did for him and his family, that he might do the work appointed him to do. I simply did what I was moved upon to do, — "Cast thy bread upon the waters; for thou shalt find it after many days."

By comparing dates, it will be seen that what has been given to the public on Oriental subjects was published years after the author's life-work was done, and the record of that work filed away, against the time when God would call it forth to vindicate the great and good men of the prehistoric past, the claims of the Mosaic record, and the Christ of the Christian church. The time, it would seem, is at hand, the conditions are ripe, for "the Primitive Revelation written in the heavens"— the "Har-Moad," the "Mountain of the Assembly," the Chronology of creation, wherein lies the religion, the philosophy, the science of all created things, expressed in the outward world in *duality, male and female* (sec. 179), but in God, joined in one androgynous unity, and hence He is the Supreme Personality, in the likeness of which man was created, and thus became the microcosm of the universe, and the *living* Temple of God.

After harvesting the crops of his newly acquired farm, he applied himself to study, and the following letter best conveys the harmony existing between us after separation.

NASHUA, N. H., January 17, 1878.

FRIEND WHIPPLE, — Yours of the 11th instant came to hand last night. I find, according to its terms, and the interpretation it gives of your previous letter, that you are indeed doing better than I asked, for which I am under much obligation to you, and hereby tender many thanks. I have no doubt that you sincerely design in the future to do better by me than I could even hope.

As for the funds for publishing the book, since you have such an interest in it, and have contributed so much to it already, and since, when published, it is to be dedicated to you, it is only proper that you should be permitted to aid in its publication ; hence, what you offer in this respect, also, is thankfully received.

I am now busy at work upon it. I shall attempt to improve it here and there ; and now that you propose to bear the expense, mostly, of publication, I shall feel that I can afford to go to considerable expense in collecting more authorities.

Thine,
O. D. MILLER.

Thereafter he revised, in part, Chapter XIII. of "the book ;" otherwise, with the addition of several notes and "more authorities," it is the same as when finished in 1876.

In 1880, he summarized the zodiacal, astronomical, and historical portions of this book ; and for the residue of time allotted to him to work, he prepared an elaborate treatise on " The Eschatology of the New Testament ;" setting forth the Bible doctrine of the *last* things, which may be accepted as the complement of the *first* things, or "the primitive revelation written in the heavens," to be hereafter published, and thus link Genesis to Revelation, and so round out the beginning and end of things temporal and eternal.

In the spring of 1886, he sold his farm in Nashua and purchased a place in South Merrimack, where his last days were spent in hope that God would gird on anew his armor and give him a life-long victory over the adverse power of the unseen world ; but alas for human hopes, his work was done, however incomplete.

And yet, I have reason to believe that he lived at times the life of God ; lived in *conscious* unity with Him and with nature, as did the primitive man ; and I have reason to further believe that it was the hope and aspiration of his life to realize in his own being the words of Christ : " The prince of this world cometh, and hath *nothing* in me " (John xiv. 30). But he could not maintain that unity against "the prince of this world ;" who, though often "cast out," would come again, and that conscious unity would

depart, to be recovered, then lost, then recovered; and so his spirit
alternated between the law of mind and the law of sin, — the flesh,
— till paralysis and the grave put an end to the conflict and brought
deliverance from "the body of this death" and the sore trials by
which "he was made perfect through suffering." He was a just
and loving spirit in a body of death; but is now transferred to
the "Mount of the Divine Presence," to the city of the living God,
to enter upon a work for Christ and his church denied him in "the
body of this death" (Rom. vii. 24).

After an absence of four and a half years I went to see him. It
was in October, 1887, and I spent with him the sixty-sixth anniversary of his life on earth.

In August, 1888, I went to see him again, wisely and well,
for the shadows of evening had begun to gather around him; but I
did not look upon his declining health and childish sleeps, in which
night lengthened into day, and day into night, other than the rest,
necessary to recuperation, newness of life, strength, and vigor, to
enable him to resume work, and carry out a long-cherished desire
to write the revelation of God in human history, *especially* in the
EXPERIENCE of INDIVIDUALS.

But he did not see another anniversary, for on the 11th of October, 1888, he slept the good sleep, to awake in the beautiful mansions of "the just made perfect," prepared for them from the foundations of the world.

And so we come at last to his life-work, and submit that work to
the candid judgment of the world; confident that whoever makes
this book a study will come to know the primitive man; come to
know the capacity of man to know God as a *living presence* in the
affairs of men; come to know the vital force of the words sincerely
uttered; "Thy kingdom come, thy will be done *on earth*, as it
is in heaven;" and so join hand and heart with the good and true
everywhere, and, with one united effort, lead man in the way of
supreme excellence; "Be ye *perfect*, even as your Father in heaven
is perfect;" and so fulfill the law of love, — love to God, love to
man, — for on this law depends the unity of all things, even Christian unity, so dear to every believing heart.

                                                    S. M. W.

December, 1891.

GENERAL CONTENTS.

BOOK III.

THE CELESTIAL EARTH.

CHAPTER VII.

CHAPTER VIII.

CHAPTER IX.

BOOK IV.

THE TWELVE STARS OF PHŒNICIA.

CHAPTER X.

CHAPTER XI.

CHAPTER XII.

BOOK V.

ZODIACAL CHRONOLOGY.

CHAPTER XIII.

CHAPTER XIV.

CHAPTER XV.

CONCLUDING REMARKS.

CHAPTER XVI.

PLATE ILLUSTRATIONS.

HAR-MOAD.

BOOK I.

CUSHITE ARCHÆOLOGY.

CHAPTER I.

CUSHITE ORIGIN OF THE SACRED WRITING, LANGUAGE, AND LITERATURE OF BABYLON.

SECTION 1. The problem whose solution is to be attempted in this chapter is one upon which I should hesitate to enter, did it not seem forcibly to present itself, as the first difficulty to be encountered, in the investigations to which the following pages are devoted. The cuneiform system of writing, together with the more ancient language and literature of which it was made the depository, is to constitute one of the chief sources of information upon the subjects treated in the present volume. It would be desirable, if possible, not only to determine who were the inventors of this palæographical system, but to fix ethnologically their character, as well as to classify their language and literature. To be able to settle these points satisfactorily would facilitate in some measure the investigations to follow, although it could not be said by any means to constitute their basis. But unfortunately these questions are involved in obscurities and difficulties, the nature and extent of which can be fully appreciated only by those who have devoted to them the most careful consideration. I am aware that, among cuneiform scholars, a theory has been adopted, being now held by the majority, probably, which affirms the Turanian, or more definitely the Ugro-Finnish character of the people, language, literature, etc., to which we refer; and that for a still more specific and local designation, the

term *Accadian*, derived from *Akkad*, Biblical "Accad," is coming
gradually into use; if not for its technical accuracy, at least for the
sake of convenience and uniformity.   At first, I adopted without
hesitation the theory and terminology here indicated.   But every
subsequent attempt to give a scientific account of them, especially
in view of the recent discussions of these questions in France, only
served to render apparent the extreme uncertainty in which the
entire subject is involved.   The very pertinent and suggestive
remarks of M. Ernest Renan, in his annual report to the Asiatic
Society of France, for the year 1875, the substance of which as
bearing upon our subject will be hereafter presented, have finally
afforded me the hint that has determined the course of investigations in the present chapter.   M. Renan has employed the term
Cushite, a title that suggests at once the theory and terminology
in perfect accord with the Mosaic account of the original settlement
of the Euphrates valley, and which ought, when properly applied,
to offer a satisfactory explanation of all the facts pertaining to our
problems.   Indeed, the hypothesis, substantially, long since proposed by the Messrs. Rawlinson, if it had been more critically and
consistently worked out, contained all the elements of a complete
solution of these questions.   I cannot hope by any efforts of my
own to supply fully the defects of the hypothesis here alluded to,
but the materials so abundantly supplied by others, together with
the suggestions afforded by the recent discussions of French Assyriologues, ought to constitute some ground of confidence in a partial
success in this direction.   I avail myself, first, of the advantage of
M. Renan's valuable and critical remarks.

SEC. 2. In the year 1874, M. Jos. Halévy had submitted a learned
paper to the Asiatic Society, in which he strongly protested against
the theory, to use his own expression, of "the pretended Turanians
of Babylon;" insisting upon the strictly Semitic character of the
population, language, literature, as well as palæographical system
appertaining to the Chaldæo-Assyrian empires.[1]   Although not himself a specialist in cuneiform studies, his familiarity with the Hamite and Semitic formations of language enabled M. Halévy not only
to expose the weaker points of the Turanian hypothesis, but to
develop some quite serious objections to it.   Dr. Jules Oppert, having been one of the founders of cuneiform science, and being one of

[1] *Journal Asiatique*, June, 1874, pp. 461–536.

the first, if not the very first, to proclaim the Turanian origin of the Babylonian civilization and culture, was naturally the one to whom all eyes were turned for the defense of a theory which had been adopted by the majority of cuneiform scholars, but which was now assailed with such power. Dr. Oppert is a man of great ability, as well as learning, and his reply was worthy of his distinguished reputation.[1] It proved, to say the least, that M. Halévy's argument was in no sense a finality. The fact that the primitive language, as it appears in the earliest inscriptions, was so different from the Semitic, or the Semitic-Assyrian, as to render the bilingual texts necessary, the translation of one idiom into another, was seen at once to be fatal to the extreme Semitic theory. On the other hand, it was to be admitted that M. Halévy had essentially weakened the ultra-Turanian hypothesis. The inquiry at this stage of the discussion almost forced itself upon the minds of scholars : whether a middle ground did not exist between the two extremes ? In other words, was it not possible to suppose here an original Hamite or Cushite formation, which had served as the basis of Babylonian civilization and culture ? It is at this point that the remarks of M. Renan should be introduced. With his habitual caution under such circumstances, this eminent critic proceeds : —

" We doubt then, even now, notwithstanding the impression made upon us by the learned authorities, and by certain facts sufficiently striking, whether the foundation of the Assyrian civilization was Turanian. We do not believe, on the other hand, that it was Semitic. We regard our position as unaffected by the proofs, according to which M. Oppert was able to show twenty years since, that the cuneiform system of writing was an *importation*, and that, in the *Assyrian* texts, properly speaking, it was applied to a language *for which it was never invented.* For *what language was this invention ?* We fear to add only new elements to a confusion worse than that of Babel, in reminding you of a class of scholars some twenty years ago, among whom I esteem the venerable Baron D'Eckstein, now too much neglected, as holding the first rank, scholars who ventured, rashly without doubt, to designate this primitive formation by the name of *Cushite,* placing it in affinity with the Hamite civilization. This Hamite and Cushite foundation, constituting the two civilizations of Egypt and Assyria, equal in antiquity, closely resembling each other, and withal anterior to the entrance of the Aryans and Semites into history, was for us a

[1] *Journal Asiatique,* May–June, 1875, pp. 442–497.

seducing hypothesis. It is easy to find a better one, without doubt, but the proof is wanting; and we still refuse to see in the most ancient civilization of Babylon the work of Finns and Lapps." [1]

In his annual report the year before, M. Renan had alluded to these questions, and had used the following language : —

" If we employ the term *Turanian* in its strict sense, if we attribute the origin of the refined civilization of Babylon to the Turks, Finns, and Hungarians, to races, in fact, who have only destroyed, but have never created a civilization of their own, we simply declare that this astonishes us." [2]

Thus, M. Renan is unable to see in the Lapps and Finns, or in any people directly related to them ethnologically, for whom it would be impossible otherwise to prove a world-historical importance, the real founders of a civilization that filled all antiquity with its renown. He would much prefer to recognize here the mental and physical activity of a race that never failed to mark its progress with powerful dynasties, with monuments of industry and grandeur. It would be very difficult for any one, I think, not to be more or less affected with similar sentiments. But as regards the hypothesis put forth by him, the author here referred to makes no pretensions to originality. On the contrary, he but asks for a reconsideration of the views of a former period, advanced by those to whom modern criticism is indebted for many of its most splendid achievements. In other words, the Cushite origin of the Chaldæo-Assyrian civilization, which is substantially the basis upon which the Messrs. Rawlinson labored long since to solve the problem before us, constitutes for him the true point of departure, if we would arrive at permanent and satisfactory results.

SEC. 3. The theory of the eminent English authorities just named, as I have already expressed the opinion, contained the principal elements of truth, respecting the subject upon which we have entered. Some defects in the method of working it out, if my estimate of their labors here is correct, were the chief cause of its being to a great extent abandoned by cuneiform scholars, and of the adoption of the extreme Turanian hypothesis as the substitute. As a basis of further criticism and progress, it will be most convenient to present in some detail the scheme which was formerly proposed by these authors. In the 11th essay published in the first volume of

[1] *Journal Asiatique*, July, 1875, pp. 39, 40.   [2] Ibid., July, 1874, p. 42.

the new version of Herodotus by Rev. Geo. Rawlinson, the preface
to which is dated January, 1858, it is probable that we should find
the views of Sir Henry Rawlinson at this period, although his ini-
tials are not attached to this paper in the American edition of the
work. At the early stage of cuneiform researches here indicated,
it would be unreasonable to demand matured opinions on such a
difficult subject, even from the founders of the science, among whom
Sir Henry is to be placed in the foremost ranks. In the essay to
which I refer, the Turanians are regarded as the primitive population, settled in the valleys of the Euphrates and Tigris. At that
epoch linguistic differences had not become very marked; but gradually the Hamite or Cushite formation made its appearance, and
out of this Semitism was developed; an event which the author is
inclined to date from about 2000 years B. C. In his table of ethnological affinities appended to the essay, the writer places the
Hamites, including the Susianians, Chaldæans, etc., in the general
category of Turanian populations. But it is clear that he does not
consider the Turanians, strictly so termed, as the inventors of the
cuneiform system of writing, and the founders of the Babylonian
civilization. It is only when the original Turanian has developed
itself into the Hamite or Cushite formation that this invention
and civilizing process definitely take place. Thus their really
Cushite or Hamite origin is the hypothesis to which the mind of
the author is evidently inclined, at this early stage of cuneiform
research. We subjoin here some extracts embodying his views,
from which our interpretation of them has been drawn: —

"The monuments of Babylon furnish abundant evidence of the
fact that a Hamite race held possession of that country in the earliest times, and continued to be a powerful element in the population down to a period but very little preceding the accession of
Nebuchadnezzar. The most ancient historical records found in the
country, and many of the religious and scientific documents to the
time of the conqueror of Judæa, are written in a language which
belongs to the Allophylian family, presenting affinities with the
dialects of Africa on the one hand, and with those of high Asia on
the other. The people by whom this language was spoken, whose
principal tribe was the *Akkad*, may be regarded as represented by
the Chaldæans of the Greeks, the Casdein of the Hebrew writers.
This race seems to have gradually developed the type of language
known as Semitism, which became in the course of time the general
language of the country; still, as a priest-caste a portion of the

Akkad preserved their ancient tongue, and formed the learned and scientific Chaldæans of later times."

" The early Babylonian language, in its affinity with the Susianian, the second column of the cuneiform trilingual inscriptions, the Armenian cuneiform, and the Mantchoo Tartar on the one hand; with the Galla, the Gheez, and the ancient Egyptian on the other, may be cited as a proof of the original unity between the languages of Africa and Asia; a unity sufficiently shadowed out in Genesis (x. 6–20), and confirmed by the manifold traditions concerning the two Ethiopics, the Cushites above Egypt, and the Cushites of the Persian Gulf. Hamitism, then, although no doubt the form of speech out of which Semitism was developed, is itself rather Turanian than Semite."

" The primitive or Turanian character of speech exhibited a power of development, becoming first Hamite, and then, after a considerable interval, and by a fresh effort, throwing out Semitism. It is impossible to say at what exact time the form of speech as Hamite originated. Probably its rise preceded the invention of letters, and there are reasons for assigning the origination of the change to Egypt." " The development of Semitism, as has been already remarked, belongs to the early part of the 20th century B. C., long subsequently to the time when Hamite kingdoms were set up on the banks of the Nile and of the Euphrates." [1]

Sec. 4. We have introduced some of the briefer extracts at the close, with a view to indicate Sir Rawlinson's general ideas, at the time, as regards the actual period to which the rise of Hamitism should be assigned. He very prudently declines to fix the date, and it would be hazardous for any one to do so even at the present day. However, as he dates the rise of Semitism in the 20th century B. C., long after the development of Hamitism, he could not assign the latter consistently to a period later than the 24th or 25th century before our era. The author's chronological estimate for the rise of Semitism is remarkably correct, even in the light of all the facts known to-day; for it would be difficult to prove the existence of this form of speech at a period much earlier. The indications are, however, that Hamitism preceded by some centuries the dates assigned to it above. According to the views of M. F. Lenormant, a Cushite development existed at Babylon long prior to the era of Urukh, or Lik-Bagas, the earliest known king of Chaldæa.[2] As to the period of Urukh's reign, Rev. Geo. Rawlinson remarks : —

[1] Rawlinson's *Herodotus*, i. p. 525, note ; and pp. 526, 533, 534.
[2] Vid. *La Magie*, pp. 295, 29?.

" We must place his accession at least as early as B. C. 2326 ; possibly it may have fallen a century earlier." [1]

These data seem to necessitate a chronology, for the origin of the Cushite development of Babylon, of at least 2500 to 3000 years B. C. But as regards the ethnological character of the language, writing, and literature of which there is here question, it is apparent that Sir Henry was strongly inclined, when he wrote the foregoing extracts, to the Hamite or Cushite hypothesis. He places the invention of letters after the development of Hamitism. But he identifies the Cushites with the Accadians, and likewise with the Chaldæans. Here are two elements of confusion, judging from the standpoint of cuneiform scholars. The Accadians are now supposed, by the majority of Assyriologues, to have been Turanians, strictly speaking; while others regard the Chaldæans of Babylon as properly Semites. The classification of the Hamites under the general category of Turanian populations, regarded from the point of view now generally adopted, is not sufficiently definite for scientific purposes, and tends to complicate the problem before us. Nevertheless, the affinities traced, and correctly too, between the sacred language of Babylon and the Galla, Gheez, etc., of Africa on one hand, and the languages of high Asia on the other, point toward a mingling of Hamite and Turanian elements in one general development, which might be designated either as Hamite or Turanian, yet too loosely for the purposes of strict accuracy. The Hamite element allies itself to the dialects of Africa, and the Turanian to those of high Asia ; there is a marked difference between them ; and it is necessary to maintain this distinction if we would attain to anything like scientific results. It was chiefly for the want of this, as it appears to me, that the theory proposed by the Messrs. Rawlinson, as a solution of the problem in question, proved to be quite inadequate, at the same time that it involved nearly all the elements of a correct solution. We pass now to a consideration of the views of Rev. Geo. Rawlinson, who devotes an entire chapter to the subject before us, in the second edition of his " Five Monarchies," etc. The author's theory may be inferred generally from the subjoined passage : —

" On the whole, therefore, it seems most probable that the race designated in Scripture by the hero-founder Nimrod, and among the

---

[1] *Five Monarchies*, i. p. 156.

Greeks by the eponym of Belus, passed from East Africa, by way
of Arabia, to the valley of the Euphrates, shortly before the opening of the historical period.   Upon the ethnic basis here indicated,
there was grafted, it would seem, at a very early period, a second,
probably Turanian element, which very importantly affected the
character and composition of the people.   The *Burbur* or *Akkad*,
who are found to have been a principal tribe under the early kings,
are connected by name, religion, and in some degree by language,
with an important people of Armenia, called *Burbur* and *Urarda*,
the Alarodians (apparently) of Herodotus.   It has been conjectured
that this race at a very remote date descended upon the plain country, conquering the original Cushite inhabitants, and by degrees
blending with them, though the fusion remained incomplete to the
time of Abraham.   The language of the early inscriptions, though
Cushite in its vocabulary, is Turanian in many points of its gram-
matical structure, as in its use of post-positions, particles, and pro-
nominal suffixes; and it would seem, therefore, scarcely to admit of a
doubt that the Cushites of Lower Babylon must in some way or other
have become mixed with a Turanian people.   The mode and time
of the commixture are matters altogether beyond our knowledge." [1]

The author does not state definitely, in the foregoing extract, to
which of the two peoples brought into view by him he would attribute the sacred writing and science of Babylon, and the original
foundation of her civilization.   But the language following seems
to imply that the Cushites were the principals in this work : —

" For the last three thousand years, the world has been mainly
indebted for its advancement to the Semitic and Indo-European
races; but it was otherwise in the first ages.   Egypt and Babylon,
Mizraim and Nimrod, both descendants of Ham, led the way, and
acted as the pioneers of mankind in the various untrodden fields of
art, literature, and science.   Alphabetic writing, astronomy, history,
chronology, architecture, plastic art, sculpture, navigation, agriculture, textile industry, seem, all of them, to have had their origin in
one or other of these two countries." [2]

[1] *Five Monarchies*, i. pp. 54, 55.

[2] Ibid., p. 60.   From the author's language cited in the text, we would naturally infer the Cushite origin of the cuneiform system of writing.   But formerly
Sir H. Rawlinson had been strongly inclined to trace the origin of this system
to the Hamites of Egypt.   Thus he observed : —

" Whether the cuneiform letters, in their primitive shapes, were intended, like
the hieroglyphs, to represent actual objects, and were afterwards degraded to their
present forms; or whether the point of departure was from the hieratic, or perhaps the demotic character, the first change from a picture to a sign having thus
taken place before Assyria formed her alphabet, I will not undertake to decide;

Sec. 5. Such are the principal points in the theory relative to the problem before us, as it passed from the hands of the author in 1871, the date of the volume from which the quotations have been made. The last extract seems to affirm the Cushite origin of the Babylonian civilization in all its characteristics. Yet the author appears in other passages to attribute all this to the Chaldæans; and by the Chaldæans he means, as expressly stated by him, the mixed population composed chiefly of Cushites from Africa and Turanians from the north, together with a less proportion at first of Semitic and even Aryan elements. As relates to our general problem, and especially with a view to definite conceptions, the term *Chaldæan* ought to be excluded from the discussion, since it is connected in the past, and even in the present state of science, with so many different and quite contradictory ideas. Hardly two writers could be named who attach to this term precisely the same ethnological value; and it is to me somewhat doubtful whether Rev. Rawlinson himself employs it uniformly in the same sense. The designation of the Babylonian civilization, therefore, as Chaldæan appears to me quite objectionable, since the true import of this title is far from being settled. Another point made by the author, involving no small degree of uncertainty, is the supposed conquest of the Cushites of Babylon at a remote period, by a Turanian race, the *Burbur* or *Urarda*, descending from the mountains of Armenia. Directly opposed to this view is the opinion of many Assyriologues, that an original Turanian population was conquered by a Semitic race. It would be difficult to reconcile these two suppositions, and much more so, to produce really conclusive evidence in favor of either of them. The ground here is debatable even at the present

but the whole structure of the Assyrian graphic system evidently betrays an Egyptian origin." — *Commentary on the Cuneif. Inscript. of Babylonia and Assyria.* London, 1850, p. 4.

This extract was penned as early as 1850. No one now would think of deriving the cuneiform system from Egypt. It is curious to note the great changes that have taken place in the views of cuneiform scholars since the foregoing extract was published. That the cuneiform system originated in high Asia, instead of in Egypt, or in a region far to the north or northeast, in relation to Babylonia, seems at the present day to be a well-established fact. As regards the original form of the characters, there seems to be no proof that they were pictures of concrete objects. What is termed their hieratic form, which rarely shows any trace of hieroglyphism, was probably primitive. There exists no proof to the contrary.

time, and it is thus an element of doubt, so far as concerns our
general problem. But the most objectionable position assumed by
the author is that which derives the primitive Cushite population
of Babylon from East Africa, or the Ethiopia above Egypt. The
Cushites of Africa, judging from the notices of them upon the
Egyptian monuments, had not assumed any importance at the commencement of the Middle Empire, dating from the twelfth dynasty.[1]
An emigration to the Euphrates before this period, therefore, is
hardly to be supposed; and one subsequently to it would not answer the conditions of the problem, for Babylon must have been
settled long before. Again, the earliest traditions of Southern
Arabia are directly in conflict with Rev. G. Rawlinson's views respecting such an emigration. According to the investigations of M.
Caussin de Perceval, followed by those of M. F. Lenormant, the
first Cushite dynasty of Southern Arabia, under the name of " The
Adites," was founded by colonists from the Euphrates, being overthrown about eighteen centuries B. C. It was followed by another
Cushite dynasty bearing the same name, which was destroyed by
the Joctanian Arabians.[2] Now, the commencement of the first
dynasty, whose overthrow is dated 1800 years B. C., must be carried
back to at least 2000 years before our era, at which period a movement of Cushite populations takes place from the Euphrates into
Arabia. It will be seen at a glance that these data do not admit of
any such emigration from East Africa, through Arabia, to Babylon,
as Rev. Rawlinson supposes. Finally, our author's position is
opposed to the usual interpretations of the Mosaic account of the
first settlement of Babylon. The *direction* of the movement of populations to the land of Shinar, under the leadership of Nimrod, is
said to have been from the *east* instead of from the west or southwest : that is to say, from the African Ethiopia.

It will be seen, from the foregoing remarks, that the solution of
the problem upon which we are engaged, as originally proposed by
Sir Henry Rawlinson, and more recently worked out by Rev.
George Rawlinson, must be regarded as defective in some important particulars ; although, as I have already expressed the conviction, it contains all the essential elements of a satisfactory theory

---

[1] Vid. F. Chabas, *Etudes L'Antiq. Historique*, p. 132.
[2] Vid. Lenormant, *Man. d'Hist. Ancienne*, etc., t. iii. pp. 256-258 ; and 261-
263.

upon the subject to which it relates. So far as I know, a really scientific value was never claimed for it ; nothing more, in fact, than a statement of the leading facts, together with such inferences drawn from them as the authors believed legitimate. For strictly scientific purposes, the theory was put forth in terms too general, and it lacked certain critical distinctions that would be desirable, so far as the state of knowledge permitted. It is quite possible that at the present writing the authors themselves have modified their views somewhat, and, indeed, I believe Sir Henry is to-day much inclined to favor the Turanian rather than the Cushite hypothesis. Be this as it may, it was quite certain that cuneiform scholars would not be long satisfied with a theory involving so many points of doubt. The attempt would be made to narrow the problem down, to reduce it to as few elements as possible ; and we pass now to an examination of the later phases assumed by these questions ; to the opinions, in fact, as at present held by the majority of Assyriologues.

SEC. 6. The two ethnic titles — admitting for the present that they have strictly an ethnological value — which appear most frequently in the cuneiform texts, relating to the ancient populations settled upon the banks of the Euphrates, are *Sumir* and *Akkad*, designating the countries, or *Sumeri* and *Akkadi*, applied to the people inhabiting them. As usually held, the first named were properly Semites, while the others were Turanians. Dr. Jules Oppert, however, maintains that the *Sumeri* were the real Turanians of Babylon, the *Akkadi* being regarded by him as a Semitic population. Not to enter here upon a discussion of this question, the opinions of the majority will be adopted. The Accadians, then, were properly the Turanians of the Euphrates valley ; and they were, as now held, the inventors of the cuneiform system of writing, as well as the founders of the Babylonian civilization. They were the Finns and Lapps to which M. Renan has alluded. Of course, he does not believe the Accadians were such ; and even the advocates of the Turanian hypothesis regard them as only distantly related to the Lapps and Finns of modern times. Mr. George Smith gives expression to the present views of most cuneiform scholars in the passage here subjoined : —

" Intimately connected with these historical studies is the question of the origin and history of the great Turanian race which first established civilization in the Euphrates valley. It is the opinion of the majority of Assyrian scholars that the civilization, literature, mythology, and science of Babylon and Assyria, were not the work of a Semitic race, but of a totally different people, speaking a language quite distinct from that of all the Semitic tribes.   There is, however, a more remarkable point than this ; it is supposed that at a very early period the Akkad or Turanian population, with its high cultivation and remarkable civilization, was conquered by the Semitic race, and that the conquerors imposed only their language on the conquered, adopting from the subjugated people its mythology, laws, literature, and almost every art of civilization.   Such a curious revolution would be without parallel in the history of the world, and the most singular point in connection with the subject is the entire silence of the inscriptions as to any such conquest. There does not appear any break in their traditions, or change in the character of the country to mark this great revolution, and the question of how the change was effected, or when it took place, is at present quite obscure." [1]

If Mr. Smith had aimed to overthrow completely the hypothesis which he explains, it would be difficult to conceive a more effectual method of doing so than by the statements contained in the last half of the foregoing extract, assuming them to be correct, which they undoubtedly are.   It is morally impossible that a series of circumstances, such as he describes, should occur in the natural history of any nation or country.   As the writer well observes, it would be " without a parallel in the history of the world."   According to the terms of the description, the Semites must have been in a semi-barbarous condition ; an ignorant, nomadic race, at the period of their conquest of the highly cultured Accadian population ; for they are said to have borrowed from the conquered people " its mythology, laws, literature, and almost every art of civilized life."   In the first place, then, for a rude, uncultured race to impose its language upon a highly civilized people, at the same time adopting its mythology, literature, science, institutions and laws, and even its mode of writing, is virtually a contradiction of terms.   If, however, such an anomalous event be supposed, for the conquerors to so obliterate all traces of a revolution thus radical that the keen eye of modern criticism is unable to detect them, and to show that the continuity of development has been interrupted, the chain of traditionary ideas broken, must be regarded, I think, as wholly incredible.   Reverse all the conditions of this statement ; say that it was a cultured race

[1] *Assyr. Discoveries*, pp. 449, 450.

that imposed its language, literature, institutions, and laws upon a people ignorant and barbarous, effacing from history all indications of the previous condition, — there is then 'nothing in the proposition that is repugnant to reason. But it would require sufficient proof of such a change as having taken place, before assuming it as a basis of scientific conclusions. According to the supposition this, too, is wanting. Regarding, then, the language of Mr. Smith as embodying the actual conditions of the problem, the only natural solution of it is that no such revolution as supposed has ever occurred in the valley of the Euphrates.

Sec. 7. The passage cited from Mr. Smith in the last section, although it seems to embody his own views upon the subject, was never intended as a formal and critical statement of the Turanian hypothesis relative to the origin of the Babylonian civilization. In fact, the ground assumed is an extreme one as compared with the opinions of some of the most distinguished advocates of the theory referred to, among whom I include especially M. F. Lenormant. This eminent French Assyriologue has been inclined heretofore to admit the existence of a powerful Cushite influence in the early history of Babylon, and since the discussions of these questions in France, to which reference has been made, he appears to be still more positively disposed in this direction. In view of the remarks of M. Renan in 1874, M. Lenormant expresses himself as follows :

" It will be seen that I am far from attributing to the Turanians, primitively settled in Chaldæa, the entire work of the high civilization of Babylon. I see here only one of its factors, and this not the most important." " In this great and learned civilization the principal and the most noble part has proceeded from the Cushito-Semitic element, from the element of which the Assyrian language became the national idiom." " Babylon in particular, at least the Babylon of historical epochs, has been always preëminently Cushite." [1]

The positions assumed in these extracts do not differ widely from the views put forth by Sir H. Rawlinson, as already presented. A blending of Turanian and Cushite influences is expressly held, and by far the greater importance is attached to the Cushite element. On the whole, it is sufficiently obvious that something of a reaction has taken place in the author's mind, mainly due to the profound

[1] *La Magie*, pp. 305, 306.

criticisms of M. Renan. Indeed, M. Lenormant is free to admit
that he may have been somewhat unguarded in his statements heretofore. We have not space here to present in detail, and in his own
language, M. Lenormant's present standpoint respecting our general
problem, but substantially his positions are the following : 1st. At
Babylon, as well as throughout Western Asia, perhaps, the really
primitive layer of population was Turanian. Not much importance
is attached to it, however, in the work of Babylonian civilization.
2d. Somewhat later, the Cushite race took possession of Babylonia
in particular, and then commenced definitely the civilizing process.
The arts and sciences were cultivated, such as astronomy and architecture, and the astro-mythology was developed. 3d. Still later,
perhaps, the *Akkadi*, related ethnologically to the Finns and Lapps,
more directly to the *Urarda* of Armenia, descended upon the plains
of Chaldæa, bringing with them the original system of writing in its
cruder form, from which the regular cuneiform of later epochs proceeded. The Accadians were agriculturists preëminently, addicted
to magic, and to the worship of the natural elements. 4th. The
Babylonian civilization as known to us was the product of the
admixture of all these elements, the Cushite or Cushito-Semitic
influence being usually in the ascendant. The population last designated appertained to the *Sumir*, perhaps the *Shinar* of the Mosaic
text ; while the *Akkadi*, occupying principally the country of Chaldæa, were Turanians strictly speaking, and properly the *Kaldi*, or
the Chaldæans, in relation to whom so much uncertainty has existed.

I think the foregoing outlines represent fairly the scheme as at
present held by M. Lenormant, whose opinions, certainly, upon subjects so familiar to him are entitled to very serious consideration.
The marked difference, in several important particulars, between this
theory and that already set forth by Mr. Smith will be readily recognized, and the variations also from the opinions of the Messrs.
Rawlinson on various points are not less notable. But, on the
whole, M. Lenormant's present views are a partial return to the
old Cushite hypothesis, with which we started upon these investigations. That in which he most essentially differs from this scheme
to-day, pertains to the original invention of the cuneiform system
of writing, which he still holds to have been a work of the Accadians, or, in other words, the Turanians of Chaldæa. This point is

to me the most obscure, and the least satisfactory, of any included in the author's theory. Were the Cushites of Babylon destitute of any system of writing, prior to the settlement of the Accadians in Chaldæa? It would be extremely difficult to sustain such a thesis, and would tend to throw many obstacles in the way of the solution of our problem. The oldest inscriptions of Chaldæa, being exclusively of the hieratic or primitive type, appertain to the period of *Urukh*, or *Lik-Bagas*, as M. Lenormant proposes to read this name, and cannot be dated earlier than 2500 years B. C. At this time, according to our author's own showing, in a passage to be hereafter cited, the Cushite civilization of Babylon had attained no inconsiderable development; and even *Lik-Bagas* himself was but an imitator of Babylonian usages and a convert to the Cushite religion. Yet during all the period prior to this date, we are to suppose the Babylonian development had proceeded without a paleographical system, being finally indebted to the Accadians of Chaldæa for one which had been an importation from some region of high Asia. But another objection to the ground here assumed is one which no man can better appreciate than M. Lenormant himself. I refer to the vital connection which exists between the sacred science and traditions of Babylon on one hand, and on the other, the very characters, the paleographic symbols, by which these notions have been recorded. This point will be more fully illustrated in the two chapters immediately following the present. Suffice it to say here, that the vital connection between the notion and written symbol is such, in a great number of instances, as to preclude the idea of their separated development in connection with the Babylonian science. Both the notion and symbol belonged originally to the scientific system under the forms in which it has been transmitted to us. For illustration take the hieratic form of the Accadian *U*, "to measure, a measure, a cubit," etc. This character appertained to the metrical system of Babylon, and to the method of land-divisions, from the moment of its origin. It was never borrowed from a different race, and afterwards adapted by the Cushites of Babylon to their sacred science. The scientific conception, the written symbol, and the regularly developed system in which both are embodied, at the very earliest period to which the inscriptions pertain, were the product of the same spiritual genius. The sacred science and the sacred writing of Babylon had a common origin. One was not Cushite, and the other Turanian or Accadian; both were Cushite, or, if not, both were Accadian.

Sec. 8. The reader is now fully prepared to realize the nature and extent of the difficulties connected with the problem upon which we are engaged. They do not afford much encouragement toward an attempt, on our own part, to establish something definite relative to the questions before us. Nevertheless, it is necessary to abandon here the course of merely negative criticism, and, if possible, to sketch the outlines of a theory that shall satisfy all the conditions of this complicated subject. Some fundamental position ought to be seized upon as a fixed point of departure; and I believe this may be found in the Mosaic account of the first settlement of the Euphrates valley by emigrants from the *East* under the leadership of the Cushite hero Nimrod. The following passages from the Biblical text embody the substance of the narrative to which we allude: —

"And Cush begat Nimrod: he began to be a mighty one in the earth. He was a mighty hunter before the Lord: wherefore it is said, Even as Nimrod the mighty hunter before the Lord. And the beginning of his kingdom was Babel, and Erech, and Accad, and Calneh, in the land of Shinar. Out of that land went forth Asshur, and builded Nineveh, and the city Rehoboth, and Calah, and Resen between Nineveh and Calah: the same is a great city." (Gen. x. 8-12.)

"These are the families of the sons of Noah, after their generations, in their nations: and by these were the nations divided in the earth after the flood. And the whole earth was of one language, and of one speech. And it came to pass, as they journeyed from the east, that they found a plain in the land of Shinar; and they dwelt there. And they said one to another, Go to, let us make brick, and burn them throughly. And they had brick for stone, and slime had they for morter. And they said, Go to, let us build us a city and a tower, whose top may reach unto heaven: and let us make us a name, lest we be scattered abroad upon the face of the whole earth. And the Lord came down to see the city and the tower, which the children of men builded. And the Lord said, Behold, the people is one, and they have all one language; and this they begin to do: and now nothing will be restrained from them, which they have imagined to do. Go to, let us go down, and there confound their language, that they may not understand one another's speech. So the Lord scattered them abroad from thence upon the face of all the earth: and they left off to build the city. Therefore is the name of it called Babel: because the Lord did there confound the language of all the earth: and from thence did the Lord scatter them abroad upon the face of all the earth." (Gen. x. 32-xi. 1-9.)

The precise character of the genealogy contained in the tenth chapter is sufficiently indicated in the last two verses. It is not exclusively a genealogy in the ordinary and modern sense, nor is it, on the other hand, entirely geographical, as some exegetes have held. It was uniformly the practice in high antiquity for a political community to trace its descent from one ancestral head, just as the Jewish nation was regarded as wholly descended from Abraham. The Mosaic genealogy, then, as here referred to, embodies the double conception: 1st. Of a literal descent from a common ancestry; 2d. Of distinct nationalities proceeding therefrom, located geographically in different quarters of the world. These two ideas are definitely embodied in the expression: "After their generations, in their nations" (x. 32). It would be decidedly erroneous, I think, to interpret the text as revealing simply a system of colonizations from one nation and geographical locality to another. Hence, in tracing a descent of Cush from Ham, and of Nimrod from Cush, Moses does not mean that Cush and Nimrod proceeded from the *land of Ham*, that is to say, from Egypt. Yet this appears to be Rev. Mr. Rawlinson's view of the matter, when he derives the Cushites of Babylon from East Africa.[1]

The Mosaic narrative, as contained in the eleventh chapter (ver. 2–9) was intended obviously to describe the very first settlements in the land of Shinar, after the deluge, and not simply a conquest of peoples previously inhabiting the country. The city of Babylon was not in existence prior to this emigration. The original founding of Babylon, therefore, and of the cities surrounding it, must be attributed to these colonists, journeying from a region eastward to the land of Shinar. These emigrants must thus be regarded as actually the founders of the Babylonian kingdom and civilization. The question arises, then, Who were these colonists from the east? Were they principally Cushites under the leadership of Nimrod, or some other people of whom we have no definite knowledge? It is obvious that the sacred writer, in the tenth chapter (ver. 10–12), intends to attribute the foundation of the Babylonian kingdom in the region called Shinar to Nimrod; and, at a subsequent period, the founding of the Assyrian kingdom in a country outside of and to the north of Shinar, to Asshur. Thus, we must identify the first settlement of Babylon and building of the tower with the beginning of Nimrod's kingdom consisting of Babel, Erech, Accad, and Calneh. It would be very difficult to establish a beginning for the Nimrodic empire at a period before or after the building of the tower; for universal tradition and the opinions generally of critics tend equally to place these events in close connection with

[1] Vid. *Five Monarchies*, i. p. 54.

each other, although they are narrated in different chapters of the
sacred text. So far, then, as concerns the Mosaic record, it is plain
that we ought to attribute the origin of the civilization of Babylon
to a people who had migrated from the east under the leadership
of Nimrod, the Cushite.

SEC. 9. According to the views of Rev. G. Rawlinson, as already
exposed, the Cushites originally settled upon the banks of the Euphrates had migrated through Arabia from East Africa, a region to
the west or southwest instead of east in relation to Babylonia.
We have objected to this opinion : 1st. That the Cushite settlement
in Africa was no more ancient, probably, than that in the Euphrates valley ; 2d. That this scheme is directly opposed by the earliest
traditionary notices relative to Southern Arabia ; 3d. That the
Mosaic text connects the original settlement of the land of Shinar with a popular emigration, under the conduct of Nimrod, from
a country situated in the east in relation to Babylon. Now, are
there any evidences proving the existence of a primitive Cushite
population, more or less advanced in civilization, inhabiting a country far to the east as regards the Euphrates valley? If a people
thus described and located were known, and could be referred to a
very high antiquity, such a fact would tend to explain and confirm
the Mosaic account of the first settlement of Shinar by a Cushite
emigration from the east, sufficiently advanced in architectural ideas
to undertake the building of an immense tower. The following
notices from M. F. Lenormant will furnish us the desired information : —

"The Bible, in the recital relative to Eden, which has been preserved by Moses under a traditional form exceedingly ancient, and
certainly anterior to Abraham, places a country of Cush upon the
borders of Gihon, or the Oxus, and a country Havilah or Khavila,
a name which is that of one of the sons of Cush, upon the Pison,
that is to say, the upper course of the Indus. Thus, we are shown
here a Cushite people inhabiting the two slopes of the Indian Caucasus, long before the Aryan development, and from whom has been
derived the name of the *Hindu-Cush.* Herodotus, who reports the
traditions which he had collected at Babylon, characterizes distinct-
ively the inhabitants of Gedrosia as Cushites, directly related to the
dark tribes of the Indus, since he qualifies them as Ethiopians. The
Baron D'Eckstein has proved that the Aryans of India originally
designated this race by the term *Sudras,* whom they had supplanted
in their rich domains included in the region called by the generic

name of *Kausikas*, preserved at a later period in certain sacerdotal families, deriving their origin from the people occupying the country before the Aryans, and who were admitted by the latter to their own rank. This term *Kausikas* is manifestly one with *Kush* (Cush)." [1]

In a word, the author shows from his own researches, following those of Dr. Lassen and the Baron D'Eckstein, that a Cushite pop-

---

[1] *Manuel Hist. Anc. Orient.*, iii. pp. 417, 418, etc.

Dr. J. Grill (*Die Erzväter der Menschheit*, etc., i. pp. 250–279) has developed quite recently a similar order of facts, derived from Hindu sources, to those upon which M. Lenormant bases his opinions. They tend to confirm the hypothesis of an extremely ancient Cushite civilization in the region to the northwest of India. Dr. Grill is, however, as it appears to me, wholly unauthorized to consider these Cushites either as Aryans or as ancestors of the Hebrews. If the Hebrew Scriptures have in any way been indebted to Cushite traditions, it was mediately through the Babylonian civilization; not in any sense directly from the Cushites of India. The author's theory, that the Mosaic account of Eden is of Aryan origin, really derives no support from the fact of this Cushite civilization of India; for the latter evidently preceded the Aryan development, and especially the entrance of the Aryans into India. The Hindus, as stated in the text, were greatly indebted to the Cushites, who were far more advanced. Another German author, Hr. Ernst von Bunsen, adopts the theory of a Cushite civilization in the region of the upper Indus at an early period (*Biblische Gleichzeitigkeiten oder Uebereinstimmende Zeitrechnung bei Babyloniern, Assyrern, Aegyptern, und Hebräern.* Berlin, 1875, pp. 53, 54).

At the last session of the International Congress of Orientalists at St. Petersburg, in September, 1876, Dr. Schmidt of Gevelsberg exhibited the first pages of a work, in which some entirely new views are presented on the subject of the origin of the Hamite civilization, especially the Egyptian. In substance, his opinions are the following :—

"Leaving aside the relation of the Chinese to the ancient Egyptians, since I have not the materials necessary to this subject, I have put forth the opinion in my work, that the Egyptians are the most ancient people among those whom I call the Armeno-Caucasians. Their civilization had its birth in Mesopotamia. From there, their civilization traversed the Persian Gulf, Arabia, and Ethiopia; and passed from thence to the valley of the Nile. After the expulsion of the dynasty of the Medo-Iranians, of Sha-suor Sakes, the Egyptian kings carried their civilization into Syria and Assyria. It was to the powerful impulsion of this civilization that Semitic Assyrians owed the development of their civilization." — *Bulletin du Congrès International des Orientalistes, etc.* p. 89.

Such views, put forth before such an august body, are entitled to great respect. Yet it is difficult to perceive by what process the author could fully sustain them. I cite them here to show the improbability of the first origin of Babylonian civilization from Africa. Certainly, as it seems to me, the people who built the brick pyramid in stages at Sakkara, in Egypt, must have emigrated from the Euphrates, rather than those who built the pyramid of Borsippa, from the Nile valley.

ulation occupied the region of the Hindu Caucasus, long before the
Aryan developments; that they had occupied India before the Aryans entered this country, extending themselves along the banks of
the Indus to its mouth, and thence upon the shores of the Indian
Ocean westward to the Persian Gulf and mouth of the Euphrates.
When the Aryans entered India the Cushites settled there were
well advanced in civilization, and became literally the teachers of
their ruder conquerors. The facts here brought to light accord precisely with the Mosaic accounts, as we have construed the narrative
relative to the first settlement of Babylon by a colony from the
east, under the leadership of the Cushite hero Nimrod. The absolutely primitive Cush or Ethiopia, as thus appears, was located
around the waters of the upper Indus, and it was from thence that
the emigration took place to the land of Shinar. By what route
the Nimrodic expedition to Babylonia proceeded, whether along the
shores of the Indian Ocean and thence up the Euphrates, or by a
more direct and northern course, it is impossible to decide.

The theory which derives the Cushites of Babylon from the east,
according to our interpretation of the language of Moses, finds definite support in the primitive traditions of the country of the Euphrates, as shown by the cuneiform inscriptions. Thus, we have
the frequent expression *Kharsak-Kurra*, for the Accadian, or " *Bit-
Kharris* of the east," for the Assyrian, designating the supposed
primitive abode of man, and the seat of a civilization prior even to
that of Babylon. A slightly different phrase is " *Bit-Kharris* of the
east, the father of countries," in which we recognize the sacred
locality from which, according to tradition, the civilizers of Babylon had departed on their westward journey. Rev. A. H. Sayce
explains the phrase *Kharsak-Kurra* by " mountain of the world," and
is inclined to locate it in the " Highlands of Elam," the modern
Susiana; but this geographical scheme for the great Asiatic Olympus is essentially erroneous, as will be shown in a future chapter.
The region designated is the high table-land of Central Asia. [1]

---

[1] I see that Dr. Schrader, in a critical paper upon the "Origin of the Chaldæans, and the Primitive Home of the Semites," inclines to derive the first Cushite
settlers of Babylon under Nimrod from Southern Arabia, thus favoring substantially Rev. G. Rawlinson's views (*Zeitschrift d. Deutsch. Morgend. Gesellschaft*,
Leip., 1873, pp. 419–422). But the primitive Babylonian tradition which embodies itself in the phrase, " Bit-Kharris of the east, the father of countries,"
cannot be construed with reference to Southern Arabia. We have, 1st. The fact

Sec. 10. We return now to the tower of Babel, and the traditions centring in and around it. It is supposed by most cuneiform scholars to be identical with the ruins of the pyramidal temple of Borsippa, whose modern name is *Birs Nimrod*, "mound or tower of Nimrod," from which Sir H. Rawlinson obtained formerly some inscribed cylinders, giving an account of the restoration of the structure by Nebuchadnezzar, after it had remained for a long period in a state of decay. The subjoined passage from one of these inscriptions, as translated by M. Lenormant, is quite important :

"The temple of the seven lights of the earth, the monument of the traditions of Borsippa, had been constructed by the most ancient king ; he had given it an area (at the base) of 42 agrarian measures ; but he had not completed it to the top. Since the days of the deluge (*ultu yum riknt*), it had been abandoned without repairing its water courses ; thus the rains and tempests had destroyed the construction in crude brick, and the facings in burnt brick were cracked, so that the mass of crude brick had crumbled into the shape of cones."[1]

There have been many versions, by different Assyriologues, of this particular passage in the text of Nebuchadnezzar, the majority of them being substantially alike except as regards the phrase, "since the days of the deluge," which Sir H. Rawlinson and other English scholars render "since the days remote," without any special reference to the deluge. The whole question depends upon the true reading of the cuneiform text in this place. If the reading is *ultu yum riknt*, as maintained by M. Lenormant, then his translation as given above is correct. But Sir H. Rawlinson's reading of the original here is *ultu yumi ruquti*, which is an ordinary expression employed in the inscriptions for "days remote," or "ancient times," without reference to any particular event such as the deluge. Certain circumstances strongly favor M. Lenormant's proposed reading, which had been adopted long before by Dr. Oppert. 1st. The pyramidal temple at Borsippa, and the one at Babylon were traditionally regarded as absolutely primitive structures in the valley of the Euphrates, being the original types of all others. 2d.

of a very early Cushite civilization in the region of the upper Indus. 2d. The Mosaic account of the migration "from the East." 3d. The Babylonian tradition pointing directly to the east. These facts perfectly harmonize with and confirm each other. It is very difficult to resist the conclusion dictated by them.

[1] Vid. *Frag. de Bérose*, p. 352.

Cuneiform scholars are for the most part agreed that the pyramid of Borsippa is to be identified definitely with the tower of Babel. 3d. Is a philological question pertaining to the two terms *rikut* and *ruquti ;* M. Lenormant contends that the latter reading involves a permutation of the sounds *k* and *q,* which is contrary to prevailing usage in the texts. As he has devoted a special consideration to this one point, it is probable that his views are correct, and as they are supported substantially by Dr. Oppert, I believe it is safe to adopt the rendering, " since the days of the deluge ; " involving thus a direct allusion to the traditionary events recorded by Moses. The name *Borsippa* is often written in the texts with the characters *Bora-sip-ki,* which M. Lenormant interprets as a mystical title having the sense of " city of the confusion of tongues," or of " the stuttering of words." [1]   It is somewhat doubtful, I think, whether this interpretation is legitimate, but as I have not seen it called in question, it may be provisionally adopted.   An ancient and mystical name of Babylon is written with the signs *Din-tir-ki,* which M. Lenormant renders " city of the root of languages." [2]   This interpretation is sustained by the ordinary meaning attached to the characters, and it points to the tradition of the original unity of language.   The more frequent mode of writing the name of Babylon in the inscriptions is with the Accadian characters *Ka-An-ra-ki,* literally, " the gate of the god of the deluge ; " but its Semitic reading would be *Bab-ilu,* " gate of Ilu," or *El,* supreme divinity of Babylon.   This is the origin of the name *Babel,* as employed by Moses.   It has been supposed by some that Borsippa was included anciently within the walls of Babylon.

The facts placed here before the reader will be seen to confirm literally the Mosaic account, considered as a faithful echo of primitive tradition respecting : 1st. The original unity of language ; 2d. The building of the tower of Babel and the confusion of tongues ; 3d. The connection of this event with Nimrod, as indicated in the modern name, *Birs Nimrod.*   All this goes to establish the hypothesis that the colonists from the east, from the primitive Ethiopia near the headwaters of the Indus, under the conduct of Nimrod, were actually the founders of the Babylonian civilization.   Our interpretation of the Mosaic text to the effect that this emigration was really from the east, is confirmed by the expressions already referred to,

[1] Vid. *Frag. de Bérose,* p. 349.   [2] Ibid.

as often found in the inscriptions, relating to the "Bit-kharris of the east, the father of countries," etc.

One other fact deserves to be noted in connection with the pyramidal temples of Borsippa and Babylon, which, as we have said, were regarded as the primitive and typical structures of this class in the country of the Euphrates. These edifices were built of brick, and in stages retreating one upon the other. It is remarkable that what is now regarded by Egyptologists as the most ancient pyramidal structure in the valley of the Nile is that of Sakkara, the primitive site of Memphis, which is constructed also in stages and of brick, according to the express statement of M. Lenormant.[1] Mr. Birch, of the British Museum, in a recent work published by him, conveys a different impression as to the materials employed; for he says: "It was constructed of calcareous stone and granite, and had seven steps like the Babylonian towers."[2] In relation to the seven stages, like the pyramid of Borsippa, it is probable that Mr. Birch is correct, but I think he is in error as to the material employed. M. Lenormant's statement is explicit and repeated, to the effect that the great pyramid of Sakkara was constructed of brick. This is confirmed by the following observation of Mr. Bayard Taylor: "As we passed the *brick pyramid* of Sakkara," etc.[3] Other pyramids in stages, much smaller, located in the same region, are built of stone; and it is probable that the error into which Mr. Birch appears to have fallen is due to this fact. As regards the primitive character of the edifice to which we refer, the opinion of M. Lenormant, as just cited, is confirmed by the remarks of Aug. Mariette-Bey: "At the Great Pyramids, we find nothing anterior to Cheops." "Sakkara, on the contrary, exhibits monuments at each step that embrace all the long period comprised between the first dynasty and the emperors." "It is, in effect, upon the northern plateau of Sakkara that are seen the most ancient monuments of Egypt."[4] From the foregoing notices, then, it is safe to infer: 1st. That the great pyramid of Sakkara was constructed in stages, and of brick; 2d. That it was absolutely the primitive pyramidal structure in the Nile valley; 3d. That it was strictly analogous in its

---

[1] Ibid., p. 363.

[2] *Ancient History from the Monuments, Egypt*, p. 25.

[3] *Journey to Central Africa*, p. 69.

[4] *Musée A Boulaq*, etc., p. 287. Cf. *Aperçu de l'histoire ancienne d'Egypte*, p. 76.

character to the pyramids of Babylon and Borsippa, these being the most ancient in the valley of the Euphrates.

SEC. 11. In the facts brought to light in the last section, we have the basis of some important conclusions, to which we invite particular attention : —

1st. The pyramids of Borsippa, Babylon, and Sakkara in Egypt, belong to the same chronological epoch; and M. Renan is correct in the opinion already expressed, that the two civilizations of the Nile and Euphrates were " equal in antiquity." But this epoch, which is, so to speak, historical as pertains to Egypt, had become purely legendary at Babylon, for the reason that the primitive monuments maintained no such state of preservation, as in the valley of the Nile. The pyramid of Sakkara is wholly exceptional in its character, so far as concerns Egypt. Nothing of the kind appears either before or after it in the country. It marks, thus, a distinct epoch, and this the primitive one. It shows that when the Hamites entered Egypt, they carried with them the same traditionary notions as those which prevailed originally at Babylon: although shortly after, these ideas took on a different material expression. Now, if we assume that the pyramidal temples of Babylon appertained to a period much later, it will be very difficult to account for the origin of the conceptions embodied in them. If the Babylonians had derived their notions from the Nile valley at a later epoch, they would have been different, since in this country the style of such structures had been essentially modified. If we would derive these architectural ideas from the far East, from Central Asia, very much depends upon the date assumed. It is probable that the pyramidal temple was absolutely primitive in all Asia, but at a very early period it had assumed the style of the pagoda in Central and in Farther Asia. Thus, everything forcibly tends to fix a high antiquity for the pyramids of Babylon and Borsippa, equal even to that of the brick structure of Sakkara in Egypt. That an original type of these pyramids in stages, built up in the mass, did exist in Eastern Asia may be inferred from the fact that, in Central America, we have the same primitive style of sacred edifices, evidently of Asiatic origin. But everything forces us to assume an immensely remote period, if we say that the temples of Babylon were derived either from Africa or Central Asia. The only easy and natural inference is, that the Hamites settled Egypt and the Cushites Babylon, at about the same epoch, both having emigrated from Central Asia, and carrying with them the same architectural notions. This supposition fully accords with the Mosaic narrative, and it would be difficult to frame a different scheme that would thus accord, and at the same time harmonize with all the essential facts.

2d. The founders of the civilization of the Euphrates and of the Nile were equally members of the great Hamite race. As M. Renan has said, they were "equal in antiquity, closely resembling each other, and, with all, anterior to the entrance of the Aryans and Semites into history." There is here absolutely no question as to the Aryans and Semites; so that the problem is narrowed down to that between the Hamites and Turanians; or, as regards the country of the Euphrates, between the Cushites and Accadians. The subjoined extracts from a recent work by M. Lenormant ought in such case to settle the question : —

"It is a very striking fact, and one already noticed by many scholars, that at the base of all the pyramidal temples of *Chaldæa* strictly speaking, at Ur, at Erech, Nipur, and Larsam, we find uniformly the name of the same king inscribed upon the bricks, which I would read Lik-Bagas (Urukh). 'In all Chaldæa,' observes Rev. G. Rawlinson, 'so far as the explorations have been at present extended, we find no trace of a sacred monument that can be assigned reasonably to a date anterior to this monarch.' His inscriptions are the earliest of any at present known; yet he belongs fully to the historical period, and it cannot be said that he opens an era, like that of Menes in Egypt. The temples in form of a pyramid, with stages retreating one upon the other, were thus *recent in Chaldæa*, as compared with those in the country of Shinar, or Sumir, where the indigenous traditions, like that of the Bible, associated the confusion of tongues with the construction of the original edifice of this class; and where, as in respect to the pyramids of Babylon and Borsippa, they did not pretend to attribute them to any king belonging to the historical dynasties, making use rather of the expression in such cases of 'the king very ancient,' or 'the most ancient king.' But in the country of Akkad (Chaldæa), instead of being a fact equally primitive and indigenous, the construction of edifices of this type is in reality only an imitation of the usages of Babylon; an imitation undertaken and pursued in all the cities at the same epoch, and by one and the same monarch, who, instead of belonging to the traditionary period, appertained strictly to the historical epoch." "Now, it is necessary to go farther, and to distinguish Babylon in the primitive ages as Cushite, from Chaldæa, which remained for a long period almost wholly Accadian, or Turanian."[1]

In what sense, then, the reader will now naturally inquire, does M. Lenormant still hold to the Accadian or Turanian origin of the Babylonian civilization? The answer must be: Mainly as regards the system of writing. Contrary to the statement as already cited from Mr. George Smith, our French Assyriologue maintains that Lik-Bagas borrowed from Babylon, not only its sacred architecture, but its mythology, its science, if not also its industrial arts, institutions, and laws. It is the paleographical system, almost exclusively, for whose Accadian or Turanian origin he still contends. Consistently with the foregoing extracts, it is difficult to perceive how he could maintain much, if anything, more than this; and it appears to me impossible now to sustain even such an hypothesis. I insist again that the Cushites of Babylon could not have remained destitute of a paleographic system during the long period prior to the time of Lik-Bagas; could not have attained to the advanced state of development supposed by M. Lenormant, without such a system. There exist to-day undiscovered in Babylonia inscriptions appertaining to a period long before Lik-Bagas; inscriptions in a character which produced the cuneiform, or in some other yet unknown; the primitive Phœnician, perhaps, or the Himyaric of a date anterior to any that scholars are at present willing to allow.[2]

[1] *La Magie*, pp. 295, 296, 298.

[2] Sir H. C. Rawlinson many years since held the following language relative to the paleography of the primitive Babylonians: "The Babylonian is unquestionably the most ancient of the three great classes of cuneiform writing. It is well known that legends in this character are stamped upon the bricks which are excavated from the foundations of all the buildings in Mesopotamia, Babylonia, and Chaldæa, that possess the highest and most authentic claims to antiquity, and it is hardly extravagant, therefore, to assign its invention to the primitive race which settled in the plains of Shinar." "It is natural to infer from the peculiar form of cuneiform writing that, in all ages and in all countries, it must have been confined exclusively to sculptures and impressions. In Babylonia and Assyria there was certainly a cursive character employed in a very high antiquity, synchronously with the lapidary cuneiform." (*The Persian Cuneiform Inscription at Behisteen.* London, 1846. Pp. 20, 31.)

M. E. Renan maintained long since, that no Semitic language ever existed in a written form that had not an alphabet peculiar to itself of the general Phœnician

M. Lenormant, in the foregoing passages, has sketched with a masterly hand the exact state of the facts according to the latest developments. Neither in the tenth chapter of Genesis, nor in the *Iz-dhu-bar* Tablets, in both which the hero-founder of the Babylonian kingdom is so often mentioned, — since *Iz-dhu-bar* and Nimrod are held to be the same,[1] — do we have any mention of the city of *Ur*, the Biblical " Ur of the Chaldees," which was the chief capital city of Lik-Bagas. It is evident that this city did not exist at the period to which the account of Genesis and the *Iz-dhu-bar* legends pertain. If, then, the inscriptions of Lik-Bagas are the earliest yet known, which is true, and notwithstanding he must be assigned to the twenty-fourth or twenty-fifth century before our era, it is obvious that these are not the origin of writing in the valley of the Euphrates. The hieroglyphic system of writing is known to date from the earliest dynasties in Egypt, and there is much reason for the supposition that the Hamites had invented this system even before the time of Menes, the first historical king. The near relationship of the Cushites to the Hamites leads to the inference that the Cushites likewise had possessed a similar system at an epoch not much later. The explorations thus far have not been by any means so thorough at Babylon and the surrounding district as at Nineveh; and it cannot be said of Babylonia, as of Chaldæa, that its earliest inscribed monuments are already known.

SEC. 12. As before remarked, the Aryan and Semitic elements are completely eliminated from our problem, since the Babylonian

type; and that such a mode of writing must have existed at Babylon at a very early period (*Histoire générale*, pp. 70–75). Thus, I have expressed the opinion in the text that either the cuneiform system of writing, or some other allied to the Phœnician, was absolutely primitive at Babylon. Its renowned civilization never grew up without a paleographic system from the start. The probabilities are that this was the primitive cuneiform.

[1] Dr. J. Oppert, at the last session of the International Congress of Orientalists, took ground against the identification of Iz-dhu-bar and Nimrod ; and he held that Nimrod was not the name of a person, but of a people (*Bulletin des Congrès des Orientalistes*, etc., p. 124). I do not think, myself, that there exist sufficient grounds for the identification of these two personages. I have merely followed the opinions of Mr. George Smith, in a matter not at all vital to my subject. Dr. Oppert presented to the Congress alluded to a new version of the Iz-dhu-bar tablets, in which he differs somewhat from the views of Mr. Smith, and which, I suppose, will appear in the regular publications of the Congress.

civilization obviously dates from a period prior to the entrance of
the Aryans and Semites into history. So far as relates to the Accadians or Turanians of Chaldæa, this element may be also excluded. for it cannot be traced to a period earlier than Lik-Bagas,
who was really a convert to the religion, and an imitator of the
usages, appertaining to the Cushites of Babylon. The question
before us now rests between the Cushites on one hand, and a possible Turanian element prior to the ancient Chaldæan dynasty
headed by Lik-Bagas. We know that there was an *Akkad* in the
valley of the Euphrates long before the earliest of the monarchs of
Chaldæa, as the name occurs in the list of cities constituting the
beginning of Nimrod's kingdom. More than this; it is probable
that the *Akkad* of Nimrod's time was not situated in Chaldæa, but
in the vicinity of Babylon itself. Mr. George Smith remarks:
" The capital of Sargon (*the ancient*) was the great city of Agadi,
called by the Semites Akkad, mentioned in Genesis as a capital of
Nimrod (x. 10), and here he reigned for forty-five years. Akkad
lay near the city of Sippara on the Euphrates and north of Babylon." " I have only recently discovered the identity of Akkad with
the capital of Sargon." [1]   In a matter of such importance. and one
upon which there has been so much doubt and discussion, it is to be
regretted that the author offers no proof of the foregoing statement.
Nevertheless, one circumstance tends powerfully to support it.   In
the symbolical geography of Sargon the ancient, which appears to
be fundamental in the mythological and astrological texts attributed
to him, *Akkad* is considered as situated at the centre of the world,
surrounded by four countries located exactly in the direction of the
cardinal points. Now we know that Agani. probably the same as
Agadi mentioned by Mr. Smith, was the capital of Sargon's empire.
It is wholly improbable that he should take an *Akkad* in Chaldæa
as centre of his symbolical system of geography, instead of his own
city or country. The *Akkad*, therefore, of the texts of Sargon was
his own capital city, and this was the *Akkad* also of Nimrod's time.
Another consideration tends to support Mr. Smith's hypothesis
above stated. According to the Mosaic text, the Biblical Akkad
was situated within the limits of Shinar. Now Shinar must be
distinguished from Assyria, on one hand. since Asshur went forth
from Shinar to found the Assyrian kingdom; and it is probable, on

---

[1] *Assyrian Discoveries*, p. 225, and *note*.

the other hand, that Shinar should be distinguished from Chaldæa. This tends definitely to fix the locality of the original *Akkad* in the vicinity of. Babylon, outside of either Chaldæa or Assyria. The facts here brought out, if one considers a moment their bearings, completely overthrow the Turanian hypothesis as regards the origin of the Babylonian civilization, if we are to identify the supposed Turanians with the *Akkadi*, or Accadians. Attribute, if you please, this civilization to *Akkad*, yet the *Akkad* of Nimrod's era was included in his kingdom; the Cushites under Nimrod were the ruling class; and it was to the ruling class, in such cases, that the entire culture, as well as civil and religious functions appertained. This remark applies as well to the paleographic system. Trace its origin to the *Akkadi*, yet the ruling class among the *Akkadi* of primitive times were Cushites, and they must have been the inventors of the system. It is probable, in fact, that these primitive Accadians were themselves Cushites. For myself, I am unable to perceive how it is possible to avoid the conclusions here deduced. Nor do these rest wholly on Mr. Smith's proposed identification of *Akkad* with the capital city of the ancient Sargon. We might have said long ago: Yes, the Accadians were the inventors of the cuneiform writing and the civilizers of Babylon; but the Cushites were the ruling element of population among the primitive Accadians of the Euphrates valley, and to this preponderating element is due the invention and civilizing process, even if the Accadians themselves were not Cushites. But the discovery made by Mr. Smith, and the other considerations urged by myself, render the basis of these conclusions much more solid and reliable. The Turanian hypothesis, then, must be abandoned, or it must be divorced from the population termed *Akkadi* in the cuneiform texts; and this is precisely that which Dr. Oppert believes to have been done already; that is to say, he claims that the *Akkadi* were properly Semites, while the *Sumeri* were the Turanians of Babylon, the founders of its civilization, and the inventors of its paleographical system.

SEC. 13. We have not been willing, in these researches, to enter upon the question raised by Dr. Oppert, which has remained principally up to the present time as between himself and M. Lenormant. Outside of France, very few, if any, Assyriologues have adopted the doctor's theory, although it has been supported on his part with great ability. As regards the *Sumeri*, it is not easy to fix any special locality for them from the texts, nor to assign to them any definite chronology. The terms *Sumir* and *Sumeri* seem to appertain, at least under their present form, rather to the historical than the legendary period. They do not occur in the Mosaic text, nor in the "Izdhubar Tablets." Nevertheless, under a different form, it is possible that *Sumir* was really very primitive. M. Lenormant supposes that *Sumir* and the Biblical *Shinar* were originally the same; and I believe that Dr. Oppert formerly favored this assimilation, being, in fact, the first to suggest it. In such case, we find here no objection to the Cushite hypothesis assumed by us, since, according to all known facts, the Cushites were the original ruling class in the land of Shinar. Again: a very ancient title assumed by the monarchs especially of Chaldæa was the Accadian *Ungal ki Engi ki Akkad*, which Dr. Oppert and Mr. Smith render "king of *Sumir* and *Akkad*." M. Lenormant maintains that *ki engi* means simply "a country," and translates the phrase "king of the *country* of *Akkad*." It is probable, however, that *Engi* is the name of a particular country, for it is preceded by the determinative *ki* in the same manner as *Akkad*. Now the word *Akkad* is supposed to signify "highland," and *Engi* is interpreted by Dr. Oppert as "true-lord." The inquiry, then, is quite pertinent: What personage, as "true-lord," has furnished thus the name of the country called *Engi?* We cannot long hesitate in saying that it must have been Nimrod, primitive ruler of Shinar; and the proposed assimilation of this Biblical name to *Sumir* on'y goes to confirm the suggestion here made. Thus, I believe it is unnecessary to devote much space here to show that the word *Sumir* affords no serious objection to the Cushite theory, to establish which has been the leading aim of the present chapter. That this hypothesis is wholly free from doubt and difficulty is not pretended; but it appears to me to offer the most plausible explanation of all the facts; and it is in perfect accord with the Mosaic narrative, of which also it affords throughout a striking confirmation.[1]

[1] In the *Journal Asiatique* for March-April, 1876, M. Halévy returns to the question of the origin of the cuneiform system of writing, and in a paper occupying nearly two hundred pages, attempts to show that its originators were Semites, that is to say, the Assyrians. In other respects, the paper is very valuable; but it will hardly convince Assyriologues that the Semites were the authors of this palæographical system. They will share rather the views of M. Renan, in his very last annual report, published in the *Journal* of July, 1876, now just at hand, namely, that it is "little satisfactory" (vid. p. 42).

CHAPTER II.

SEC. 14. That the original character of the systems of writing prevailing among the ancients was hieroglyphic, not excepting that from which our own alphabet was derived, is the hypothesis now generally and tacitly received among the learned; and according to the opinion usually held by Assyrian scholars, the cuneiform system constitutes no exception to this rule. I have been surprised at times at the very slight indications and, as it appeared to me, quite insufficient proofs upon which writers have applied the hypothesis named in particular cases, among which I include especially the ancient Phœnician and the Accadian or Cushite systems. It will be hazardous, perhaps, to call in question the fundamental accuracy of the hieroglyphic principle, in its application to the paleographic systems of antiquity. But the more I have investigated the subject the firmer has been my conviction that this principle has been too hastily adopted in some instances, and too exclusively applied in all. That the first system of writing known to men consisted simply of pictures of concrete objects, or in other words of pure hieroglyphs, appears to me extremely doubtful, notwithstanding the general tendency among the authorities to support this view. But I would hold the advocates of the hieroglyphic theory strictly to their definitions and doctrines. When a concrete object has acquired a symbolical character, and this fact has constituted the obvious motive of selecting an image of it for paleographic purposes, we have then, not hieroglyphic, but symbolical writing. Take for example the Hebrew letters *Aleph*, "ox;" *Beth*, "house, temple;" *Teth*, "serpent;" *Kaph*, "hand;" *Nun*, "fish;" *Ayin*, "eye;" and *Tau*, "cross;" admit that these characters were originally pictures of concrete objects, and represented phonetically their names; yet this does not prove that they were hieroglyphs in the strict sense of the theory. Every one of them had acquired a symbolico-religious character, and as such was celebrated throughout all antiquity. It could be readily shown that the religious notions attached to them were far more ancient than any known written monuments. The selection of the images of these objects, therefore, as palæographic signs, was no mere picture writing, but was rather, and in the proper sense, symbolical. This method having been once established, an extension of it subsequently to objects destitute of any symbolic import would be quite natural; and I believe that such was the case among the Chinese and Egyptians. In regard to the Accadian or Cushite system, from which the cuneiform was derived, the hieroglyphic theory has been quite too hastily adopted. The recent important admission of M. Lenormant, that the existing hieratic type of the characters was probably the primitive one, must be considered as fatal to this theory as applied to them.[1] Undoubtedly, we have here many concrete objects represented as the original basis of the characters employed; but it would be easy to show that the objects themselves had acquired a symbolic import, and were for this reason selected for palæographic purposes. In general, then, my hypothesis is as follows: symbolic writing was the most ancient among the cultured nations of antiquity; but subsequently, and with certain peoples, the original method was extended to objects to which no æsthetical ideas were attached. This theory accords perfectly with the universal tradition respecting the invention of writing, at the same time that it accounts for all the facts now known. I do not take into consideration here the crude attempts at picture writing existing among certain savage tribes or semi-barbarous races, for the simple reason that there exists not the slightest proof that they were the most ancient known to man.

SEC. 15. The archæological studies proposed for the present and next succeeding chapter will be based to a considerable extent upon various cuneiform characters, under their most ancient or hieratic form; these being taken for the most part at their ideographic value. For this reason it has seemed necessary to insist at the outset that these characters are not mere figures of concrete

---

[1] Vid. *Etudes Accadiennes*, t. i. pt. 3d, p. 7. The author's repertory of Accadian signs, contained in this part of the work cited, will constitute our authority upon the values of the signs, except in cases specially noted. It is, so far as I have knowledge, the latest and most complete treatise upon the subject.

PLATE I.

objects, but that they attach to themselves an exalted and instructive symbolism, highly important to be considered, and constituting a proper subject for archeological investigation. The reader finds in our first plate several groups of paleographic signs, to which reference will be had in these studies, the greater portion of them belonging to the cuneiform system of writing; while the others, either appertaining to other systems, or strictly to the art monuments, will be explained in their proper connection. The characters are arranged in couplets or triplets, as the case may be, having the hieratic form on the left and the corresponding modern form on the right; the small letters being introduced for convenience of reference. For the value of these characters, I shall follow chiefly the excellent treatise by M. Lenormant, which was cited in the last section (note 1). The signs marked *a*, *b*, and *c* of the first group are only so far different as would be necessary to distinguish readily between them, and it is natural to infer that they appertained originally to a class of conceptions not less intimately connected. They constitute the objects to which our first attention will be directed.

The second character, or the one marked *b*, appears to involve the central idea of the whole, and its Accadian values are as follows: *Ni*, "to sweep, to scrape, to clean; shovel, hearth, God;" *Kisal*, "altar, sacrifice;" *Zal*, "joint, vicinity, or neighborhood" (Rep. 142). It cannot be denied that the *shovel* is a concrete object, and one sufficiently ordinary and humble; but who can fail to admire the artistic skill that has raised this common utensil of the household to the rank of a symbol of the primitive worship of mankind — that of the *hearth!* The hearth, and the divinity of the hearth, constituted the focus of all the ancient civilizations. It was around the firesides of primeval humanity that those elemental organizations were formed, those social and semi-political customs instituted, and religious conceptions and sentiments nurtured, which subsequently, by the simple process of expansion and reduplication, developed themselves into tribal and national institutions. The hearth was the family altar, and the cheerful blaze kindled upon it was the symbol of the divinity who presided over the destinies of the household, seeming to share in its fortunes, receiving the grateful remembrance and adoration of its members. The altar of the tribe was its hearth, and the national altar was the national hearth; the same notions and customs were transferred from one to the other, being

modified only so far as necessary to adapt them to the changed circumstances. The God of the hearth was really the paternal head of
the household, and its members were his family. So, too, when the
same divinity had been transferred to the national altar, he became
the father of the nation, while all the members of the commonwealth
were his children. The entire territory belonging to the state was
a common patrimony, and the state itself an organization of brothers
and sisters, whose focus was the national hearth or altar. It was impossible for a stranger to become a member of the commonwealth
except by regular adoption into some family. This principle of
adoption was an important feature of the patriarchal institutions.
By a fiction of law, a stranger could be made a regular member of
the household, and admitted to the family sacra or worship, being
thence regarded as a blood relation, descended from the same ancestral head. The custom of *adoption* had a religious as well as political significance, even from its origin ; and this explains the fundamental importance attached to it by the Jewish and Christian
sacred writers. It was only upon the principle of adoption that the
Gentiles could be admitted into the family of Jehovah, and be
accounted as regular descendants in the ancestral line of Abraham.
It was only in this way that they could be admitted to the divine
sacra, and be permitted to feast upon the great Sacrifice, which
was Christ.

SEC. 16. When it is found that the Accadian character whose
usual reading is *Ni* had the meaning of *hearth, altar, God,* this is
sufficient to demonstrate that the God *Ni* was the hearth-divinity
of the Accadian or Cushite race. The cuneiform texts afford but
little information respecting the special character and primitive worship of this divine personage, except the one great fact which will
be developed hereafter. The general absence of this name from
the ordinary lists of divinities favors the supposition that the worship of *Ni* had been very ancient, but had fallen into neglect at the
period to which the existing monuments pertain. But the peculiar
nature of the house-gods of antiquity may be learned from the
classic nations, particularly the Romans, or Etrusco-Romans. It is
now known that the Etrusco-Roman civilization was derived, in a
great measure, from the valley of the Euphrates. Thus, it is probable that we may find in the Roman cultus of the *Penates* and
Lares of the Latin nation a reflex of the religious conceptions and

customs centring in the primitive Cushite god of the hearth. Dr. William Smith has the following notice of the Roman house-divinities : —

"*Penates*, the household gods of the Romans, both those of a private family and *of the state* as the *great family of citizens*. Hence we have to distinguish between private and public Penates. The name is connected with *penus*, and the images of those gods were kept in the *penetralia*, or the central part of the house. The Lares were included among the Penates; both names, in fact, are often used synonymously. . . . The Lares, however, though included in the Penates, were not the only Penates; for each family had usually no more than one Lar, whereas the Penates are always spoken of in the plural. . . . Since Jupiter and Juno were regarded as the protectors of happiness and peace in the family, these divinities were worshiped as Penates. Vesta was also reckoned among the Penates; for each hearth, being the symbol of domestic union, had its Vesta." "Most ancient writers believe that the Penates of the state were brought by Æneas from Troy into Italy, . . . and were preserved first at Lavinium, afterward at Alba Longa, and finally at Rome. . . . At Rome they had a chapel near the centre of the city in a place called *Sub Velia*. As the public Lares were worshiped in the central part of the city and at the public hearth, so the private Penates had their place at the hearth of every house, and the table also was sacred to them. On the hearth a perpetual fire was kept up in their honor, and the table always contained the salt-cellar and the firstlings of fruit for these divinities. Every meal that was taken in the house thus resembled a sacrifice offered to the Penates, beginning with a purification and ending with a libation, which was poured either on the table or upon the hearth. After every absence from the hearth, the Penates were saluted like the living inhabitants of the house; and whoever went abroad prayed to the Penates and Lares for a happy return, and when he came back to his house, he hung up his armor, staff, and the like, by the side of their images." (Class. Dic., art. *Penates*.)

It is observed by another author that : "In general, and as principal tutelary divinities, the Penates bore the name of great gods (*magni dii*, θεοί μεγάλοι, δυνατοί). It was doubtless from this fact that they were identified with the great gods carried from Accadia to Samothrace" (Bernard. Dic. Myth., art. *Penates*). On the subject of the expansion or reduplication of the family organization into that of the tribe and nation, Mr. H. S. Maine remarks : —

"In most of the Greek states and in Rome there long remained the vestiges of an ascending series of groups out of which the state

was at first constituted. The family, house, and tribe of the
Romans may be taken as the type of them, and they are so described to us that we can scarcely help conceiving them as a system
of concentric circles, which have gradually expanded from the same
point. The elementary group is the family, connected by common
subjection to the highest male ascendant. The aggregation of families forms the gens or house. The aggregation of houses makes
the tribe. The aggregation of tribes constitutes the commonwealth. Are we at liberty to follow these indications, and to lay
down that the commonwealth is a collection of persons united by
common descent from the progenitor of an original family? Of
this we may at least be certain, that all ancient societies regarded
themselves as having proceeded from one original stock, and even
labored under an incapacity for comprehending any reason except
this for their holding together in political union. The history of
political ideas begins, in fact, with the assumption that kinship in
blood is the sole possible ground of community in political functions." [1]

It rarely happened, however, that all the members of a commonwealth, or of a single tribe even, were actually descended from the
same ancestor. They were considered as such on the principle of
adoption only, as Mr. Maine has explained. The reader will not
fail to recognize, in the foregoing remarks, some of the most important facts and conceptions underlying the two religions of the Bible;
and, indeed, all the ancient religions and civilizations were, in a
great measure, founded upon these principles.

SEC. 17. We comprehend now the character of the divinity
of the hearth; and the full significance to be attached to the paleographic symbol *Ni*, designating the hearth, the altar, the deity.
But the most important fact of all is, that the Accadian or Cushite
God *Ni* was one and the same personage with *Yahveh*, or Jehovah,
of the Old Testament. If the cuneiform inscriptions contain but
few notices respecting the hearth-god in question, they at least
afford abundant proof of the statement just made. In the Syllabaries the Accadian *Ni* is repeatedly equated to the Assyrian, that
is to say, to the Semitic *Ya-hu* (Nos. 685–687). In the text here
cited, the term *Ya-hu* is put for the three values of our Accadian
sign, namely, *Ni*, *Zal*, and *Ili*, the latter being correctly interpreted "a god" by Mr. Norris.[2] That the Accadian *Ni* is translated by the Assyrian or Semitic *Ya-hu*, and that the latter is put

---

[1] *Ancient Law*, pp. 123, 124.   [2] *Assyr. Dic.*, ii. p. 476.

CUSHITE ARCHÆOLOGY. 37

for a divinity, are points upon which it is impossible to raise a doubt.
Two other texts cited by Mr. Norris, by comparing them together,
show that *Ya-hu* was considered a name of divinity. Thus, a king
of Hamath is mentioned, whose name is written *Ilu-bi'h-di* in one
instance, and *Il Yahu-bi'h-di* in a different text.[1] *Il* or *Ilu* was
the supreme divinity of Babylon, being one with the Hebrew *El*.
In the second example given above, *Ya-hu*, preceded by *Il* as characteristic of divinity, takes the place of *Ilu*, name of the Babylonian deity, in the first example. Here are two distinct proofs
that *Ya-hu* is taken as a name of God. First, we have the substitution of *Ya-hu* for *Ilu ;* and secondly, *Ya-hu* is preceded by the
determinative of divinity. In view of these facts, Dr. Schrader
has well expressed the conclusion that, as *Ilu* equals the Hebrew
El, so *Ya-hu* must be one with the Hebrew *Yahu*, *Yahveh*, or Jehovah ; especially as the Jewish Scriptures consider Jehovah and El
as the same personage.[2] If now any doubt remains respecting the
identity of the Assyrian *Ya-hu* with the Biblical *Yahveh* or Jehovah, it is removed by the fact of the occurrence in the texts of two
kings' names, *Jehu* and *Jehoahaz*, both containing the Hebrew
element *Yahu*, *Yaho*, or *Yahveh* in composition. Thus, *Jehu* is
written in the cuneiform by *Ya-hu-a ;* while in the other case we
have *Ya-hu-ha-zi* for *Jehoahaz*.[3] These examples show that it
was customary with the Assyrian scribes to write the name of the
Hebrew national divinity in the manner here indicated, and that
they considered the Assyrian *Ya-hu* equivalent to the Hebrew *Yahu*
or *Yahveh*.

[1] *Assyr. Dic.*, p. 482.
[2] Vid. Schrader, *Keilinschrift. u. d. Alt. Test.*, pp. 3–5, where the same texts
are cited, and a like inference drawn to that set forth by us, but without reference to the Accadian God *Ni*.
[3] Norris, ii. pp. 476, 477. M. James Halévy, in a recent critical paper relating to the cuneiform writing, affords a direct confirmation of our position, that
the god *Ni* was a divinity of the hearth. To the character *Ni* he attaches the
senses of "sojourn, abode, the god *Jahu*," identifying *Jahu* with the Greek 'Aós,
answering to the god *Hea* (*Jour. Asiatique*, March-April, 1876, pp. 260, 266).
This proves that *Ni* was primitively associated with the house, the family. As
for the god *Hea*, M. Lenormant has well interpreted the name itself as signifying
"abode," "dwelling," etc. The resemblance of *Hea* to Jehovah, as manifest
especially in Christ, is illustrated in various facts to be hereafter developed.
Indeed, *Hea* takes the title *Aur Kinue* (הוא כינא), in the texts, which signifies
Existent Being ; the first element being the same Semitic radical from which the
name *Yaveh*, or Jehovah, is formed.

The Accadian term *Ni*, then, is not *identical* with the Assyrian
Ya-hu, for they belong to different languages. But the Assyrian
word must be taken as a *translation* of the Accadian, both terms
being put for one and the same divine personage. The result is,
from the data that have been now submitted, that the national God
of the Jews was originally one with the ancient Accadian or Cushite divinity of the hearth. The fact thus brought to light is of
very great importance, though it will be received with some hesitancy among Biblical scholars. Nevertheless, the proofs are direct
and positive, and I entertain no doubt of the correctness of the
conclusion to which the data have conducted me. But it would be
quite illegitimate to infer, from the assimilation here established,
that the Jehovah of the Old Testament was originally regarded as
a divinity of inferior rank, like some of the house-gods of antiquity.
It has been seen that the Accadian God *Ni* was considered the same
personage as the Semitic *Ilu*, *Il*, Hebrew *El*, whom the sacred
writers identify with the Jehovah of the Jews. As before observed,
El was the supreme divinity of Babylon, and it is well known that
He was held primitively in the highest estimation by the entire
Semitic race. In addition to this, we have seen that, among the
Romans, the highest divinities were worshiped as Penates, such as
Jupiter, Juno, and Vesta, and that the Penates themselves received
the title of "great gods." Vesta was characteristically a divinity
of the hearth, yet she received the highest honors. Dr. Smith has
the following remarks relative to her character and worship : —

" *Vesta*, one of the great Roman divinities, identical with the
Greek *Hestia*, both in name and import. She was the goddess of
the hearth, and therefore inseparably connected with the Penates;
for Æneas was believed to have brought the eternal fire of Vesta
from Troy along with the images of the Penates ; and the prætors,
consuls, and dictators, before entering upon their official functions,
sacrificed, not only to the Penates, but also to Vesta at Lavinium.
In the ancient Roman house, the hearth was the central part, and
around it all the inmates daily assembled for their common meal
(*cœna*) ; every meal thus taken was a fresh bond of union and
affection among the members of the family, and at the same time
an act of worship of Vesta, combined with a sacrifice to her and the
Penates. Every dwelling-house, therefore, was in some sense a
temple of Vesta ; but *a public sanctuary* united *all the citizens of the
state into one family.*" (Class. Dic., art. *Vesta*.)

It has been already shown that the divinity of the hearth was transferred successively to the altar of the tribe, and thence to that of the nation, which was thus the national hearth. In those instances where the national divinities were really different from the primitive house-gods, the former must have been a later conception ; for the family was the original unit of society, from whose expansion or reduplication the tribe and state were subsequently formed. The divinity of the hearth was thus not only primitive, but was the exclusive object of worship in the first ages of the world. Such was the Accadian God *Ni*, identical with *Yahveh* of the Hebrew Scriptures.

SEC. 18. A comparison of the chief attributes of the Jehovah of the Old Testament, and of the essential relations He sustained to the Israelitish people, with the principal features of the Cushite divinity *Ni*, as interpreted by the notions and customs of the Etrusco-Romans, will contribute materially, not only as additional evidence of the original identity of the two divine personages, but as an important key to the underlying conceptions of the Mosaic religion, which, if they were insisted upon by the writers of former periods, have fallen into general neglect, and have been often called in question by more modern authorities. With a view to such general comparison, I introduce here some lengthy extracts from the learned Dr. Cudworth, which will admirably serve our purpose : —

" In like manner, I say, the eating of sacrifices, which were God's meat, was a federal rite between God and those that did partake of them, and signified there was a covenant of friendship between him and them ; for the better conceiving whereof, we must observe that sacrifices, beside the nature of expiation, had the notion of feasts, which God himself did, as it were, feed upon, which I explain thus : when God had brought the children of Israel out of Egypt, resolving to manifest himself in a peculiar manner present among them, he thought good to dwell amongst them in a visible and external manner, and therefore, while they were in the wilderness, and sojourned in tents, he would have a tent or tabernacle built, to sojourn with them also. This mystery of the tabernacle was fully understood by the learned Nachmanides, who in few words, but pregnant, thus expresseth it ; that is, the mystery of the tabernacle was this, that it was to be a place for the Shechinah, or habitation of the divinity to be fixed in ; and this, no doubt, as a special type of God's future dwelling in Christ's human nature, which was the *true Shechinah.* But, when the Jews were come into their land, and had there built them houses, God intended to have fixed a dwelling-house also ; and therefore his movable tabernacle was to be turned into a standing temple.

" Now the tabernacle or temple being thus as a house for God to dwell in visibly, to make up the notion of a dwelling or habitation complete there must be all things suitable to a house belonging to it. Hence, in the holy place, there must be a table and a candlestick, because this was the ordinary furniture of a room, as the fore commended Nachmanides observes : He addeth a table and a candlestick, because these suit the notion of a dwelling house. The table must have its dishes, and spoons, and bowls, and covers, belonging to it, though they were never used, and always be furnished with bread upon it. The candlestick must have its lamps continually burning. Hence also there must be a continued fire kept in this house of God's upon the altar, as the *focus* of it, to which notion, I conceive, the prophet Isaiah doth allude (xxxi. 9), which I would thus translate : Who hath his fire in Sion, and his focus in Jerusalem.

" And besides all this, to carry the notion still further, there must be some constant meat and provision brought into this house, which was done in the sacrifices that were partly consumed by fire upon God's own altar, and partly eaten by the priests, which were God's family, and therefore to be maintained by him. That which was consumed upon God's altar was accounted *God's mess,* as appeareth from the first chapter of Malachi (v. 12), where the altar is called *God's table,* and the sacrifice upon it *God's meat :* ' Ye say, the table of God is polluted, and the fruit thereof, his meat, is contemptible.' And often in the land the sacrifice is called God's (לחם) bread or food. •

" The sacrifices, then, being God's feasts, they that did partake of them must needs be his *guests* (convœ), and in a manner *eat* and *drink* with him. And that this did bear the notion of a federal rite in the Scripture account, I prove from that place (Lev. ii. 13) : ' Thou shall not suffer the *salt of the covenant* of thy God to be lacking ; with all thine offerings thou shalt offer salt.' Where the salt that was to be cast upon all the sacrifices is called *the salt of the covenant,* to signify that as men did use to make covenants by eating and drinking together, where salt is a necessary appendix, so God by these sacrifices, and the feasts upon them, did ratify and confirm his covenant with those that did partake of them, inasmuch as they did in a manner *eat* and *drink* with him. For salt was ever accounted amongst the ancients a most necessary concomitant of feasts, and condiment of all meats. . . . And therefore because covenants and reconciliations were made by eating and drinking, where salt was always used, salt itself was accounted among the ancients a symbol of friendship (*amicitiæ symbolum.*)" [1]

SEC. 19. It will be impossible not to recognize in the foregoing extracts the various notions and customs, modified to adapt them to the purposes of a commonwealth, that originally pertained to the divinity of the hearth, before families had reduplicated into tribes, and tribes into nationalities. To complete the comparison, however, it will be necessary to introduce here a brief explanation of the two cuneiform signs *a* and *c*, in the group upon which we have been engaged. For the first we have the values: *Rû*, "to make, to construct;" *Kak*, "to make, to complete, all;" *Pâ*, "to make, to construct." The character marked *c* stands for *Ir*, "fruit, embryo, fœtus;" *Sukal*, the same (Rep. 141, 143). The typical constructions were the house and the temple, primitively considered as one; and the typical fruits were the products of the field, and the first-born child, through whom the inheritance of the family was to be perpetuated.

These conceptions, together with the various facts that have been now collected into view, will enable us to transport ourselves, so to speak, to the hearth-stones of primitive humanity. He who illumes both sun and star, and kindles the fires upon creation's hearth-stone, had thus early in the history of our race taken up his abode in human habitations. He who is enthroned in the heavens, who issues from the gates of morning with beams of light, that fall on the world in golden showers, had sought an abiding-place with his rational creatures. But *He was not a guest* beneath the humble roof which He had chosen for his temple, and where men and women first learned to worship, to love, and to obey. They were *His guests, His people, His sons and daughters.* That was *His house, His table, His fireside.* The consecrated hearth, whose mystic flame was the symbol of his own divine existence, was *His focus, His altar.* It was there that the institutions and civilizations of the ancient world were cradled, and it was He who had forged their nerves and sinews with his own hands. It was from the hot bosom of the domestic hearth, under the watchful care of its presiding divinity, that those giants leaped forth who were the first founders of religions and of states, and it was the divine artisan, with the chimney-corner for his *smithy*, who welded those

[1] *Intellect. System,* etc., ii. pp. 536–539.

bonds of human society which were destined to unite all the families and kindreds of earth in one brotherhood.

But it was not as a simple taskmaster that the Deity thus early selected the family circle for his favorite abiding-place. He knew that, if anywhere on earth, *there* would be *love*, between father and mother, brother and sister. It was his nature to love, and only in the circle of loving hearts could He find a home. The world without was beautiful, the heavens were peopled with shining hosts, and the earth, from her mountain peaks to her ocean depths, was alive with the living forms which He had created. He could dwell on those heights where the thunders and the lightnings have their birth; in those deep watery caverns whose floors are studded with pearls; or beneath the shady oaks and pines where the zephyrs play and the birds sing. But it was man alone whom He had created in his own image, and whom He loved with a father's affection. It was thus with the sons and daughters of men that He desired to dwell, and into their habitations that he wished to be received. He would share their lot and destiny, would be their provider, protector, their friend and their God, if they would only love Him. From the bright morning when the bridal pair first invoked his presence and blessing upon the hearth, through all the long years of toil and struggle, till the frosts gathered upon their heads, and finally the crimson sea had frozen over in their hearts, He would be with them, and abide with them, and be their God forever. The first-fruits of the harvest and the first-born of the household should be his, and every feast and joyous festival should be sacred to Him, as a pledge of his friendship, as a covenant of salt between Him and them![1]

Such was one of the original conceptions of the Jewish theocracy. Earlier than the time of Abraham, earlier than the tower of Babel, and while the Hamite and Semite, the Turanian and Aryan, were yet as one family, this grand idea had been taught the world. We have the proof in the existence of the Accadian or Cushite *Ni*, a term that, while it designated the God of the hearth, proved to have been one with *Yahveh* of the Hebrew Scriptures, was at the same time a suffix pronoun of the Cushite tongue, and thus appertained to the primitively developed stages of this ancient language.

[1] [What goes before was condensed into an eight-page article and published in *The American Antiquarian and Oriental Journal*, July, 1882. S. M. W.]

As a pronoun it involved the notion of *possession*, being put for the possessive case. The hearth as a possession, and so the first-born child, were co-related and typical ideas; and the connection of the God *Ni* with *Yahveh*, as divinity of the hearth, calls forcibly to mind the joyful exclamation of the first mother, when she said : *I have gotten a man from Yahveh* (Gen. iv. 1). Thus early invoked at the firesides of our race, on the multiplication and expansion of families into tribes, and of tribes into nationalities, the divinity of the hearth was transferred from one to the other, presiding thus at the inauguration of states and of national religions, the Hebrew among the rest. The notion that God *dwells*, *inhabits*, the same as man, was everywhere fundamental ; and it was for this reason that the national temple was considered *God's House*, and its altar the *National Hearth*.

Sec. 20. To bestow upon any object an individual name is in so far to distinguish it from the mass, and actually to raise it to a higher rank in our conception. If a race of men were discovered, of which the individuals had no personal names, this fact alone would be conclusive as to the exceedingly low order of development of the race itself. For a domestic animal to receive a particular name, learning to recognize and answer to it, is really to distinguish it from the generality and in a measure to elevate it. With the ancients, much importance was attached to personal names, these having usually a symbolical import, which could be etymologically explained. The true title of the Deity, considered as a personal name, instead of a general or generic one, was held in the highest veneration, being supposed not only to express the nature of God, but to be in some sense the Deity himself in his external manifestation. To call upon this *name* of God was to enter into a concrete personal relation with him. In fact, where the Deity is habitually addressed in worship by generic titles, or by those that have become such practically in conception, it is evidence that men have ceased, in a measure, to enter into that intimate, personal relation with the Divine Being that those of antiquity believed themselves to do. This leads us to our second group of cuneiform characters.

The signs marked from *a* to *e* of the second group, in their hieratic form, present a striking analogy in their construction, leading to the inference that the notions attached to them must be also fundamentally related. For the character *a* we have the values: *Mu*,

"to give, to call, name, memorial, year" (Rep. 24). The sign *b* is
put for *Gi*, "flame; to found, foundation ; to deport, to transport ; to
restore" (Rep. 85). The next in order is *Zi*, "to live, life, soul,
person, spirit; regulation, rule, law" (Rep. 84). Then follows *Ri*,
"to heap up, column, to elevate, elevation ; to rise, appearance, as a
star" (Rep. 87). Finally is *Ar*, whose meaning is not given by M.
Lenormant, although it must signify "vineyard, palm-tree," according to the texts which will be cited hereafter. The Accadian *Mu*
is found ordinarily in the Assyrian texts as the monogram for *Sam*,
"name," and *Sanat*, "year." The character *Gi*, "flame," has been
recognized by Sir H. Rawlinson as a monogram for the Scythic or
Accadian *Fire-god*. The sign *Ri* is often employed as a monogram
for the goddess *Is-tar*, or Babylonian Venus. The last character,
Ar, is strictly composite, being constituted of the sign *Ri* and the
Accadian *Si*, "eye, face, presence, prospect, country" (Rep. 359).

It will be evident, even from a superficial view of these hieratic
symbols, that they represent material objects, either natural or artificial, and it will be necessary, if possible, to ascertain what they
are. The second hieratic form for *Mu*, if placed uprightly before
the eye, seems to show for itself that it was intended for a *bush* or
tree. The other form, although constructed somewhat differently,
must represent the same object. The two figures, then, being simple variants of the characters before us, must be taken for a bush
or tree ; one presenting this object approximatively in its natural
form, while the other shows an artificial tree, similar to those which
appear so frequently upon the ancient art monuments. If now we
compare these two forms with that marked *b*, and thus with each
one of the series, it will be seen that the gradations from one to the
other are perfectly natural, affording only sufficient variations to
distinguish between them as paleographic symbols ; and the conclusion becomes obvious that the sacred tree, so celebrated in the reli
gions of antiquity, formed the original basis of all these characters.
This supposition will become more and more apparent as we proceed with these investigations.

SEC. 21. If the hieratic form of *Zi*, signifying "life, soul, spirit,"
etc., represented a tree, this must have been no other than the
"tree of life;" and as the traditions pertaining to this one object
evidently formed the staple element in all the religious conceptions
of this class, so the character *Zi* may be considered naturally the

centre of the group to be studied. The Scriptures afford but few notices of the tree of life, serving to convey a definite idea of it, but these few are very significant, and, taken in connection with other facts now known, it will be possible to deduce some conclusions of a nature quite important and reliable. The Revelator alludes to this subject in the following terms : —

" And he shewed me a pure river of water of life, clear as crystal, proceeding out of the throne of God and of the Lamb. In the middle of the street of it, and on either side of the river, was there the tree of life, which bare twelve manner of fruits, and yielded her fruit every month : and the leaves of the tree were for the healing of the nations." (Rev. xxii. 1, 2.)

In addition to the foregoing, and uniformly interpreted by exegetes as relating to the tree of life, is the language of Ezekiel : —

" And by the river upon the bank thereof, on this side and on that side, shall grow all trees for meat, whose leaf shall not fade, neither shall the fruit thereof be consumed (exhausted); it shall bring forth new fruit according to his months, because their waters they issued out of the sanctuary ; and the fruit thereof shall be for meat, and the leaf thereof for medicine." (Ezek. xlvii. 12.)

The data upon which both these passages substantially agree are, 1st. The tree is planted upon both banks of the sacred river, this having its source under the sanctuary or beneath the throne of God. 2d. It produces a fruit-harvest each month, there being twelve during the year. 3d. The fruits serve the purpose of food, and the leaves that of medicine. Professor Moses Stewart has some very judicious remarks upon the first text cited above, which I subjoin :

" The writer conceives here of the river as running through the whole city ; then of streets parallel to it on each side ; and then, on the banks of the river, between the water and the street, the whole stream is lined on each side with two rows of the *tree of life*. The phrase (ξυλον ζωῆς) is generic, and means something equivalent to our word *grove*. *Producing twelve fruit-harvests*, not (as our version) *twelve manner of fruits*. In order to afford an abundant supply for all the inhabitants, it bears twelve crops in a year instead of one." [1]

In short, the Revelator depicts here the highest ideal of an Oriental city, with its luxurious gardens, abundantly supplied with pure water and food-trees, like the sacred river and tree of life of the traditional paradise ; and we know now, not only that these rich gardens of the Eastern monarchs were termed " paradises," but that they were often expressly designed as imitations of the first

[1] *Commentary on the Apocalypse* in loc.

abode of humanity on earth. The question arises here, whether any of the sacred trees of antiquity, known to us as such, answers exactly, or nearly so, to the Biblical description of the " tree of life." I think the Orientalist will hardly hesitate in naming the palm-tree, especially the date-palm. The following, from the pen of M. F. Lenormant in reference to Chaldæa, bears directly upon the point before us : —

" It was the palm, the tree that furnished the major portion of food to the inhabitants, and from whose fruit a fermented and exhilarating beverage was derived, the tree to which they attributed, in a song mentioned by Strabo, as many blessings as there are days in the year ; — it was the palm, we say, which was regarded in this country as the sacred tree, the tree of life." [1]

In another treatise, the same author speaks of Arabia : " I have shown from the testimony of the monuments that the palm was the tree of life, the sacred tree *par excellence* in a portion of Chaldæa. It was the same in many localities of Arabia ; this was the tree to which they devoted most frequently their adoration." [2] But that which tends to exclude all doubt in reference to the point in question is the statement of Professors Roediger and Pott, in the critical paper cited below. They say : " A branch of the palm served as a symbolical designation of the year, in the Egyptian hieroglyphic writing, because the palm engenders each month, or twelve times during the year." [3] It is a characteristic of this celebrated tree that it throws out new blossoms every few weeks, so that it is not unusual to behold ripe fruit and new blossoms at one and the same time. I had hesitated to consider the statement of Professors Roediger and Pott as wholly reliable, having been made at a time when the facts were not so well known as at the present day. But all doubt is removed by an Assyrian cylinder, from which a cut is given in a work just published by Mr. George Smith, entitled " The Chaldæan Account of Genesis." The scene represented is the battle between Bel and the Dragon (p. 99), having a cosmical import, but at the same time a reference to the calendar. On either hand is shown a palm-tree, each having six branches of fruit, three on one side and three on the other side of the trunk, suspended from

---

[1] *Frag. de Bérose*, p. 330.   [2] *Lettres Assyriologiques*, ii. p. 104.
[3] *Kurdische Studien*, von E. Roediger u. A. F. Pott; *Zeitschrift für die Kunde des Morgenlandes*, von Chr. Lassen, B. vii. H. 1, Bonn, 1846, pp. 104, 105.

beneath the outspreading branches. Here, then, are the *twelve-fruit-harvests* literally represented to the eye. A much superior engraving from this cylinder was published many years since by M. F. Lajard, but the twelve bunches of fruit are not quite so fully represented.[1] That this cylinder had a cosmical import is shown from the facts developed by Mr. Smith, and its reference to the calendar is proved, not only by the twelve fruits of the palm, a recognized symbol of the year, but by the crescent exhibited in the field, with the three projections from the outer surface, denoting obviously the three phases of the moon. The demonstration, therefore, is complete ; the palm considered as a sacred tree answers precisely to the Biblical description of the tree of life. If it was not regarded by the sacred writers as the tree of life itself, there can be no question that it had been selected as a type of it.

SEC. 22. In connection with the palm, whose monthly harvests not only yielded an abundance of food, but from which also a fermented beverage was manufactured, it seems proper to introduce here a brief consideration of the cuneiform *Ar*, whose signification of " vineyard " and " palm-tree " has been already suggested. In the New Syllabaries (No. 125), a second equation of the Accadian *Ub*, " region " (Rep. 266), has *Ar* in the first column, and *Karmu* in the third, or Assyrian column. As for *Karmu*, it is assimilated by Dr. Delitzsch to the Hebrew *Kerem* (כֶּרֶם), " a park of noble plants, a garden," especially a " wine-garden, or vineyard." [2] The connection of *Karmu*, in the sense of " vineyard," with the Accadian *Ub*, " region, cardinal point," is perfectly natural, since the vineyards of antiquity were considered as a species of *templum*, being laid out with especial reference to the cardinal regions. But the primitive application of the word *Karmu*, Hebrew *Kerem*, was doubtless as a designation of the palm, particularly the date-palm, from the fruit of which, instead of from the grape, it is probable that the ancients first manufactured wine. Hr. Leo Reinisch has developed a class of facts tending to establish the declaration just made, from which he draws the conclusion as follows : " The most ancient artificial drink known to our primitive ancestors was thus the palm-wine ; but subsequently, when the manufacture of wine from the grape was discovered, the same name was applied to the

---

[1] Vid. *Culte de Venus*, plate iv. No. 12.

[2] *Assyrische Studien*, Heft i. p. 134.

new product."[1]   In fact, it results from the investigations of this
author, and from those of Professors Roediger and Pott, in the place
already cited, that the term *Karmu*, as designating the palm, thence
put for palm-wine and vineyard, and under the various modified
forms of *Kerem*, *Kurma*, *Karma*, *Khorma*, *Karm*, etc., etc., prevailed
from Armenia in the north to Middle Africa in the south, so that
the equation of the Accadian *Ar* to the Assyrian *Karmu* seems a satisfactory indication of the senses which I have attached to it.   It is
probable that it designated a watch-tower also, the ancient vineyards
being provided with such structures from which to guard the products from depredations.   As before stated, *Ar* is a composite sign,
consisting of *Si*, " eye," and *Ri*, "column."   Thus, just as *Si + e*,
eye + temple or tower, means " astronomical observation, observa-
tory " (Rep. 363) ; so *Si-ri*, forming the character *Ar*, might well
be put for " the watch-tower of a vineyard," although it appears primarily to have designated the tree from which the products of the
garden were derived.   It is worthy of note in this connection that
the Mosaic account of Noah's " vineyard " has the Hebrew *Kerem*
in the original, leading to the conjecture, at least, that it was not a
grape-garden, but a palm-garden, which was planted by Noah after
leaving the Ark.

We have seen that the tree of life was regarded preëminently as
a food-tree, and it was for this reason that other fruit-bearing trees,
especially the oak, certain species of which afforded an esculent product, came to be considered sacred, and were held in great veneration.   It is a remarkable fact that the term most frequently denoting the oak, as it appears under various forms in the Aryan languages, was derived from the same original theme, which produced
another class of words relating to the process of eating : thus indicating a very early association of the conceptions of food and of
eating with the oak, considered as a food-tree.   Probably it was
owing to its usefulness in this respect that the oak was venerated as
a sacred tree by so many and so widely separated branches of the
Aryan race.   But there are other important ideas connected with
the primitive Aryan root just referred to, quite essential to our
present researches ; and it will be desirable to group them together
with those already noticed, in the natural order of their development.   We have, then, 1st. The root *Bhag*, " to allot, to impart,

------

[1] *Einheit. Ursprung d. Sprachen d. Alt. Welt.*, i. p. 342, note.

to apportion, especially food to be eaten." 2d. The substantive form, masculine, *Bhaga*, or *Baga*, literally "the apportioner, he who allots, a portion;" then "bread-lord, lord of bread;" and so "Lord" in general, "divinity;" employed at an early period as name of the Sun-god. 3d. From the idea of abundance of food is derived that of "luck, good fortune, well-being," and thence the "God of fortune, Lord of destiny." 4th. From the notion of food again proceeds the feminine form *Bhâga* or *Bâga*, denoting the oak-tree, considered especially as a food-tree, corresponding to the Latin *Quercus*, Greek *Phēgos*, both derived originally from the root *Bhag*. 5th. Those derived forms signifying "portion, portion of food, meal-time," also "to enjoy, to eat," etc.[1]

SEC. 23. In the cuneiform texts appertaining to the Achæmenian period, *Baga* occurs frequently as the equivalent of the Accadian *An*, Babylonian *Ilu*, that is to say, as highest divinity like the Persian Ormuzd, but at the same time as God in general. When written phonetically, we usually find it with the characters *Ba-ga;* but I have noticed the forms *Bak* and *Bak-da* (*ba-ak* and *ba-ak-da*), in composition with other names. In all these cases, it is evident that *Baga* represents a male divinity, although the term is often applied to female divinities in Aryan mythology. As a female divinity, or goddess, *Bâga* must have been known to the Chaldæans from the earliest period; for we have the phrase *An Bagas*, the "Goddess Bagas," in the name of the monarch who heads the list of the ancient Chaldæan kings. That is, the name which was formerly read *Urukh* is probably, according to M. Lenormant, to be more correctly read *Lik-Bagas*, or *Lik An Bagas*. In an inscription of *Dun-gi*, son of this ancient monarch, as published in the last volume of the "Cuneiform Inscriptions," the father's name is written phonetically *Ba-ga-kit*.[2] The element *kit* is obviously here a simple adjunct, the same as in the name *Bil-kit*, "Lord of the Abyss," according to which *Ba-ga-kit* would mean "Baga of the Abyss." We have likewise the phrases *Ba-ga-ra* and *Ba-ga-ra-kit*, going to show that *Baga*, or *Bagas*, is really the true reading, and that *kit* is merely an adjunct.[3] Thus, if we adopt the reading *Ba-ga*, justified by the

---

[1] Fick, *Woerterb. d. Indog. Sprachen*, i. pp. 154, 687. Cf. Curtius, *Grundzüge*, p. 298, etc.

[2] 4th Rawl. Pl. 35, No. 2. Cf. Lenormant, *Etudes Accad.*, t. i. pt. 3d, p. 76.

[3] 3d Rawl. Pl. 67, No. 2, lines 48, 49; and 4th Rawl. Pl. 5, col. 2, lines 42, 46.

facts just noticed, its identity with the Aryan *Baga* seems quite apparent. The final *s* results doubtless from the Accadian value of *khas* (Rep. 2), the six-rayed star inclosed in a square, constituting the monogram for *Bagas ;* but it is proved non-essential by the phonetics *Ba-ga-kit* of the inscription of *Dungi*, already cited. The six-rayed star, having the value of *khas*, "to strike violently, to cut," etc., answers precisely in meaning and form to the "wheel," ordinary mythological symbol of the Goddess of fortune, or of destiny; another indication of the connection of the Accadian *Ba-ga* with the Persian *Baga*, "God of fortune." That *An Bagas* is a female. divinity, a goddess, appears from the fact that the same characters occur in several texts, as the title of such a person-age.[1]

But we find *Baga*, evidently as a title of divinity, in a connection still more unexpected than that with which we were last occupied. According to Dr. D. Chwolsohn, the Haranite Sabæans, living in the midst of a Semitic population, and evidently Semites them-selves, celebrated the mysteries of *Shemal*, or Samael, in an under-ground room like a cave, which was termed by them the "House of Bogdariten;" and the initiates were called the "Sons of Bogdari-ten." This name *Bogdariten*, as the author states, proceeds from *Bogdariun*, plural of *Bog-dar*, in which the Slavic *Bog*, one with the Persian *Baga*, appears at a glance. Among the ceremonies of these mystics was that in which they partook of bread prepared in the manner of the shepherds; another in which they partook of food and wine; and still another custom was the preparation of cakes consisting of meal, kneaded with the boiled flesh of a male child offered in sacrifice ; and these cakes served as a mystical bread during the entire year. The sanctuary itself being a cave, and the custom of preparing a kind of bread after the manner of the shep-herds, lead to the conclusion, as Professor Chwolsohn thinks, that these mysteries pertained to a very early epoch, and were probably founded on the worship of the cave-dwellers themselves.[2] Consider-

---

[1] 2d Rawl. Pl. 54, 3; Obs. lines 17, 18 ; and 3d Rawl. Pl. 59, No. 1; Obs. l. 25. *An Bagas* is here put for mother of *En-ki-ga-kit*, or "Lord of the region of the Abyss." *Enkigakit* is explained in other places as the God Hea. Mr. George Smith has a different reading for the name of this goddess, which is probably cor-rect. Yet for the Accadian I prefer the one here adopted, as it appears to be well supported.

[2] *Ssabier u. d. Ssabismus,* ii. Excursus to chap. ix. pp. 319–364.

ing the connection of this sanctuary with the Persian *Baga*, " Lord of bread," together with the mystical food prepared in the manner stated, it is probable that we should see here a reference also to the oak, denoted by the term *Baga*, considered as a sacred tree, especially a food-tree. The God *Shemal*, Hebrew *Semol*, who constituted the central point of these mysteries, was the great divinity of the north, the word *Semol* being an ancient Semitic term put for the north, and North Star. Seven great gods were associated with Shemal, probably the seven planets; also seven genii, supposed to be the seven stars of the constellation of the northern Dipper. In relation to these seven stars, Shemal as polar-star constituted the *Eighth*, calling to mind the Eighth Cabiriac divinity; and the highpriest, or hierophant, who presided over the initiations, was called *Kabir*, that is to say, one of the Cabiri.[1]

Like the Aryan *Bâga*, the Semitic *Ilu*, Hebrew *El*, is closely connected with the oak-tree. The term *El* (אל) signifies the " Strong One," thence put for the Almighty. According to Dr. Fürst (Heb. Lex. *sub voc.*), *El* is equivalent to *a-yil* (איל), also the " Strong One," but otherwise denoting the " tall, strong tree," particularly the oak. We read that " Abraham planted a grove in Beer-sheba, and called there on the name of the Lord (*Yahveh*), the everlasting God " (*El olam;* Gen. xxi. 33). The original for " grove " in this passage is *Eshel* (אשל), which is singular, though some exegetes maintain that it has a generic sense, and may properly denote a " grove " instead of a single tree. It occurs in two passages only besides this (1 Sam. xxii. 6; and xxxi. 13), and in the last the idea of plurality is absolutely excluded. In fact, it is rendered " tree," the singular, in both instances. Gesenius, Rosenmuller, and others regard the *Eshel* as a *Tamarisk;* but Professor Bush observes that " Among the ancient versions some render it by *oak* or *oak-grove*, and others, like the English, simply *a grove* " (notes *in loc.*). But the *tamarisk* is a species of *oak*, so that the reference would be much the same in either case. It is far more probable, I think, that Abraham planted a single tree, an oak or tamarisk, calling there upon the *name Yahveh*, identified with *El* as the everlasting God. This was in accordance with prevailing custom, which, in the time of Abraham, had not been perverted to idolatrous purposes.

SEC. 24. We proceed now to the especial consideration of the

---

[1] Ibid.

character *Mu*, having the sense of " to give, memorial, name, year."
As already stated, the Accadian *Mu* is the ordinary Assyrian monogram for *Sanat*, " year."   Thus, if the character itself represents
a tree, the sacred tree, and especially the palm, is obviously intended.
We have seen that the palm, from the fact of its engendering each
month, producing twelve fruit-harvests during the year, was chosen
as a symbol, and even paleographical symbol, of the year.   While
the Egyptians selected a simple branch of the palm for this purpose,
the Accadians or Cushites represented the tree entire, sufficiently
contracted, however, for convenience in writing.   Do we not have
here a direct indication, not only that this paleographic symbol was
in use, as such, before the separation of the Hamite race, but that
the Egyptian system of writing was at first symbolical?   Be this as
it may, the proof here afforded that the hieratic character *Mu* was
intended for the sacred tree must be considered, I think, quite satisfactory.

The sign *Mu* occurs also in the texts as the ordinary Assyrian
monogram for *Sam* or *Sum*, Hebrew *Shem* (שׁם), " name ; " and it
may be interpreted as *memorial name*, since this character attaches
to itself both these significations.   Another fact not yet noticed,
but quite important in the present connection, is that *Mu* is usually
employed in the Accadian as a personal pronoun, first person, singular, denoting thus the *Person*, the *Ego*, or the *I am*, humanly
speaking ; and it shows that the two conceptions of " name " and
" person " were intimately associated.   In view of these considerations, and of the fact that the hieratic form of this paleographic sign
evidently represents a *bush* or *tree*, it is probable we should see here
some relation to the remarkable circumstance, which is thus recorded
in Scripture : —

" Now Moses kept the flock of Jethro his father in law, the priest of Midian :
and he led the flock to the backside of the desert, and came to the mountain of
God, even to Horeb.   And the angel of the Lord appeared unto him in a flame
of fire out of the midst of a bush : and he looked, and, behold, the bush burned
with fire, and the bush was not consumed."   " And God said unto Moses, *I am
that I am :* and he said, Thus shalt thou say unto the children of Israel, *I am* hath
sent me unto you.   And God said moreover unto Moses, Thus shalt thou say unto
the children of Israel, The Lord God of your fathers, the God of Abraham, the
God of Isaac, and the God of Jacob, hath sent me unto you : this is my *name*
for ever, and this is my *memorial* unto all generations."   (Ex. iii. 1, 2; and 14,
15.)

The definite association of the name of God, the Divine Word,
with the sacred tree, particularly the *Fire-tree*, must have been very

ancient and widely prevalent. In the primitive epochs, it was customary to generate the fire, whether of the hearth or of the altar, by means of the friction of two pieces of wood. As the hearth was the primitive altar, and its flame the natural symbol of the hearth-divinity, whose character doubtless constituted the basis of all the fire-gods of antiquity, this method of generating fire naturally attached to itself a peculiar significance and sanctity; and it was in the midst of the flames thus generated that the Divine Being was believed to manifest himself to man, and to communicate with the world. The two pieces of wood employed for this purpose were called *Arani* by the Aryans. M. Léon Carré cites two texts from the Gâthas, constituting the most ancient portions of the Zend-Avesta, in which *Sraosha* or *Serosh*, the personification of the Word of Ormuzd, and through whom the Good Spirit communicated to men, is directly referred to in connection with the *Arani*: —

"Thou didst command me not to create, without having first received a revelation; not till Çraosha should come with majestic truth; — he who deigns to reveal the wisdom by means of the two pieces of wood destined to generate the fire." "May the instructed man, who thinks only of truth and of the two lives, O Ahura! may his language be free, and may they hear his words of truth, promulgated by thy brilliant and beneficent fire, produced by the friction of two pieces of wood."[1]

Rev. Mr. Rawlinson has the following in relation to the general attributes and office of Serosh: —

"Armaiti, however, the genius of the earth, and Sraosha or Serosh, an angel, are very clearly and distinctly personified. Sraosha is Ormazd's messenger. He delivers revelations, shows men the paths of happiness, and brings them the blessings which Ormazd has assigned to their share. Another of his functions is to protect the true faith. He is called in a very special sense, 'the friend of Ormazd,' and is employed by Ormazd not only to distribute his gifts, but also to conduct to him the souls of the faithful, when this life is over, and they enter on the celestial scene."[2]

According to the views of Professor Theod. Benfey, *Baga* was conceived also as the Divine Word, and nearly allied to Serosh. Re-

---

[1] *L'Ancien Orient*. ii. p. 319, and note. The author quotes from Langlois' version. The second-hand translation given above is probably faulty, but that which relates to the connection of *Serosh* with the *Arani* is doubtless sufficiently accurate.

[2] *Five Monarchies*, ii. pp. 326, 327; cf. pp. 336–339.

ferring to the Greek names *Bagarazos* and *Bagaios*, this writer observes: "They appear to me derived from the frequently occurring term *Baga*, 'the Sacred Word.'" Upon the name *Bigtan*, a loan word from the Persian occurring in the Bible, the same author says: "This appears to me to be *Baga-tanu*, 'the Word having a Body;' in a similar signification we have *tanumanthra* applied to the Ized Çraosha."[1]

SEC. 25. In connection with the fire-tree, it will be most convenient to introduce here some remarks relative to the character *Gi*, signifying "flame, foundation, to found, to transport, to deport, to restore." In the foundation of a new state, nothing was esteemed more important by the ancients than the transportation of the sacred fire from the altar or hearth of the parent state to that of the new colony. It seems probable that there is some reference to this custom in the notions of "flame, transportation, foundation," etc., appertaining to the sign *Gi*. Mr. Norris states the fact that *Gi*, preceded by the determinative of divinity, is believed by Sir H. Rawlinson to be the Scythic, or Accadian fire-god; and that it is sometimes substituted for the Accadian characters *Ne* and *Iz-bar*, ordinarily put for the fire-god.[2] The sign *Iz*, in the name *Iz-bar*, has the meaning of "wood, tree" (Rep. 233); while *Bar*, among others, has the sense of "to burn" (Rep. 69). All this goes to show that the primary notion of flame, heat, fire, and thence of the fire-god, was intimately connected with the tree, from which fuel was derived. The term *Iz-bar*, "tree, to burn," as name of the fire-god, shows for itself a primitive association with the hearth and the altar, originally one and the same thing; but the sun, as primary source of all heat, was doubtless at a very early period considered as a fire-divinity. A well-known title of this god was *Bar-sam*, that is, "*Bar* by *name*," or whose name is *Bar*, "to burn," forcibly calling to mind the *Name Yahveh*, proclaimed from the burning bush of Mount Horeb. We see, then, from the data here established, not only the connection of *Gi* with the sacred tree, especially considered as the fire-tree, but the definite association of the Divine Name with it. Connected with the facts having a similar bearing, presented in the previous section, it will be difficult not to admit the habitual association of the Divine Word or Name,

[1] *Monatsnamen*, pp. 198, 199; cf. p. 67.
[2] *Assyr. Dic.*, i. p. 342; Fürst, Heb. Lex. art. זָבַר

on the part of the ancients, with the sacred tree, and especially the fire-tree. Not only this, but it is evident that all these notions must be referred primitively to the hearth, the family altar, where the first flames were kindled, the first bread consecrated, and the first offerings presented to God ; where, in fact, the Sacred Name was first invoked by mortal man. But upon the connection of the Divine Name with the sacred tree in general, and especially upon its reference to the external manifestations of the Deity, considered as his Other, his Face, that is to say, his Feminine Form, I wish to present some extracts from the pen of M. De Vogüe, which, although quite lengthy, are too important not to appear here : —

" Like the supreme divinity of the Egyptians, *Baal* was not absolutely distinct from created nature, at least in the epochs of history accessible to our researches; as early as we are able to penetrate into the annals of the Canaanitish populations, we find the worship of Baal associated with certain trees and stones, considered as abodes of the Divinity (ביתאל), or *Beth-el.* In other words, they adored in God the hidden spring of nature, the principle of life that animates the material."

" The first formula that we recognize is that so often repeated in the inscriptions of Carthage, in which the goddess Tanit is called *Phan-Baal* (פנבעל). This expression signifies properly, *facies, persona Baalis,* and M. de Sauley has very happily rendered the first by 'Manifestation of Baal.' M. Zotenberg has demonstrated that it contains also the idea of conjugal union. Tanit, then, does not differ essentially from Baal; this is, so to speak, only the subjective form of the Primal Deity ; a second Divine Person, sufficiently distinct to be associated conjugally, yet no other than the Deity himself in his external manifestation."

" The second formula is more explicit; Astarte, the goddess of Sidon, associated with Baal of Sidon in the inscription of Eshmunazar, is qualified as *Sam-Baal* (שםבעל), *Nomen Baalis.* The abstraction is more pure here than in the preceding example. At Carthage the goddess was a divine person ; but here she is only, so to speak, a theological locution. This is Baal less under another aspect than under another name, although the personality is enough distinct for them to be designated as male and female divinities ; the author of the inscription employs the plural; he calls them the *gods of the Sidonians.* Astarte, then, is the personification of the Divine Name ; of that *Name* to which all the religions of antiquity have attributed a mysterious power. This is as if *Sam-Yahveh* (שםיהוה) had taken a body. Already in the Bible this expression is found employed in an active acceptation, that relates more to the *numen* (power) than the *nomen* (name); it is applied to the external manifestations of divine power. It is by virtue of the *Sam* (שם) that the angel works, charged with communications to men; it is the *Sam* also who resides in the temple of Jerusalem. But while the Jews preserved the abstract value of this expression, the Phœnicians had given it a distinct existence; they made of it a special divinity by an operation similar to that according to which they deified the *Face* of their God. One will not deny, however, that an analogy exists between the two phrases *Sam-Baal* and *Phan-Baal*. Gesenius had traced the connection of *Sam-Yahveh* with *Phan-Yahveh*, at an epoch when the Phœnician inscriptions were wholly unknown or imperfectly explained." [1]

The substantial accuracy and great value of the views expressed in the foregoing passages have been adverted to frequently by European scholars. We see here how the sacred name of divinity gave rise to the conception of a distinct goddess, who was adored often under the form of sacred trees or stones, with which this divine name was usually in such cases associated. A striking example of this is the term *Semiramis*, a title of the Assyrian Venus, and which Drs. Movers and Gesenius substantially agree in explaining as " the most exalted name." [2] We have seen that Abraham planted a tree in Beer-sheba, calling there upon the *name Yahveh* (שם־יהוה), identified with *El* as the everlasting God. The Patriarch Jacob likewise, in memory of his vision of the mystical ladder, set up a stone, and poured oil upon the top of it, calling the name of that place *Beth-el* (בית־אל), " the House of El " (Gen. xxvii. 18, 19). We see that these notions and customs were universal at this early period, and it is evident that it was only much later that they became associated with polytheistic and idolatrous conceptions. The Hebrews, as correctly observed by M. De Vogüe, preserved the abstract value of the terms *Sam* and *Phan*, applied to the Deity, denoting thereby the external manifestations of divine power; while the other Semitic races, for the most part, gave to these attributes a distinct existence under the form of a goddess conjugally related to the primal divinity. As M. Zotenberg has shown, these

---

[1] *Mélanges d'Arch. Orient.*, pp. 52-55. Cf. *Syrie Centrale*, p. 53 note, for a substantial confirmation of the author's views.

[2] See Dr. K Schlottmann, *Die Inschrift Eschmunazars*, etc., pp. 75-79 and 142-146; Movers, *Phœnizier*, i. p. 634; Gesenius, *Heb. Lex.* art. שם־יהוה.

expressions primarily involve the idea of conjugal union, and it was but natural that they should give rise, sooner or later, to the conception of a female divinity more or less distinct from the primal Deity. With the Hebrews, this dualism applied more to the relations between the divine and human, God and the Church.

Sec. 26. Of the group constituting the subject of our present study, the only remaining character is that of *Ri*, "to heap up, column, to rise, appearance of a star." It has been already observed that *Ri* is an ordinary monogram for the Goddess *Is-tar*, the Babylonian Venus; and we find this character thus employed in its hieratic form in the very earliest texts now known. The conical stone (for which was often substituted a heap of stones, implied in the sense of "to heap up," in other words, the *Betyle*, the *Beth-el*), appertained to the Goddess *Ri*, or *Is-tar*, as is now generally understood by Oriental scholars. But the *Asherah* (אֲשֵׁרָה), from the root *Ashar* (אָשַׁר), "to be blessed, to be happy," to which frequent reference is made in the Hebrew Scriptures, was also dedicated to Venus, particularly as the goddess of fortune, proceeding from the idea of happiness, of well-being, expressed in the root *Ashar*. The Asherah, however, was usually made of wood, consisting of a tree or trunk of a tree artificially fashioned, and thus briefly described by Dr. Gesenius: "A statue, image, of Asherah, made of wood; a wooden pillar of great size, which on account of its height was fixed or planted in the ground." "Of the ancient versions some render this word *Astarte*, others *a wooden pillar*, others a *tree*;— by which they seem to have understood a sacred tree" (Heb. Lex. Art. אֲשֵׁרָה). In a list of stars contained in an Assyrian tablet, Venus is styled the *Star Izlie*, "the star of fortune," as interpreted by M. Lenormant and Dr. Oppert. The name *Iz-li-e*, since the first syllable means "a tree," shows that the planet Venus derives this title from some kind of a sacred tree; and from M. Lenormant's transliteration of the term (עֵכֶל), I am persuaded that it should be identified with the Hebrew Eshel (אֶשֶׁל), "the *oak* or *tamarisk*," planted by Abraham at Beer-sheba.[1] This assimilation of *Izlie* to *Eshel*, as a species of oak, calls to mind again the term *Bâga*, under its feminine form, frequently applied to the oak; and thus also the goddess *Bâga*, as mistress of fortune, the same as Is-tar or Venus.

[1] *Lettres Assyriologiques*, ii. p. 162. Cf. 3d Rawl. Pl. 57, 6, l. 51. Mr. Norris cites various texts in which *Izli* occurs as a tree (*Assyr. Dic.*, i. p. 346).

Another circle of ideas deserves a brief notice in this connection.
We have seen that *Baga*, as title of the lord of bread, thence as
god of fortune, ruler of destiny, proceeds from the root *Bhag*, " to
allot, to award, to apportion," especially food or bread to be eaten.
Along with *Gad* (לד) lord of fortune, supposed to be the planet
Jupiter, the majority of the Semites of Western Asia adored a
female divinity, *Meni* (מְנִי), mistress of fortune, usually assimilated
to the planet Venus.   Now *Meni* is derived from *Mā-nāh* (מָנָה), " to
divide out, to allot," involving thus the same primary conceptions
as the Aryan *Bhag*, from which *Baga* is derived.   There is also the
obsolete root *Mānan* (מָנַן), having the same sense of " to allot, to
divide out ; " and from this comes the word *Manna* (Heb. מָן), the
name of the food upon which the Israelites subsisted in the wilderness.   It is remarkable that the *Manna* of Arabia and other countries of Western Asia is derived from the tamarisk, a species of the
oak-tree.   We have here a complete, and in every particular an exact
parallelism, between the Aryan and Semitic conceptions, proceeding
primarily from roots signifying to allot, to apportion, thence applied
to food derived from the oak, and finally to the idea of god or goddess of fortune.   The manna is a kind of gum, having a sweet taste
like honey, that exudes from the leaf or bark of the tree, on being
punctured by a species of insect.   The Arabians call this the gift
or food of heaven.   Many exegetes still maintain that the manna
upon which the Israelites subsisted was different, and was miracu-
lously provided.   But the fact that a food in all respects similar,
and having the same name, is produced even to-day in the same
region of country tends strongly in favor of the naturalistic view.

No additional facts are required to establish the point that the
Accadian *Ri* represented originally a form of the sacred tree, and
that it appertained definitely to the Asiatic Venus.   Nor does it
seem necessary to adduce further proof by way of substantiating
our general theory, not only of the highly symbolical character of
the entire group of signs forming the subject of these studies, but
of their direct primary reference to the sacred tree.   Nevertheless,
I desire to cite one more illustration from the art monuments, that
seems to embrace in one all the conceptions which have been developed thus far, relating to the subject before us.   I refer to the eagle-
headed man from the Nimrod sculpture, represented in a plate
opposite page 102 of Mr. Smith's " Chaldæan Account of Genesis,"

already cited in these researches. These man-eagles, so often appearing upon the Assyrian monuments, are the especial guardians
of the sacred tree, and correspond remarkably with the *Garuda* of
the Hindu tradition. Upon the sacred basket held in the left hand
of the figure here alluded to, the mystical tree appears under a form
more simple than is usual to find it; and it admits of a ready explanation of the symbolism attached to it. We have here a plain
trunk planted uprightly in the ground, like the Asherah described
to us by Dr. Gesenius. From this are seen twelve arms branching
off at right angles to it, six upon one side and six upon the other.
It is plain that these represent the calendar, the two halves of the
year, six months in each. This explains to us fully the Accadian
Mu, Assyrian *Sanat*, "year." The rays of light proceeding from
the top of the column, intended for seven, or the seven planets, as
will be seen by a comparison with the other cuts of the sacred tree
given by Mr. Smith, not only confirm the notion of an astronomical
reference, but give us the additional idea of flame, of the fire-tree,
answering to the cuneiform sign *Gi*. Finally, the simple fact that
this tree is guarded by the man-eagle proves its reference to the tree
of life, to which corresponds the character *Zi*, "to live, life, soul,"
etc. An indirect relation to the sign *Ar* might be traced through
its signification of the palm, shown to have been a recognized symbol of the calendar.

SEC. 27. During the investigations of the present chapter, the
materials have seemed constantly to multiply upon our hands; and
though much of importance has been excluded, these pages will
doubtless appear to the reader too much crowded with individual
facts and details; and to avoid the confused impression that might
otherwise be left upon the mind, it will be necessary to recapitulate
briefly the chief points upon which we have been occupied, and
which we have labored to establish. Undoubtedly the character of
the Jehovah of the Old Testament presents individual attributes
that closely resemble the fire-gods of antiquity. This fact has been
frequently noticed by Orientalists, and has been often made the
subject of critical research. Dr. Movers has pointed out the analogy existing between Jehovah considered as a destroying power
and the Moloch of the Canaanitish worship. M. Obry has traced
the resemblances between the national God of the Jews and Agni
of the Hindu religion. The data that have been presented in the

present chapter, relating to this point, all tend to the conclusion
that these striking analogies are chiefly due to the fact that the
primitive object of the worship of mankind was unquestionably the
divinity of the hearth. For the populations of ancient Italy and
Greece, it would be easy to verify this last statement. It is well
known that Agni of the Hindus was primitively a hearth-god.
We have shown that the Accadian *Ni* was such, and that this personage was one with Jehovah of the Hebrew Scriptures. I am confident that the same general statement could be shown to apply to
all the cultured nations of antiquity. Although conceptions more
or less distinct, yet similar in character, were applied to the sun as
the ultimate source of all heat, I believe that the fire-god was first
worshiped at the hearth-stones of primitive humanity.

But little doubt can be entertained, I think, that the conceptions
centring in the Aryan *Baga* or *Bâja*, lord or mistress of bread,
and thence of fortune and destiny, to which those connected with
the Semitic *Mann* or *Meni* present in all respects exact correspondences, — but little doubt can be entertained, I say, that all these
ideas, and the customs originating from them, appertained primitively to the hearth and to the divinity of the hearth ; to those
early stages of society when the food-tree furnished the staple products for the support of human life. Applied at first, perhaps, to
the master and mistress of the house, as apportioners of food to the
inmates, they were naturally transferred to the divinity presiding
over the well-being and fortunes of the family, who then became in
a mystical and religious sense the apportioner of food, the lord of
bread. As it was here that the first covenants of salt were ratified
between God and man, so here the first allotment of bread was
shared between them, in token of perpetual friendship. The judicious remarks of Dr. Cudworth which have been cited, relating to
the feast upon the sacrifice considered as a federal rite between
God and man, receive here a striking confirmation, the origin of
such customs being now clearly demonstrated. Their significance
as fundamental conceptions of the Mosaic religion has been already
adverted to, and will not fail to be appreciated by the reader.

As the altar and the hearth were originally one, and as the flames
had occasionally to be rekindled upon the hearth, the *Arani*, or two
pieces of wood destined to generate the fire, became thus a necessity
of every household, and the veneration paid to the fire so mysteri-

ously produced, so appropriate a symbol of the Deity himself, was naturally transferred in a measure to the fire-tree from which the *Arani* had been derived. Both the tree that furnished the fuel for the hearth and the food for the table would attach to themselves a sacred character. They afforded the necessary sustenance of life, as well as the means of life's chief conveniences and comforts. As the generous tree from whose branches they plucked their daily food seemed so essential to the continuance of mortal existence, the transference of their ideas to the tree of life *par excellence*, from whose products the life beyond the tomb was sustained, was but natural to those with whom the present and future existence appeared to be merely a continuance, one of the other, although under changed and improved circumstances.

The veneration of certain trees, considered as abodes of divinity, was at the first no such vulgar superstition as has been supposed by some modern writers. M. De Vogüe has very correctly observed that, with the men of high antiquity, God was not conceived as wholly distinct from created nature. On the contrary, nature was considered as the Face, the Name, the external manifestation of divinity. As something purely universal and abstract, far removed from the work of his own wisdom and power, the Deity was almost wholly unknown in the first ages of humanity. The house which He had built was also his own habitation. The Divine Mind was everywhere present in the outward world, and everything that had life was a symbol of God. The process was therefore wholly logical and normal by which a natural object, distinguished from the mass by its remarkable characteristics, or by some consecrating ceremonies, was conceived as a dwelling-place of God, who might be thus approached by his rational creature man, and his Sacred Name invoked. Things *in general* were as chaos, as nonentity, to the apperceptions of the first men. Everything appeared to them in the *concrete*. Nor was this by any means an incorrect view. Modern physical science demonstrates to-day the great truth, announced many years since by Mr. J. B. Stallo as an axiom of nature-philosophy : " Life only then appears when the whole energizes in a part," and the immortal Goethe had long before said : " If you would appreciate the whole, you must recognize the whole in the small-est." [1] God is universal, infinite, but not as an abstraction. He

---

[1] *Philosophy of Nature*, pp. 14 and 15, note.

alone is really universal who is present in his fullness in each and
every particular. If the Infinite dwells not in this tree and in this
stone, He is nowhere. If it is only beneath temple domes that the
Deity takes up his abode, primitive humanity was without a God,
for then there were no temples.

SEC. 28. But that which many modern writers have failed most
essentially to appreciate, in the religious conceptions and customs of
high antiquity, regards the one great idea fundamental in all the
facts which have been grouped together in the present chapter,
and which has been already insisted upon, namely, that God
dwells, inhabits, the same as man. It is mind alone that dwells,
and it is matter alone that constitutes the dwelling. To conceive
of mind without a dwelling is to regard it as an abstraction instead
of a reality. It is impossible for the human soul to enter into intelligent, personal relation with the storm-cloud, the whirlwind, or
with empty space. If God would communicate with man, it must
be through the concrete, through the particular; the Deity must, to
this end, subject himself to the limitations of time and space, must
unite the universal in the individual as one; — in a word, God must
dwell somewhere, the same as man. This eternal law of all mind,
which philosophy, science, and speculative theology have now well-
nigh forgotten, is that alone which gives meaning to the phrase,
" House of God." Whether it be, therefore, the burning bush of
Horeb, the oak of Beer-sheba, or the unhewn stone of Bethel;
whether it be the temple on Mount Sion, the church of St. Peter,
or the humble chapel by the wayside; if the Infinite Mind be not
conceived as dwelling there, neither does the finite mind enter into
intelligent, personal relation with him. The same principle applies with nearly equal force in reference to the sacred name of
God. With the ancients, if the Deity was not known by his
personal name, He was as a stranger to them. To invoke the
Divine One by merely generic titles was to them as if we should
say Thou Man! or Thou Woman! to a near and dear friend. As
already observed, to those of the earlier ages, things *in general*
were as chaos, as nonentity. That which manifested itself in the
concrete was to them the truly existent: and it was only under a
personal name that they knew how to commune with God. Practically, according to their philosophy, to conceive the Deity as merely
the Universal, the Infinite, and under terms having a like general

import, was to banish Him from the human soul, from the sanctuary, and even from creation itself.

It is certain that the divinity was not thus conceived in the primitive worship of mankind. The Deity is not too great to be good, too powerful to aid the weak, too infinite to exclude him from the humble hearth where tried souls wrestle with adversity. We obtain a glimpse of the hard lot and sadness, as well as the difficulties, attending the ordinary life of man in the first ages of the world, and while as yet the earth was unsubdued, in the language of Lamech at the birth of Noah: "And he called his name Noah, saying, This same shall comfort us concerning our work and toil of our hands, because of the ground which the Lord hath cursed" (Gen. v. 29). A comfortable home, a cheerful blaze upon the hearth, a table supplied with food for the support of life, and above all the gift of sons and daughters to love, were the chief temporal blessings that Providence could bestow upon those with whom the terrible calamity that had befallen the first human pair was still fresh in mind. But the unsettled state of society, the increasing wickedness of mankind, with even the brute creation and rude elements of nature to contend with, rendered everything precarious and uncertain; and it was then, if ever, that the rational creatures of God needed his protection, and his personal, immediate presence with them. The exiles from the peaceful bowers of Eden had not been deserted by the Divine One in whose mercies they still trusted. It was now their loss and not his anger that He remembered. He had created them that He might love them, and in the midst of their misfortunes He would mingle the tokens of his kindness with the bitter fruits of their transgression. It was thus that He became a dweller in human habitations, and was invoked as the God of the hearth, as the Lord of bread. The flames kindled there, generated from the sacred fire-tree, and the table which was there spread with the products plucked from the sacred food-tree, were the most significant symbols of God's friendship for man, and of man's dependence upon the bountiful Provider. Humble indeed were these first tabernacles in which the Infinite and Finite mind covenanted together, and simple was the repast in which the Divine and Human ratified the pledges of mutual fidelity. But immortal destinies were being moulded there, and the Lord of destinies was present to superintend the process. The blaze kindled upon those

hearths had been brought, not from Troy, but from the sun; and
its forked tongues were the tongues of prophecy, proclaiming the
results of the divine economy on earth.

But time's noiseless shuttles flew rapidly to and fro; generations,
like clouds, flitted across the mirror of life, and they who had consecrated the ancient hearths with their tears became the progenitors
of powerful tribes, the ancestral heads of flourishing states. The
God of the fathers has now become the national divinity, and He
whose mess was once the humble product of the food-tree receives
the united offerings of a prosperous commonwealth, bound together
by the ties of a common lineage. The traditions, the ideas and
customs, associated with the primitive worship, may be traced in
the solemn rites and imposing ritual constituting the national cultus.
Thus it was that He who had been invoked at the hearth-stones
of the Hebrew patriarchs was worshiped on Mount Sion by the
assembled tribes of Israel, whose destinies He had wrought out amid
so many fiery ordeals.

But the divine name which had been proclaimed from the burning bush of Horeb was yet to take a body; and the sacred tree,
whose fruits were "for meat, and the leaves thereof for medicine,"
was yet to be transplanted, not upon Mount Sion, but upon Cal-
vary. — "I am that bread of life" (John vi. 48). This, then, is the
Lord of Bread; of that Bread gathered from a Food-tree whose
fruits can never be exhausted. This, too, is now the Divinity of
the Hearth; He who dwells to-day in the habitations of men.

CHAPTER III.

THE CABIRI.

SEC. 29. In the legend upon the signet cylinder of *Lik-Bagas* (for we prefer this reading to that of *Urukh*) who heads the list of ancient Chaldæan kings, so far as known, this monarch assumes the title of *Pa-te-shi*, an expression often occurring in the cuneiform texts, having the sense of "master, pontiff, sovereign-pontiff, or priest-king."[1] Lik-Bagas was literally a priest-king, like Melchizedek, who was king of Salem, and at the same time a priest of *El-elyon*, or the "Most High God" (Gen. xiv. 18). Abraham must have recognized at once the exalted character of Melchizedek, for he was born in "Ur of the Chaldees," which was the chief capital of Lik-Bagas, and the order of priest-kings was well known at Babylon. At the earliest period of which the inscriptions afford us any knowledge, the country of Assyria also was governed by sovereign-pontiffs, or *Pa-te-shi*, though they seem to have been tributary to Chaldæa or Babylon. The term *Pa-te-shi*, for which Assyriologues often substitute the reading of *Pa-te-si*, must be regarded as a technical expression, designating a class of personages universally recognized, and in whom the kingly and sacerdotal characters were united. Coming down to the era of Nebuchadnezzar, we find that this monarch applies to the god Marduk, the patron divinity of Babylon, the phrase *Pa-te-shi tsi-ri;* and Nebuchadnezzar himself frequently assumes the same title, from which it appears that the locution was employed in reference equally to certain deities and to men in high official stations.[2] But it is, I think, satisfactorily shown, from the critical investigations of Rev. A. H. Sayce and M. Grivel, that the god Marduk, whose name very frequently appears

[1] 1st Rawl. Pl. 1, No. 10. For the meaning of this term consult Lenormant, *Etudes Accadiennes*, t. i. pt. 1st, p. 49, and pt. 3d, p. 77.

[2] 1st Rawl. Pl. 53, col. 1, l. 4, 5. Cf. Oppert, *Etudes Assyriennes*, pp. 26, 27.

66 HAR-MOAD.

in the inscriptions under the Accadian form *Amer-ud*, was in point
of fact no other than the Biblical Nimrod, Cushite founder of the
primitive Babylonian kingdom ; although Mr. George Smith, in his
recently published work, objects to this view.[1] He urges that the
god Marduk was regarded as creator of the world, like the Babylonian *Bel* or Belus, and could not have been at first merely a deified mortal. The author forgets, however, that *Bel*, as creator, and
Marduk, as patron deity of Babylon, were originally two distinct
personages, whose characters were blended only at an epoch much
later, as long since established by Sir H. Rawlinson, and more
recently by M. Lenormant.[2] The philological and other evidences
adduced by M. Grivel and Rev. A. H. Sayce should be regarded as
conclusive, as it appears to me, until proofs of an equally critical
character are discovered, tending to destroy their force. But to
return ; it is obvious that the entire phrase *Pa-te-shi tsi-ri* was
employed technically, also, the same as *Pa-te-shi* when used alone.
As the latter designated the entire class of sovereign-pontiffs, the
former must have been limited in its application to those of a high
rank. In this sense it is probable that we should interpret the full
phrase, when assumed by Nebuchadnezzar as his own title, and
likewise when bestowed upon the god Marduk, shown to have been
one with Nimrod. The connection of Nimrod with the temple
structures of Babylon and Borsippa, to which the inscription of
Nebuchadnezzar just cited largely relates, was fully established in
the first chapter. It will be interesting, then, to study briefly the
particular notions inherent in the two expressions, *Pa-te-shi* and
Pa-te-shi tsi-ri, assumed as official titles by the ancient rulers of the
country of the Euphrates.

We have first the sign *Pa*, explained by M. Lenormant as follows : " to anoint, royal unction, power ; " also *Khat* and *Kun*, " the
rising day, the dawn" (Rep. 234). Then follows *Te*, "foundation,
base, duration," etc., with the Assyrian value of " corner-stone,"
(Rep. 355). The third character is *Shi*, " horn, to strike with the
horn, blow, side ; to accomplish, to fill " (Rep. 125). Of the term
Tsi-ri, the first element has the sense of " to see, view, appearance,

---

[1] Vid. the critical papers of Rev. A. H. Sayce and M. Grivel, in *Trans. Bib.
Arch. So. London*, ii. pp. 243–249 ; and iii. pp. 136–144. See Mr. Smith's objections in *Chaldæan Account of Genesis*, p. 180.

[2] Rawl., *Herod.*, i. pp. 483, 484, 512, 513. Lenormant, *Frag. de Bérose*, p. 67.

to rise, as a star " (Rep. 195). The sign *Ri*, " column," etc., is already familiar to the reader. A very frequent monogram for the god *Nabu*, assimilated to the planet Mercury, being thus the same as the Egyptian *Thoth* and the Greek *Hermes*, was the Accadian *Pa*, designating him as the patron of religious and civil institutions, as he who anoints both king and priest, bestowing upon them the divine unction and power. The same symbol in the sense of the opening day, the dawn, refers to Mercury as morning star, as he who foretells the sunrise. In this phase of his character, Nabu is the instructor and prophet *par excellence;* accordingly one of his titles is *Dun-pa-uddu*, " peaceful prophet of the rising sun." The Accadian *Te*, Assyrian *Temin*, is particularly employed with reference to constructions, especially of a sacred character, as temples or towers, and denotes the base, the foundation, the corner-stone ; but it is necessary always to bear in mind that at Babylon, and generally throughout the country of the Euphrates, the materials for such structures were rather *brick* than *stone ;* hence the language of Genesis : " And they had brick for stone, and slime had they for morter " (xi. 3). The great tower at Babylon is termed by Nebuchadnezzar, " the temple of the foundation (*Te*) of the earth." [1] The element *Shi* or *Si* was highly symbolical, the same as the two characters just explained. The horns of the taurus and buffalo were types of strength and impetuosity ; then there was the horn of the altar, of the ship, the crescent ; finally, the horn of the tower and of the sacred mountain. In a hymn to Marduk, the tower at Babylon is termed the " horn of the habitation," alluding to the upper sanctuary ; and in another hymn to the great mountain of Bel, the scribe speaks of its " horn as a ray of the luminous sun, as the star of heaven that announces the day, completing its effulgence." [2] The mountain peak catches the first beams of the sun, and thus foretells the approaching light, the same as the morning star. In the term *Tsi-ri* there is also a latent symbolism, since the first element relates to the appearance, rising of a star, while *Ri*, " column," also " rising of a star," has frequent reference to Venus, Babylonian *Is-tar*. All the indications, then, inherent in the characters composing the two expressions being studied, seem to direct the mind toward the east, as the region where the great rising takes

[1] 1st Rawl. Pl. 54, col. 3, l. 15.
[2] Vid. Lenormant, *Premières Civilisations*, t. ii. pp. 174, 175.

place, from whence the light approaches; that region where wisdom was first revealed and the first foundations laid. That which completes the view is the fact that the entire phrase, *Pa-te-shi tsi-ri* is explained by Dr. Schrader as "*Sublime Master*."[1]

I have already insisted upon the idea that the expression here referred to is not simply laudatory and general, but technical; and this results not only from what has been before stated, but from the assimilation of the term *Pa-te-shi* by Dr. Schrader, in the place just cited, to the Hebrew *Pat-tish* (פַּטִּישׁ), "hammer," a well known symbol of the *Cabiri*, whose mysteries were celebrated throughout antiquity. In fact, a term denoting the "hammer" was the name of one of the chief Cabiriac divinities. A still further confirmation of our view here, amounting almost to a demonstration, is found in the term *Patæci*, habitual designation of the pigmy images of the Cabiri. Much doubt has existed as to the true etymology of the word *Patæci*, but Dr. Movers is probably correct in deriving it from the Greek verb *Patassō* (πατάσσω), "to beat, to pound, to knock," from whence *Patæci*, "a hammer."[2] The sense of *Patassō*, "to pound, to knock," answers precisely to that of *Shi* or *Si*, "horn, to

---

[1] *Keilinschrift. u. d. Alt. Test.* p. 276. The Assyrian scholar might well object to the analysis given in the text of the phrase *Pa-te-shi tsi-ri*, if it is to be taken as a purely *Semitic* expression. But Dr. Oppert has well observed, that it is a "designation from a source non-Semitic" (*Études Assyriennes*, p. 27). It is evidently Accadian, or Cushite, in its origin, the characters composing it being employed, not merely phonetically, but ideographically. My conjecture is, that *Pa-te-shi* especially appertained to the Cabiriac craftsmen, originally designating the *Gavel*, perhaps made of "horn," and taken as symbol of the highest authority. In such case, the relation of this term to the Hebrew *Pattish*, "hammer," and to *Patæci*, applied to the Cabiriac images, would be easily explained. I venture to submit another conjecture, relative to the qualifying word *Tsi-ri*; or *Tsi-ru*, if it is really to be taken as Assyrian. Both elements composing it refer among other things to the appearance, rising of a star, or celestial orb, in the east. The application of this term to Nimrod as *Sublime Master* finds its explanation in the fact that *Amer-ud*, name of the god *Marduk*, and from which that of *Nimrod* is derived, literally signifies "the circle of the sun;" thus, *Ud* "Sun " + *Amer* " to go in a circle." The meaning of "ridge, back," given to this word by Dr. Delitzsch (*Assyr. Studien*, p. 17, note 2), might well arise from the sun's appearance above the mountain ridges of the east. The notion of "sublimity" would naturally proceed from the same phenomenon. But the technical, and even mystical, import of the phrase in question is that upon which I especially insist.

[2] *Phœnizier*, i. p. 653.

strike with the horn, blow," etc. We assume without hesitation, then, the original identity of the two terms *Pa-te-shi* and *Patæci*, as denoting primarily " a hammer," and thence as symbolical designation of the Cabiri, the reputed sovereign-pontiffs, or priest-kings, like those of the valley of the Euphrates. We have, 1st. Their exact phonetical equivalence ; the reading *Pa-te-si*, often adopted by cuneiform scholars, can hardly be distinguished from *Pa-tæ-ci*, so frequently applied to the Cabiriac images. 2d. The assimilation of *Pa-te-shi* by Dr. Schrader to the Hebrew *Pat-tish* " hammer " accords perfectly with Dr. Movers' derivation of *Patæci* from *Pa-tassö*, denoting thus " a hammer." 3d. Both terms were unquestionably employed with reference to a priest-class, in whom the civil and sacerdotal functions were united. The data thus briefly presented must go far to establish the conclusion, not only that the priest-kings of the Chaldæo-Assyrian empire were. Cabiri, technically so designated by the term *Pa-te-shi* so often applied to them, but that the chief builders of the tower of Babel, identified with that of Borsippa, were also Cabiri, among whom the Cushite hero Nimrod, under the title of *Pa-te-shi tsi-ri*, was recognized as a *Sublime Master*.

SEC. 30. It will be necessary now to call attention to the third group of characters, extending from *a* to *g*. as exhibited in the first plate. The couplet marked *a* shows the Accadian *Tag*, " to extend, to dispose in layers, a layer of brick ; to prepare, to complete ; sentence, augury ; " *Tuk*, " fear " (Rep. 173). It will be seen that the hieratic form of the next couplet is only a variant of the one just explained, and it has actually the same phonetic power, *Tag*, signifying " a stone " (Rep. 163). It is probable that the same original symbol, denoting " a stone," was slightly modified at a subsequent period, when brick were substituted for stone in the Euphrates valley, in order to designate this new material used for the same purpose. In the third character we find the two forms of *Te*, Assyrian *Temin*, " foundation, corner-stone," etc., as set forth in the previous section. The similarity in form so apparent in these three signs is strictly paralleled by the direct analogy existing between the conceptions involved in them. The triplet marked *d* and *d'* is an important one, with the value *Ak*, " to make, to labor, to build, to superintend, to create " (Rep. 91). Then follows the composite sign *e*, whose value is *Akka*, " to elevate, to exalt, to sustain, to favor, to delight in," having the Assyrian reading *Ram* with the same sense (Rep. 279). The two sets of parallel lines marked *f* belong to the Egyptian system of writing, with the phonetic power of *Sesun*, signifying "eight ; " and the couplet *g*, concluding the group, affords two examples from the "eight *Kuas*," so called, constituting a primitive mode of Chinese writing. The forms marked *f* and *g* will be more fully explained in their proper connection.

The Accadian *Tag*, "stone," must have been a very primitive word for "mountain," characteristically the region of stone ; and this term occurs frequently in the names of mountains in Central Asia, as, for example, the *Belur-tag*. The two forms *Tag* and *Tak* are only variants of each other, the sounds *k* and *g* constantly interchanging in the inscriptions. The form *Tak* is obviously the primary one, both being compound syllables, which may be resolved into either *Ta-ka* or *Ta-ak ;* and instances might be cited where *Tak* is written phonetically according to both methods. But *Ta-ak* alone affords us an intelligent etymology of the word. The element *Ak* signifies " to make, to build," and *Tak* denotes the material for constructions, whether brick or stone. The term *Akka* is found very frequently preceded by the determinative of locality, as *Ki-akka*, place, exalted ; that is, " sanctuary, temple." For this reason it may be inferred that *Ak*, "to build," etc., not only helps to form the word *Tak*, material for building, but *Ki-AK-ka*, the building itself. According to the opinion of M. Lenormant, *Akka* forms the ground element in the name *Akkad ;* thus *Akka*, " to elevate," and *Akku*, " very high, supreme," conduct to *Akka-d*, "elevated, mountainous country," from which the author infers that the *Akkadi*, or Accadians, were originally " mountaineers, highlanders." [1] As already suggested, the term *Tag* or *Tak*, the same as *Akkad*, must have denoted primitively a mountain, as in the name Belur-*tag ;* so that the fundamental relation of all these words, with the element *Ak* for their common base, is a matter in relation to which not much doubt can be entertained.

The sign *Ak* constitutes another monogram for Nabu, the Babylon Mercury ; and its sense of " to make, to build, to create," shows that, originally, Mercury attached to himself a definite cosmical character among the Babylonians, the same as with the Egyptians.

[1] *Etudes Accadiennes*, t. i. pt. 1st, p. 39.

It is necessary to connect with this Accadian or Cushite symbol the two sets of parallel lines marked f and appertaining to the Egyptian writing. In an excellent and critical treatise on the Nomes of Egypt, Jacques De Rougé has the following in reference to Hermopolites : —

" The ancient nome was *Un*, chief place *Sesun*, later Hermopolis. The term *Sesun* in the Egyptian language designates the numeral *eight*. This number relates to the eight gods who assisted Thoth (Mercury) in his character as creator of the world. Thoth, the god of intelligence, the inventor of writing, compared by the Greeks to Hermes, had his principal cultus in the city of *Sesun*." [1]

If now we compare the Accadian *Ak* with the Egyptian *Sesun*, it will be seen at once that a direct relation exists between them, not only in respect to the form of the two paleographic symbols, but also as regards the conceptions attached to them. First, and with reference to the form of the two characters, the two sets of parallel lines, four in each, giving rise to the conception of " eight," is very prominent in both examples. In the cuneiform sign, we have a middle equatorial line, answering to the primary division of the cosmos ; while the same idea is expressed in the Egyptian character by the two distinct groups of lines. The cuneiform symbol is sometimes constructed with four squares above and four below this middle line ; showing that the reduplication of four into eight is fundamental in the conception of this figure, and that it must have appertained originally to it, the same as to the *Sesun* of the Egyptian writing. Secondly, and as regards the notions attached to these figures, both are connected with the god Mercury, under the various names of Nabu, Thoth, Hermes, etc. ; and in both cases the notion of fabricating, building, creating, is fundamental. As the Egyptian *Sesun* definitely related to the cosmical character of Mercury, there can be no doubt of a similar reference involved in the Accadian *Ak*, taken as a monogram of this divinity. A comparison of the two symbols demonstrates, it seems to me, not only a common origin of the characters themselves, but of the ideas associated with them.

With the two figures just studied, it is necessary to connect the specimens of Chinese writing marked g in our group. It is remarkable that the same numeral " eight " reappears here, and that the

---

[1] *Monnaies des Nomes de l'Egypte*, p. 25.

fundamental reference is to the cosmos, the creation of the world. The basis of the Chinese characters is the continuous line denoting " unity," and of the broken line symbol of " duality " as generated from the primal unit. The two then give birth to *four*, and the four to *eight*. In the *Yi-king* we have the following statement of the cosmical doctrines attached to these symbols: " The great *summit* engenders the two *principles ;* the two principles engendering the four *forms ;* and the four forms producing the *eight kuas*." [1] It should be remarked that the number eight is not represented in any character, as in the Egyptian *Sesun ;* but by eight distinct groups of parallel lines, variously arranged, and derived from the two fundamental principles. But this number is just as prominent, nevertheless, and the reference to the cosmos is undoubted ; so that the Accadian *Ak*, Egyptian *Sesun*, and Chinese *Kuas* exhibit at a glance their direct relation to each other, both in form and in the ideas attached to them. They may be traced respectively to the earliest historical period of the populations employing them ; showing that they were not derived the one from the other, but had a common origin, probably outside the countries occupied subsequently by these nationalities.

SEC. 31. From the Cushites of Chaldæa, the Hamites of Egypt, and the Turanians of China, we pass to the Aryans of Central Asia, and the races diverging from this common centre. The Aryan radical *Ak*, "to penetrate, to pierce, to enter with force, to reach," etc., constitutes the theme of a multitude of words under different forms, some of them obviously related to our subject. 1st. *Akana*, " stone, whetstone," from which is the Sanskrit *Asna*, "stone, sling-stone," together with various other Aryan words denoting implements obviously made of stone, such as the " spear, javelin," or anything sharp, pointed. 2d. The forms *Aku*, " sharp, pointed, summit, peak," evidently of a mountain ; *Akra*, " sharp-point," and *Akri*, " angle, corner," like that of a corner-stone. 3d. *Akman,* " stone," and " heaven ;" the Sanskrit form denotes " a stone," while that of the Zend signifies " heaven." To the same belongs the Greek *Akmon* or *Acmon* (ακμον), " father of *Uranos* " (heaven), also " anvil." The Greek *Akmon*, " anvil," was another name applied to one of the chief Cabiri, to be compared

---

[1] Vid. Dr. G. Schlegel, *Uranographie Chinoise*, pp. 246–261, where the eight Kuas and their reference to the *cosmos* are fully treated.

with the Hebrew *Pat-tish*, "hammer." 4th. *Aktan*, "eight," and *Aktama*, "the eighth;" to which belong the Greek and Latin *Okto* and *Octo*, "eight."[1] The derivation here shown of the Aryan word for the numeral "eight" is quite important, and it merits a particular attention. From its theme *Ak*, "to pierce, to penetrate," together with its cognates, such as *Akana*, "stone," *Akra*, "sharp-point," and *Akri*, "corner, angle," it is plain that the material object, constituting the basis of the notion of "eight" in this case, is no other than *a stone* with its sharp angles and corners. But it is necessary to conceive a definite and limited number of these angles or corners, corresponding to the numeral itself. In a word, *a dressed stone with eight corners*, having thus a cubical or oblong form, constituted the original symbol of the notion involved in *Aktan*, the Ayran numeral "eight." But I wish to add here another radical from the same family of tongues, together with some of its derived forms. The root *Tak*, varied to *Taks*, signifies "to hew, to make, to work," etc. It is obvious that the material to be hewed, or wrought upon, might be either wood or stone; though the derivatives contemplate for the most part only wood as the material. 1st. *Taksan* or *Takshan*, "worker in wood, hewer of wood, carpenter." 2d. *Takstan*, "maker, master-work-man;" also Zend *Tashtar* and Sanskrit *Trashtar*, "master-work-man."[2]

I think the most rigid linguistic criticism ought to admit an original and direct relationship between the Aryan radicals *Ak* and *Tak* and the Accadian terms *Ak* and *Tak*, together with the derived forms respectively appertaining to them. The simple element *Ak*, whatever its origin, seems to constitute the essential base of all. We proceed to point out briefly some of the more obvious connections existing between the two groups of words.

1st. We have the name *Tashtar*, varied to *Trashtar*, the form *Takstar* being primary to both, with the signification of "master-workman" common to all. The radical element is *Tak*, "to hew, to make," phonetically identical with the Accadian *Tak*, "stone," in which the syllable *Ak*, "to make, to build, to create," is obviously involved. But *Ak* is the monogram for the god *Nabu*, Egyp-

[1] Fick, *Wörterb. de I. G. Sprachen*, i. pp. 4, 5, 7. Cf. Curtius, *Grundzüge*, pp. 130, 131.

[2] Fick, ibid., pp. 86, 87. Curtius, ibid., p. 219.

tian *Thoth*, whose cosmical character has been already established. It is remarkable that *Tashter* is likewise recognized as creator, or a cosmical divinity, in Aryan mythology.[1] It is impossible not to see here a primitive connection between Tashtar and the Babylonian and Egyptian Mercury in the character of worker, builder, creator, and especially of cosmical agent.

2d. The Aryan *Aktan*, "eight," to be compared with the paleographic symbols, namely, the Accadian *Ak*, the Egyptian *Sesun*, and the Chinese *Kuas*, in all of which the notion of "eight" and a distinct reference to the work of creation are to be regarded as fundamental. We have shown that the material basis of the conception of eight as denoted by the term *Aktan*, was an eight-cornered stone dressed in the form of a cube; significant type of constructions in general, and especially of the fabrication of the world by the great Master-Builder.

3d. The form *Akman*, "stone," symbol also of "heaven," is of especial importance. The corresponding Greek form, *Akmon*, constitutes the title of the father of *Uranus*, "heaven;" but it signifies likewise an "anvil," and is the name of one of the chief Cabiriac divinities. It is a very singular and seemingly inexplicable circumstance that the same Aryan term should signify "a stone," then "heaven," then again "an anvil," being finally appropriated as the name of a Cabirus. Nevertheless, the facts which we proceed to notice in some detail will tend not only to explain these apparent contradictions, but to afford a better stand-point for contemplating the various affiliations of ideas previously noticed.

In the Phœnician mythology the eighth son of *Sydik* is called *Eshmun*, whose name signifies the "eighth," and he was thus reckoned the eighth Cabirus, in relation to the other sons of *Sydik*. *Eshmun* represented "heaven," that is, the heaven of fixed stars, regarded as the eighth celestial region in relation to the seven planetary spheres, assimilated to the seven brothers of *Eshmun*.[2] It is probable that the Phœnician *Eshmun* and the Aryan *Akman*, with the softened form *Asman*, were primitively put for the same conception at least, if they were not the same word; and they contribute much to explain each other. *Eshmun* represents "heaven," and

[1] Lenormant, *Frag. de Bérose*, pp. 177, 278.

[2] Vid. Movers, *Phœnizier*, i. pp. 227-236. Cf. Lenormant, *Frag. de Bérose*, pp. 382, 383.

Akman signifies " heaven." *Eshmun* signifies the " eighth," being
put for the eighth celestial region, and *Akman*, since it denotes
" heaven," and is radically akin to *Aktan*, " eight," evidently referred
to the same celestial region. Again, *Akman* denotes a " stone," as
well as " heaven," in which case it is evident the stone is a
symbol of heaven. But a rough, unhewn stone would never be
taken as such symbol; hence, it is almost necessary to conceive here
an eight-cornered stone, a cube. Finally, we know that *Eshmun*
was a title of the eighth Cabirus, and that the Greek *Akmon* was
actually the name of a Cabiriac deity. It is thus quite certain
that this Greek Cabirus was the eighth, and that the ancient form
Akman had a similar reference. It is impossible not to admit the
common origin of these various conceptions, and their primary reference to the cubical stone as symbol of the eighth region of heaven.
But the Greek form means also "an anvil." This may be explained
by the fact that the first anvils were meteoric stones, or masses of
iron that had fallen from heaven. M. Lenormant shows that the
first workers in this metal employed the meteoric iron, and not that
produced from ores, and that the Greek for iron, *sideros*, is related
to the Latin *sidus, sideris*, " a star." [1]

Sec. 32. We thus account readily, and probably correctly, for
the transfer of the Aryan *Akman*, " a stone," to signify a meteoric
stone, and thence " an anvil." But does this prove that *Akman*
originally designated a meteoric stone? I think not; but this is
an important inquiry to be now considered. In the first place,
then, it is only under the Greek form, and only in the Greek language, if my impression is correct, that this Aryan term ever signifies " an anvil," and this appears to be, therefore, the only reason
for supposing that it denoted, primarily, a meteoric stone; — a reason quite insufficient. Secondly, all the analogies derived from its
root *Ak*, and its cognate terms, such as *Akra, Akri*, especially
Aktan, tend to the conclusion that the ordinary stone was originally intended. Finally, we are to consider here that an inclosed
cubical space, like the stone dressed in this form, usually represents
heaven in ancient architecture. The superior sanctuary of the
temple of Borsippa, dedicated to Mercury, and forming the eighth
stage, was in the form of a cube, as represented in the cut in Rev.
Mr. Rawlinson's " New Version of Herodotus " (ii. p. 482). As the

[1] *Premières Civilisations*, i. pp. 88, 89.

other stages symbolized the seven planetary spheres, the superior or eighth stage must be taken for the eighth celestial region, the heaven of the fixed stars, corresponding to that to which *Eshmun* was assimilated. The Holy of Holies in the Hebrew tabernacle was in the form of a cube, and was put for the heaven, as will appear in a future study. The stone, therefore, to which the term *Akmon* referred was evidently a dressed stone with eight angles, and it is quite certain that no meteoric stone had been given this form, for the purpose of serving as such symbol.

As *Akmon*, put for the " anvil," was the name of a Cabirus whose symbol was the anvil, so *Akman*, denoting a dressed stone, was evidently the name of a Cabirus who had such stone for his symbol. This personage must have been conceived as *the stone par excellence*. In both instances the symbol related to the eighth celestial region, being that to which the Phœnician Eshmun was assimilated, a name which is probably only a Semitized form of Akmon. The Cabiri were evidently associated with this symbol, whether as denoting the cubical stone or the anvil ; and it follows that the Cabiriac fraternity were originally workers in stone, instead of in iron and the metals generally. This is the point to which we have wished to conduct the reader, by the most rigid analysis of the facts and ideas constituting the data before us. The steps by which we have arrived at this result may be re-stated briefly as follows : —

1st. The Aryan term *Akman* designated primitively " a stone ; " at the same time, and under its Greek form Akmon, it was appropriated as a name of one of the Cabiri. Under this Greek form, it denoted also the anvil, symbol of the Cabirus thus called. Everything indicates that the anvil was at first a meteoric stone, and it is easy thus to explain the transfer of the term *Akman* from " a stone " to " an anvil." All depends now upon the question, whether *Akman* designated primitively the meteoric stone as a symbol of heaven, or a stone having the ordinary composition of this class of materials ? If the meteoric stone was the primitive conception involved in the term, then the Cabiri were workers originally in metals, the meteoric iron being one of them. But the indications are very strong, and I presume it could be fully demonstrated, that *Akman*, " stone," meant primitively a stone of the ordinary composition.

2d. But a rough, unhewn stone, being similar to all others and quite ordinary, would never be selected as a symbol of heaven ; and since the cube was usually taken to represent the heaven of the fixed stars, it is safe to infer that a dressed stone having a cubical form was literally that which was designated primitively by the word *Akman* in its double sense of " stone " and " heaven." This supposition fully explains the origin of the notion of " eight," evidently " eight corners or angles," involved in the term *Aktan*, and it would be impossible to offer any other rational explanation of such origin.

3d. The simple existence of a stone thus wrought, with which were connected symbolical ideas of the nature already indicated, presupposes the existence also of a regularly organized craft of workers in stone ; for among laborers promiscuously associated no such ideas would be likely to prevail. This organized craft could be no other than the Cabiriac fraternity; and the evidences tend to show that the very term *Akman* was the name of a Cabirus. Recall here the Greek form *Acmon*, title of a Cabiriac divinity, and the Phœnician *Eshmun*, a term obviously related to *Akmon* in its softened form *Asman;* a word also appropriated as the name of the Cabirus who represented heaven, the same region otherwise denoted by the two forms of the Aryan word.

4th. We must conclude, then, that the Cabiri were originally workers in stone ; that as such exclusively, they belonged to the period before the discovery of the art of working metals, that is to say, to the stone age. Subsequently, they became workers in metals likewise, when the Aryan word primitively denoting a dressed stone with eight corners was transferred to a meteoric stone used as an anvil.

5th. But a dressed stone, with notions of a symbolical and sacred character connected with it, supposes a sacred edifice, a temple, for which it is designed as material for construction. The Cabiri were thus originally an organized temple-craft ; and the symbolical conceptions connected with the material thus wrought and employed by them presuppose the existence of certain esoteric ideas peculiar to their organization ; in a word, it is necessary to admit here the existence of a *Traditional Doctrine of the Templum.*

SEC. 33. The statement last made must be considered as immensely important, if it is to be admitted as correct. The data upon which we have proceeded thus far do not appertain to the later epochs of antiquity, but to the very earliest periods, even to the night of ages, when the light of the most primitive civilizations known to us was just dawning upon the world. If, then, we show the existence of a regularly organized priest-class, in whom the civil and sacerdotal functions were united; if we show that this organization was in fact a temple-craft, including within itself different grades from the priest-king, the *Sublime Master*, to the superintendent of constructions, and even the hewer of stone and maker of brick; if we show finally, that certain esoteric ideas, peculiar to this craft and relating to constructions of a sacred character, had been handed down, amounting, in fact, to a *Traditional Doctrine of the Temple;* the importance of such facts once fully verified cannot be overestimated. Of course it is to be expected that statements of such a nature will be received with extreme hesitation: and my efforts will be to place the matter in the clearest possible light.

We have introduced already some proofs, of a nature quite conclusive, that the Cabiri, as an organized priest-class, were the chief constructors of the pyramid of Borsippa, usually identified by cuneiform scholars with the original tower of Babel. In addition to these proofs, it will be regarded as significant, if we find that the Cabiriac worship was actually connected with this very structure. In his account of the repairs of the pyramid of Borsippa, which had been left in an unfinished state, Nebuchadnezzar says that he completed it to the top according to the original design. The restoration of it, as represented by Rev. G. Rawlinson, already cited on this point, shows seven stages retreating one upon the other, whose different colors are interpreted by Assyriologues as denoting the seven planets. Above all, constituting the eighth stage, was the sanctuary of a cubical form, and this was dedicated especially to Nabu, or Mercury. The exact agreement in conception of this eighth stage, representing obviously the heaven of the fixed stars, with the character of the Phœnician Eshmun assimilated to the eighth celestial region, has been noticed by Dr. Movers and M. Lenormant, who see here an indication of the connection of the Cabiri with this temple structure.[1] This conjecture is confirmed by the fact that Mercury himself, to whom the superior sanctuary was dedicated, was reckoned as one of the Cabiriac divinities. Drs. Movers and Gesenius

---

[1] Movers, *Phœnizier*, p. 528 ; Lenormant. *Frag. de Bérose*, p. 388.

have shown that Mercury, or Hermes, under the names *Cadmilus* and *Casmilus*, was included among the Cabiriac deities adored at Samothrace, whose mysteries were so celebrated in antiquity.[1] But the indications are even more positive, as we shall see : —

"The pyramidal temple of the Chaldæans," observes M. Lenormant, "was as an imitation, an artificial reproduction, of the mythical 'mountain of the assembly of the stars,' the *Har-Moad* of Isaiah (xiv. 4–20), which sacred tradition placed in the north, and of which there is yet question in the sacred books of the Mendæans."[2]

The author alludes here to the diluvian mountain, identified traditionally with the sacred mount of paradise, first abode of humanity. It was directly from this diluvian mountain that the founders of Babylon had journeyed, when they came to the plains of Shinar, and undertook the building of the tower identified with that of Borsippa. Now this pyramid of Borsippa was mystically called *Bit-Zida*, "temple of the right hand." It was thus, according to M. Lenormant just cited, an artificial *Mount Ida*, "mountain of the hand;" and it was with the Mount Idas of antiquity that the Cabiriac worship and mysteries were especially associated. The same author describes minutely an ancient cylinder, upon which a pyramid in stages is represented, with a colossal hand erected upon the upper stage, around which are grouped eight personages, obviously intended for the eight Cabiri, who, according to the mystical idea involved, are born from the hand. The author has no doubt of the reference of this cylinder to the tower of Borsippa ; and it affords a complete explanation of the phrase Bit-Zida, " temple of the right hand," applied to it."[3]

The fact, then, of the primitive association of the Cabiri and the Cabiriac cultus with the pyramid of Borsippa, and consequently with the tower of Babel itself, is here clearly demonstrated. That the chief personages engaged in this construction were a temple-craft, and that they possessed certain esoteric doctrines relative to the temple, is quite apparent from the circumstances now familiar to us, and the data already established render it difficult to entertain serious doubts upon the matter. The seven stages are the seven degrees of the temple, corresponding to the seven planetary

[1] Movers, *Phœn.*, i. pp. 520–522; Gesenius, *Mon. Phœn.*, pp. 404, 405.
[2] *Frag. de Bérose*, p. 358.   [3] Ibid., pp. 381, 382, 385–387.

orbits; and these are expressly compared by M. Lenormant and Dr.
Movers to the seven sons of *Sydik*, whose Cabiriac character is
well understood. Then the eighth stage or degree, answering to the
eighth celestial region, the heaven of the fixed stars, is not only
dedicated to Mercury, one of the Cabiriac deities, but its direct connection with Eshmun, the eighth son of Sydik and eighth Cabirus,
has been already established. The pyramid itself is styled by
Nebuchadnezzar " the temple of the seven lights of the earth ; " a
phrase whose mystical import, in connection with the seven degrees
of the structure itself, is apparent at a glance. In addition to these
facts, we should call to mind here the connection of Nimrod with
this temple-structure, to whom Nebuchadnezzar applies the phrase
Pa-te-shi tsi-ri, or " Sublime Master; " appropriating often the
same title himself. The technical application of the term *Pa-te-shi*
to the priest-kings of Babylon on the one hand, and its identity
with the word *Pataci* on the other, an ordinary designation of the
Cabiriac images, are points with which the reader is already familiar. Another circle of conceptions previously developed has an
obvious connection with the subject matter now before us. Nabu,
or Mercury, is a Cabiriac divinity, and the eighth stage of this
tower, representing heaven, is especially dedicated to him. One of
the monograms for Nabu is the sign *Ak*, " to make, to build," etc.,
whose relation to the Accadian *Tak*, " stone " or " brick " as a substitute for stone, whose relation also to the Aryan *Ak*, from which
are derived *Akman*, " stone, heaven," and *Aktan*, " eight," *Aktama*,
" the eighth," are points that have been fully illustrated. We see
here certain mystical ideas, evidently originating in Central Asia,
and around the " mountain of the assembly of the stars," to which
M. Lenormant alludes ; ideas brought by the Cushite emigrants from
the east to Babylon, and these reëmbodied in an artificial mountain of degrees, an imitation of that from which they had journeyed.
These ideas have obvious reference to the temple ; and they constitute a *Traditional Doctrine of the Templum.* The various considerations that have been presented leave no doubt, as it appears to
me, upon this important point.

SEC. 34. Another proposition is to be stated here. Not only had
there been inherited by the Cabiri a *Traditional Doctrine of the
Temple*, but this *Doctrine had for its basis their theory of the Cosmos and of the Creation of the World.* In the first place, the

pyramid of Borsippa was designedly and strikingly a representation of the cosmos, or world. Its summit, or eighth stage, was a symbol of the heaven of the fixed stars; the seven stages descending from thence represented the seven planetary spheres; while the foundation was put for the earth. In this foundation was the sanctuary of the god *Anu*, who has the mystical title of *Susru*, "the founder;" that is, the founder *par excellence*. In artistic conception, then, this tower "united the heavens and the earth," just as the sacred mountain of tradition was conceived to do, and of which it was an imitation. We have here an explanation of the really double sense involved in the singular expression of the Mosaic record applied to this tower, "whose top may reach unto heaven" (Gen. xii. 4). This rendering is sufficiently correct, but the reference is probably to the fact that this structure was artistically conceived to unite heaven and earth. An equally correct translation adopted by some exegetes would be: "whose top may represent the heaven." We have seen that the superior sanctuary in the form of a cube, constituting the eighth stage, was held especially to represent the heaven of the fixed stars. But it is hardly necessary to add further illustration, that the tower of Borsippa was the material expression of certain traditionary ideas in reference to the temple, and at the same time relating to the cosmos, the creation of the world.

Again, the Cabiriac divinities themselves were usually conceived as engaged in the work of creating the world. M. Lenormant takes particular notice of this fact.[1] As shown by his monogram *Ak*, "to make, to build, to create," the Babylonian Mercury was primitively regarded a cosmical divinity. The Egyptian Mercury, or Thoth, was assisted by the eight gods in superintending the work of creation. Thus, Mercury was unquestionably a cosmical agent, and was at the same time a Cabiriac deity. The Aryan Tashtar was conceived as creator, then as master-workman, and again as one of the Cabiri.[2] These examples will suffice to establish the cosmical character of this circle of divine personages. We see, therefore, that cosmical doctrines and doctrines of the temple were blended together; the temple itself being designed as an image of the cosmos. The three phases of character, as creative powers, as

[1] *Frag. de Bérose*, pp. 382–385.
[2] Lenormant, *Premières Civilisations*, t. i. pp. 87, 140.

priest-kings, and as a temple-craft, are in reality so blended in the Cabiri that it is often difficult to distinguish between them. They were preëminently the founders in every sense : founders of the world, of civil and religious institutions, and of temples and sacred edifices generally. Their esoteric doctrines, traditionally inherited, appertained alike to all these subjects. But that which applies equally to all, and confirms all, is the number " eight," obviously relating at one time to the cosmos, then to the temple, then again to the Cabiri, reappearing constantly under various and most striking symbols. We have the eight lines or squares in the Accadian sign *Ak*, in the Egyptian *Sesun*, the Chinese *Kuas ;* we have. too, the eight-cornered stone. the eighth stage of the temple, inclosing a cubical space ; and finally the eighth Cabirus, and the eight gods assisting Thoth in the work of creation. M. Lenormant develops the fact that an inclosed cubical space as the temple, and a dressed stone, either square or cubical, as image of the Deity, were traditional ideas among the ancient Arabians, in conformity with which they constructed their sacred edifices. The Chaaba at Mecca was of this form, and the legends affirmed that it was of heavenly origin, having been adored by the angels even, being finally brought from heaven to earth by Adam, the first man. The celebrated black stone of the Chaaba was supposed also to be of heavenly origin. Originally its color was a pure white, but it had turned black in consequence of the sins of mankind. The Arabians were passionately fond of this stone. On being restored to its place, after having been at one time taken away by enemies, the people bestowed the most lavish affections upon it, caressing and kissing it even, and condoling its dark color as attributable to their own transgressions and the sins of the world.[1]

Thus, among the populations the most widely separated in antiquity, from the Mongolians of the distant east to the Pelasgic nations of the west, and from the Hamites of Africa, the Cushites of Chaldæa, to the Aryan races of high Asia, we trace the existence of certain traditionary ideas pertaining to creation and the cosmos, to the temple and to constructions of like sacred character ; ideas so characteristic in their nature and in the symbols by which they were expressed, as to render it impossible, on one hand, not to identify them at a glance, and on the other hand, to account for their

[1] *Lettres Assyriologiques,* t. ii. pp. 123. 124. 140, 150, 151.

singular resemblances wherever found, either upon the ground of mere accident, or of natural and normal development; ideas, therefore, whose community of origin anterior to the existence of these diverse nationalities is necessary to be supposed; for we have traced them even beyond the historical period in every instance. The Accadian *Ak* and *Tak*, the Egyptian *Sesun*, and the Chinese *Kuas*, and so the Aryan radicals, *Ak* and *Tak*, the Phœnician *Sydik*, and thus the various symbolical and legendary conceptions connected with them, all appertain to epochs prior to history, so that the propagation of these notions and symbols from one country to another at later periods is entirely out of the question. Their origin must be assigned, in fact, to the first ages of humanity, and from the same primitive era must date the existence of that mysterious class of personages through whom these doctrines were transmitted to subsequent ages. These personages, whoever they were and by whatever name we call them, were the founders of the ancient civilizations, the first prophets and teachers of mankind; they built the first temples, they made the first scientific discoveries, and they were the first inventors of the useful arts.

Sec. 35. We proceed now to the consideration of the Cabiri as fire-gods and workers in metals. As such they undoubtedly pertain to an immensely remote epoch, for Tubal-Cain, who is usually identified with this class of personages, lived a thousand years before the deluge, being " an instructor of every artificer in brass and iron " (Gen. iv. 22). Notwithstanding this high antiquity, I still maintain the hypothesis that the Cabiri were originally workers in stone. Modern science establishes the fact that the stone age, as a distinct era, preceded that of metallurgy. This accords perfectly with our theory and with the data already introduced in this discussion. Everything indicates, to my mind, two distinct characteristic phases in the history of the Cabiri, although investigators, so far as my knowledge extends, have never made this a special point of inquiry. The name *Enoch* occurs twice in the antediluvian genealogies of Genesis, the first as applied to the son of Cain, and then, in the line of Seth, to the son of Jared. The meaning of this name as often given by exegetes is " the initiated," and is derived from the verbal root *Khānak* (חָנַק), " to straiten, to choke, to suffocate," also, " to be narrow, strait, close." In relation to Enoch Dr. Gesenius observes: " The later Jews, founding a conjecture on the etymology of the name, make him out to have been not only the most distinguished of the antediluvian prophets, but also the inventor of letters and learning, and have forged in his name a spurious book (comp. Jude v. 12). These fables are current also among the Arabs, by whom he is called *Idris*, i. e., " the learned " (art. חֲנוֹךְ). There must have been, in my view, something more than a simple etymology at the foundation of the traditions to which Dr. Gesenius alludes. Dr. D. Chwolsohn cites a large list of authorities, tending to show that Enoch, or Idris, was considered identical with Hermes, was a renowned prophet, the founder of states and religions, inventor of letters, author of sacred books, etc., etc.[1] The traditions to this effect appear to have been widely pre-valent ; too much so, in fact, to be easily traced to the later Jews. But even the etymology of the word points directly to Thoth, or Hermes, whose recognized symbol was the serpent. The term *Khānak*, " to choke, to straiten," is compared by Dr. Gesenius himself to the Latin *Angere*, derived from the same theme which gives *Anguis*, " serpent." Dr. Chwolsohn alludes likewise to the tradition that Enoch, Idris, identified with Hermes, was the father of *Tat, Ssa, Aschmun*, etc. (i. p. 788) ; and the last must be the same personage as the Phœnician *Eshmun*, whose symbol was also the serpent. Admit that, so far as we are now able to trace them, these traditions belong to later epochs ; yet there is such a consistency in them with the religious ideas of Western Asia that it is impossible not to attach to them considerable force. The idea that certain sacred writings, attributed to Seth, or Enoch, or to both, were pre-served from the destruction of the deluge, and transmitted to the after world, has been usually discarded by modern writers ; but such a tradition is expressly supported by Berosus in his account of the deluge, and it will be shown in another chapter that certain Babylonian monarchs not only credited it, but caused diligent search to be made for the corner-stones of ancient temples, with which copies of sacred writings were believed to have been deposited. For the rest, Mr. George Smith recognizes the name *Enoch* in the in-scriptions, under the form *Emuk* or *Enuk*, " wise." [2]

We have seen that the Dioscuri were reckoned among the Cabi-riac divinities of Samothrace. The reference is to Castor and Pol-

---

[1] *Ssabier*, i. pp. 787-792.

[2] *Chald. Account of Genesis*, p. 296.

lux; that is to say, to the zodiacal constellation of Gemini. These personages have always a mixed character, partly divine and partly human. They were typically *the brothers*, and it was on account of their fraternal attachment that Neptune rewarded them with the control of winds and storms, for which reason they became the protector divinities of the sailors. In their chief character they were the powerful helpers of man, and thence termed *Anakes*, from *Anak*, "master of the house, lord, thence king." Their images were placed upon the prows of Phœnician ships, and were like the *Patæsi*, the term which we have found employed as general designation of the Cabiri, and which has been traced to the Hebrew *Pat-tish*, "hammer," and finally to the Accadian *Pa-te-shi*, "master, pontiff, priest-king," etc. Always bearing in mind that at Babylon "they had brick for stone," we may trace a direct connection of Castor and Pollux, or of the constellation Gemini, with constructions in brick. The Accadian *Uku*, "brick" (Rep. 529), is the monogram for the month answering to Gemini, denoting the "month of constructions in brick." We have thus another indication of the association of the Cabiri with the tower of Babel, and of the technical use of the term *Pa-te-shi*, one with *Patæei*, amongst them.

SEC. 36. It is the place here to take note of the fragmentary inscriptions recently translated by Mr. George Smith, and supposed by him to refer to the building of the tower of Babel. Fragmentary they are indeed, for hardly a single sentence appears entire.[1] Only the first column, consisting of sixteen lines, many of them broken up, affords direct evidence of a reference to the building of some sort of structure, probably a tower. The lines 8–14 are tolerably perfect, and are thus rendered: "(small) and great he confounded their speech. Their strong place (tower) all the day they founded; to their strong place in the night entirely he made an end. In his anger also word thus he poured out: (to) scatter abroad he set his face, he gave this? command, their counsel was confused" (pp. 160, 161). A reference to Babylon going immediately before, and still earlier to a personage who "against the father of all the gods was wicked," help to fix the allusion to the tower of Babel. Other expressions occur tending to the same conclusion. Every one will regret with Mr. Smith that the inscrip-

---

[1] *Chaldœan Account of Genesis*, pp. 160–163.

tion is so sadly defaced and broken, and until further discoveries
are made nearly everything in relation to it must be considered as
conjectural. Even its reference to the tower of Babel is by no
means certain, but it is probable that such is the case. Not having
"The Egypt of Herodotus" at hand, in which Mr. Kenrick treats
more at length upon the subject before us, I place below some quotations from an "abstract" of his remarks by another author, who
notes especially the following points : —

"The existence of the worship of the Cabiri at Memphis under a
pigmy form, and its connection with the worship of Vulcan (*Phtha*).
The coins of Thessalonica also establish this connection ; those
which bear the legend 'Kabeiros' having a figure with *a hammer*
in his hand, the *pileus* (*cap, symbol of freemen*) and *apron* of Vulcan, and sometimes *an anvil* near the feet." "The Cabiri belonged
also to the Phœnician theology. The proofs are drawn from the
statements of Herodotus. Also the coins of Cossyra, a Phœnician
settlement, exhibit a dwarfish figure with the hammer and short
apron, and sometimes a radiated head, apparently allusive to the
element of fire, like the star of the Dioscuri." "The fable that *one
Cabirus had been killed by his brother or brothers* was probably a
moral mythus representing the result of the invention of armor."
"The worship of the Cabiri furnishes the key to the wanderings
of Æneas, the foundation of Rome, and the war of Troy itself, as
well as the Argonautic expedition. Samothrace and the Troad
were so closely connected in this worship that it is difficult to judge
in which of the two it originated, and the gods of Lavinium, the
supposed colony from Troy, were Samothracians. Also the Palla-
dium, a pigmy image, was connected at once with Æneas and the
Troad, with Rome, Vesta, and the Penates, and the religious belief
and traditions of several towns in the south of Italy. Mr. Kenrick
also recognizes a mythical personage in Æneas, whose attributes
were derived from those of the Cabiri." [1]

There appears much disagreement among ancient authors as to
the exact number of the Cabiri, and I desire to consult here one or
two modern authorities upon this point. Dr. Wm. Smith's notices
of the Dactyli, identified with the Cabiri, afford valuable information upon this and also other matters : —

"*Dactyli*, fabulous beings, to whom the discovery of iron and the

[1] *Occult Sciences*, by Smedley, Thompson. etc., pp. 161–163. Connect the
two relations of Æneas to the Cabiri, as above, with the facts stated by Dr.
Wm. Smith, respecting the Penates, the augurial temple, etc., showing in this
way that the Cabiri were a temple-craft and the Roman augurs Cabiri.

art of working it by means of fire was ascribed. Their name, Dactyls, that is, Fingers, is accounted for in various ways: by their number being five or ten, or by the fact of their serving Rhea just as the fingers serve the hand, or by the story of their having lived at the foot of Mount Ida in Phrygia as the original seat of the Dactyls, whence they are usually called Idæan Dactyls. In Phrygia they were connected with the worship of Rhea. They are sometimes confounded or identified with the Cusites, Corybantes, Cabiri, and Telchines. This confusion with the Cabiri also accounts for Samothrace being in some accounts described as their residence. Other accounts transfer them to Mount Ida in Crete, of which island they are said to have been the original inhabitants. Their number appears to have been originally three : *Celmis* (the smelter) *Damnamenus* (the hammer), and *Acmon* (the anvil). Their number was afterwards increased to five, ten (five male and five female), fifty-two, and one hundred." (Class. Dic., art. *Dactyli*.) [1]

The connection of the Cabiri with the Mount Idas (*mountains of the hand*) of antiquity has been already noticed. Authors usually agree, also, that they were the same as the Cusites, Corybantes, and Telchines. Their number is thought to have been originally three by Dr. Smith, and he gives their names, among which *Acmon* is familiar to us. For *Celmis*, " the smelter," I am unable to suggest an etymology ; and as for *Damnamenus*, " the hammer," it hardly merits one. The term *Patœci*, assimilated to the cuneiform *Pa-te-shi*, and the Hebrew *Pat-tish*, " hammer," as heretofore set forth, may be regarded as a more original and probably primitive designation of this Cabirus. But the number three, in my view, refers simply to the chief Cabiri, and does not include the entire class. The subjoined remarks by Sir G. Wilkinson are valuable as relating to this question : —

" Most authorities agree that they varied in number, and that their worship, which was very ancient in Samothrace and Phrygia, was carried to Greece from the former by the Pelasgi. Some believe them to have been Ceres, Proserpine, and Pluto, and others add a fourth, supposed to be Hermes, while others suppose them to have been Jupiter, Pallas, and Hermes." " The name Cabiri was doubtless derived from the Semitic word *Kabir*, 'great,' a title applied to Astarte (Venus), who was also worshiped in Samothrace." " The eight great gods of the Phœnicians, the offspring of one great

---

[1] Connect what Smith says of Ida, Rhea, etc., with what M. Obry shows of Ida, Rhea, and also Dr. Grill, as connected with Chaos, Meru, etc., etc., and cuneif. *Ria.*

father, Sydik, the 'just,' were called Cabiri, of whom Eshmoun was
the youngest, or the eighth, as his name imports, the *Shmoun*,
'eight' of Coptic, and the *theman* or *saman* of Arabic, and *she-
monah* of Hebrew. This Eshmoun was also Asclepius." " Herodotus mentions the Egyptian Cabiri at Memphis, whose temple no
one was permitted to enter except the priest alone ; they were said
to be sons of Vulcan or *Phtha*, and, like that god in one of his characters, were represented as pigmy figures. It is not impossible that
the Cabiri in Egypt were figured as the god Phtha-Sokar-Osiris,
who was a deity of Hades, and the three names he had agree with
the supposed number of the Cabiri of Samothrace.   The number
eight might also be thought to accord with that of the eight great
gods of Egypt. *Ashmounayn*, the Coptic and modern name of
Hermopolis in Egypt, signifying the ' two eights,' was connected
with the title of Thoth or Hermes, ' lord of the eight regions.' " [1]

It would accord perfectly with the genius of ancient mythologies
to suppose the existence of a Cabiriac triad consisting of three
chief personages; and that they were often limited to this number
considered as divinities cannot be doubted. Nevertheless, if we look
to Egypt, Phœnicia, and to Chaldæa, evidently the more primitive
seats of the Cabiriac worship, we find the number eight almost
always connected with these divinities. There were the eight gods
assisting Thoth in the work of creation, the eight sons of Sydik,
and the eighth stage of the tower of Borsippa dedicated to Mercury,
one of the Cabiri. It is probable, therefore, that this number was
in some sense a typical one. It was sometimes, however, increased
to *nine*. The Phœnician Sydik, father of the eight Cabiri, was not
unfrequently included with them, being thus the *ninth*. So the
Egyptian Thoth, assisted by the eight great gods in the work of
creation, was himself a Cabirus, and would constitute the ninth.
Nevertheless, as cosmical agents, and as denoted by their various
symbols already studied, it is obvious that the numeral eight is to
be regarded as paramount in its application to them.[2]

SEC. 37. The extracts introduced into the last section are numerous and quite lengthy, but it is believed the reader will see in them
a value sufficient to entitle them to a place in connection with our
subject. They serve to complete the view of the personages, and
of the ideas associated with them, of which there has been question,

---

[1] In connection with *Phtha*, cite Marcite, *Mother of Apis*, and Maspero on
Epistolary Class of Texts, which I have otherwise referred to.
[2] Vid. Rawl. *Herod.* ii. p. 82, note 9.

as well as to substantiate various statements heretofore made, which had been left for the time being without direct proof. More than all else, I prefer to state important facts in the language of some leading and recognized authority in such matters, when it is convenient to do so. But the most recent treatment of this subject, and one thoroughly scientific in its character, is that of M. F. Lenormant, to which reference has been already made in the foregoing pages. Several points are quite clearly brought out by the author, which have not as yet received a special notice in our own researches. Insisting, as we have done, upon the extreme antiquity, even prior to all the ancient civilizations of which we have any knowledge, to which primitively these organized communities, workers in metals, appertained, the author proceeds to trace definitely their origin, ethnologically and geographically considered. They were in his view Turanians, and their original seats were the high table-lands of Central Asia, near the great plateau of Pamir; the precise locality, according to the traditions of the Aryan and Semitic races, which constituted the first abode of humanity.[1] It was thus around the common centre, from which subsequently the various races of men diverge, that the Cabiriac fraternities were first formed ; and in this particular the Mosaic allusion to Tubal-Cain, in connection with the art of working metals, accords perfectly with the results of modern research. But M. Lenormant quotes largely from the previous investigations and writings of the Baron D'Eckstein, whose name every French scholar mentions with reverence. This great Orientalist thus alludes to the special characteristics of these ancient patrons of industrial art : —

" On one side are the races addicted to the magical cultus of the gods of metallurgy, and on the other side we find certain mystical and secret corporations, whose chiefs superintend their labors, serving as their pontiffs and sacerdotal *confrères*, traditionally illustrious. The Vedas, the Zend-Avesta, the mythology of the Thracians, of the Pelasgi, Celts and Germans, overflow with traditional notices of the affiliations of these divine workers, of a doubtful character, but parallel to the *deimones* of the ancient classical authors. Inventors, instructors, magicians, benefactors, and malefactors in one, when the memory of these corporations becomes effaced, they remain engraven upon the minds of men as dreaded and unpropitious powers."[2]

---

[1] *Premières Civilisations*, t. i. pp. 98–126.   [2] Ibid., pp. 139, 140.

The nature of the occupation of these ancient smiths, the scenes of their labors in the deep mountain gorges, or in the bowels of the earth, and everything, in fact, connected with their art, would give rise naturally to a thousand vagaries and images of a frightful character, tending to associate them with the regions of darkness and wicked dæmons. Not only this, but their superior knowledge and civilization, as compared with the surrounding nomadic tribes, and even with the shepherds and agriculturists, rendered them an object of dislike and hatred on the part of the populations surrounding them. At a very early period they were driven out of their original seats, and the Baron D'Eckstein states that they were actually persecuted at times, being often forced to change their localities. This author remarks : —

" There was an end to this primitive influence of the civilizing fraternities, an eclipse of these races of men more advanced in knowledge than the shepherds, than the Aryan and Semitic races even; and hate succeeded to the recollection of their superior culture. Above all, the Aryans of Bactria and of India distinguished themselves by their aversion for these nefarious corporations, for these worshipers of the serpent-gods, for these priest-kings, who had the flaming dragon for their emblem; this Azdehak of Afghanistan and the anti-Iranian Media; this type of the royalty of dragons, of mythical Aztahak, so termed by the Armenians, of Astyages, as styled by the Greeks. Indeed, wherever the Aryan deities appear, their heroes, priests, warriors, shepherds, and laborers all carry defiance to the serpent-gods and serpent-men; they combat these robbers, these tradesmen, these sons of the Chthonian Hermes, god of highways, etc.: they pursue them throughout the three worlds; they expel them from heaven, from the atmosphere ; and finally to exterminate them they descend even into the abyss below." [1]

Yet it was from these corporations, so detested and persecuted, that the Aryans, especially of India, received their first knowledge of the arts and sciences, and it seems that some of their members were even admitted into the Brahmanic priesthood.[2] The dragon as an emblem of royalty was primitive among the Chinese, but was probably derived from these Cabiriac corporations. This will

[1] Vid. *Premières Civilisations*, p. 141. Here these corporations are connected with Hermes, Mercury, or Cabirus.

[2] Vid. Ibid., p. 100. Cf. Lenormant, *Manuel d'hist. de l'Orient*, t. iii. pp. 432–437, etc.

appear all the more reasonable when we compare the circle of ideas attached to the *eight kuas*, previously alluded to, relating directly to the cosmos, with the fact that the dragon was also a fundamental character in the Chinese cosmogony. However this may be, we obtain here a glimpse of the varied fortunes of these first civilizers of the world, of the hatred with which they were regarded by the ignorant populations surrounding them, and of the distorted representations of their character transmitted to the after-world by those very races who had so profited by their industries, superior culture, and civil and religious institutions.

SEC. 38. We return here briefly, upon the question of the priority of the Cabiri considered as a temple-craft and as workers in stone, to those who were the inventors of metallurgy, constituting a kind of metal-craft. It is probable that after the discovery of the art of working metals, the two orders, or better, the two branches of one and the same order, existed side by side, and so continued even to a late epoch. But the temple-craft was evidently the most ancient, and was, so to speak, the parent of the metal-craft. We have a plain indication of this fact in the later modification of the characters of some of the Cabiriac deities, to correspond with the new occupation and the circle of conceptions connected with it. The Aryan *Tashtar* affords us here a striking example. M. Lenormant, in the learned paper so often cited, thus alludes to this personage in his association with metallurgy: " The invention appeared so marvelous and so beneficent that the popular imagination conceived it as a gift of the gods. Thus the pretended inventor was usually the mythological personification of fire, the natural agent of this class of labors; such is the *Trachtri* (*Trashtar*) of the Vedas, the *Hephæstos* of the Greeks, the *Vulcan* of the Latins." [1] The etymology of the name *Tashtar* from *Tak* or *Taks*, " to hew," from which come the substantive forms *Takstar* and *Tashtar*, " masterworkman," proves that this personage had no connection originally with metals, since the process of " hewing " relates to wood and stone, not to metallic substances. That Mercury was a Cabiriac divinity has been sufficiently established ; and his most frequent symbol was the dressed stone, either a square or cube, known under the name of " Hermes." His connection with sacred edifices, especially temples, is also to be considered. These facts show that the

[1] *Premières Civilisations*, pp. 86, 87.

character of Mercury, as one of the Cabiri, appertains rather to
temple-structures than to fabrications in metals. Yet the fire-god,
specially worshiped by the metallurgists, is expressly assimilated
to Mercury in some very ancient texts. Such assimilation has its
exact correspondent in the case of the Aryan *Tashtar*, but it could
have occurred only as a secondary and later phase of Mercury's
character, as a divinity appertaining to the temple-craft. The
Dioscuri were included among the Cabiriac deities at Samothrace.
The reference here is to Castor and Pollux, assimilated to the constellation Gemini. The monogram for the month answering to
Gemini, in the Babylonian calendar, is the cuneiform sign denoting
a *brick*; from whence is the phrase, " month of the brick," or
" month of constructions in brick," applied to this zodiacal division.
This circumstance plainly indicates that, in the Euphrates valley,
the Dioscuri were primitively associated with the temple-craft;
and it has been already observed that they were typically *the
brothers*. These considerations, in addition to those heretofore presented, must be regarded as conclusive, I think, that the Cabiri
were originally workers in stone, perhaps in both wood and stone;
and that as such they constituted a regularly organized temple-
craft, whose chiefs united in themselves the kingly and sacerdotal
functions. The same functions were united also in the principal
personages of the metal-craft. Whether as belonging to one or the
other order, these were really the founders of the ancient civilizations, the primitive instructors of mankind in the useful arts and
sciences, as well as in religious knowledge.

Such, we are now fully authorized to say, in view of all the facts
presented, was the origin of *priest-craft* and *king-craft*, to use here
the form of expression frequently adopted by some modern writers,
applied as opprobrious epithets to those who endeavored to conduct
the affairs of mankind before these writers were born. Doubtless,
these ancient constructors and fabricators were men, inheriting the
frailties common to humanity in all ages. On the other side, the
facts now before us tend unmistakably to the conclusion that, while
the races surrounding them were in a state of semi-barbarism, if
not of actual savagism, these primitive " corporations," so properly
termed by Baron D'Eckstein, were engaged in the pursuits of regular industry, practically laying the foundations of social order and
of human progress. It was they who adjusted the first corner-

stones, who built the first forges, who founded the first dynasties. If civilization is to be esteemed a blessing, if civil and religious institutions are of any value to man, the priest-kings of antiquity are entitled to the credit of having originated them, or the teachings of universal tradition are at serious fault. The oldest historical records presuppose their existence and labors, all the mythologies exhibit more or less distinct traces of them, and the earliest monuments known to the world were but the material expression of their doctrines. Whether as *Pa-te-shi* or *Cabiri*, whether as *Dioscuri* or as *Dactyli*, these mysterious personages were those who kept the records of the unknown past, who preserved for us the sacred tradition and science, relating alike to the beginning and to the end of things. It is probable that their doctrines were but illy understood by their less cultured contemporaries, and the distorted representations of their character, on the part of their enemies, their conquerors and persecutors, in fact, render it all the more necessary that we should discriminate in our judgment of them, and, above all, refuse to become ourselves the inheritors of the prejudices of those who were often the most indebted to them for their own advancement.

From the data which have been placed before the reader in the present chapter, nothing appears more obvious than the fact that the foundations of the world's civilization were laid in wisdom, and not in ignorance; were laid by men of intelligence, and not by benighted savages. Not one of the ancient civilizations was of spontaneous growth, having its origin in the crude notions and customs of savage tribes; not one of them but had its genealogy that can be traced back into the darkness of ages prior to its own existence, for the fundamental ideas upon which it was based; ideas so profound, that the wise ones of antiquity considered it an honor to have been instructed in them. Show us, then, the first temple constructed upon the earth: we will show that it had a model, and that this model was the universe itself; the house that God built! Show to us the first civilized community among men: we will prove that this also had its model; that it was conceived as an image of the heavens; of the order and harmony in which the heavenly bodies move; and, compared with the surrounding chaos and barbarism, it literally was such! But we are anticipating here, to some extent, the results of future investigations. What we do show, however,

is the existence of an ancient order of priest-kings, of an origin so
remote that even the earliest traditions afford us no notice of it, of
which, indeed, it might well be said that it was without beginning,
and the question remains to us, whether it has had an end?

Sec. 39.   There are no documents known to the world where
the allusions to the symbols common to the Cabiri are so frequent,
and where the ideas associated with these symbols are so fundamental, as in the books of the Old and New Testament.   These
striking analogies, which will be more definitely set forth in a few
moments, naturally conduct to a standpoint from which to view the
religions of the Bible, especially that of the Hebrews, which has
been rarely adopted, but which is becoming more and more necessary in the light of modern investigations, demonstrating the prior
existence of so many ideas, heretofore regarded as peculiar and essential to the Mosaic religion.   I adopt here, with only slight qualifications, the opinions of A. W. von Schlegel, as cited by Dr.
Movers : —

"The more I investigate the history of the ancient world, the
more firmly I am convinced that a pure worship of the Supreme
Being was primitive among the cultured nations; but that the
magical power of nature over the human imagination at this period
gradually gave rise to polytheism, so that in the end the spiritual
element in the popular belief was completely obscured; while the
sages alone preserved the ancient mysteries in the sanctuary." [1]

I should account for the rise of polytheism upon other grounds
than those indicated by Von Schlegel; and, indeed, M. Mariette-
Bey for Egypt and M. De Vogüe for Western Asia generally have
definitely traced the process which resulted in this lamentable corruption of religious ideas.   However, that the truth was once known
among men, and that subsequently it was obliged to seek refuge in
the sanctuaries and in the mysteries, accords perfectly with the results of my own investigations, and I had long since adopted this
view.   The same, also, as applied to the opinions of Dr. Movers
himself, expressed in the following : —

"Neither from the historical nor from the theological standpoint
can I regard the Mosaic religion as a development out of paganism,
but rather, and in the sense of the sacred writers themselves, as a
restoration of the pure worship of a more primitive epoch; a worship which, at different periods, among the Israelites and their fore-

[1] Vid. Movers, *Phænizier*, i. p. 313.

fathers, had become more or less corrupted; first, and according to Biblical accounts, by Abraham's ancestors in Chaldæa." [1]

Since Christianity is, in respect to many of its fundamental ideas, but a higher statement of the Mosaic religion, I think Dr. Movers' views, with some qualifications, should be applied to this system also; and it is principally upon these grounds that I interpret the facts already developed, and yet to be developed, in the present treatise.

The frequent reference by the sacred writers to the dressed stone, as a symbol of the Messiah and of his church, is a matter familiar to every student of the Scriptures. We cite here a few instances. 1st. Jacob's significant allusion to Joseph, a recognized type of the Saviour : —

"From thence is the shepherd, the stone of Israel" (Gen. xlix. 24). "The stone which the builders refused is become the head stone of the corner" (Ps. cxviii. 22). "Behold, I lay in Zion for a foundation a stone, a tried stone, a precious corner stone, a sure foundation" (Is. xxviii. 16). "And are built upon the foundation of the apostles and prophets, Jesus Christ himself being the chief corner stone ; in whom all the building fitly framed together groweth unto a holy temple in the Lord ; in whom ye also are builded together for a habitation of God through the Spirit" (Eph. ii. 20–22).

Even more frequent are the figures drawn from the process of metallurgy, and from the element of fire as a purifying agent : —

"Whose fire is in Zion, and his furnace in Jerusalem" (Is. xxxi. 9). "The house of Israel is to me become dross : all they are brass, and tin, and iron, and lead, in the midst of the furnace ; they are even the dross of silver. . . . As they gather silver, and brass, and iron, and lead, and tin, into the midst of the furnace, to blow the fire upon it, to melt it, so will I gather you" (Ezek. xxii. 18, 20, etc.). "And I will bring the third part through the fire, and will refine them as silver is refined, and will try them as gold is tried" (Zech. xiii. 9). "For he is like a refiner's fire, and like fullers' sope : and he shall sit as a refiner and purifier of silver : and he shall purify the sons of Levi, and purge them as gold and silver" (Mal. iii. 2, 3). "He shall baptize you with the Holy Ghost, and with fire" (Matt. iii. 11).

Undoubtedly, we are to attribute somewhat in these examples to the ordinary poetic spirit; much less, however, as regards the symbol of the dressed stone. But in both cases, the figures so exactly accord with the character of Jehovah, and with the under-lying conceptions of the two systems of the Bible, that it is difficult not to see here a natural and direct inheritance, on the part of the sacred writers, of the customary parlance of the ancient temple-craft on one hand, and that of the metal-craft on the other. What

[1] Ibid., p. 315.

language could point any more significantly to such a conclusion than the following?

"For we are *labourers* together with God : ye are God's husbandry, ye are God's *building.* According to the grace of God which is given unto me, as a wise *masterbuilder*, I have laid the *foundation*, and another *buildeth* thereon. But let every man take heed *how* he *buildeth* thereupon. For other *foundation* can no man *lay* than that is *laid*, which is Jesus Christ. Now if any man *build* upon this *foundation, gold, silver, precious stones,* wood, hay, stubble ; every man's *work* shall be made manifest : for the day shall declare it, because it shall be revealed by *fire;* and the *fire* shall *try* every man's *work* of what sort it is. If any man's *work* abide which he hath *built* thereupon, he shall receive a *reward.* If any man's *work* shall be *burned*, he shall suffer loss : but he himself shall be saved; yet so as by *fire.* Know ye not that ye are the *temple* of God, and that the Spirit of God *dwelleth* in you?" (1 Cor. iii. 9–16).

Here are not less than a score of words, whose mystical intent might well be insisted upon, and their reference primarily to the "corporations" of which there has been here question.

Sec. 40. Whatever may be thought of the connection of the order of priest-kings to which Christ belonged, which was certainly historical, and which was recognized by Abraham in the person of Melchizedek; whatever we may say of the connection of this mysterious order with the one whose character and history we have attempted to trace, it must be admitted that the sacred writers regarded the Messiah as the Great Master, as the Rock of Ages, the stone *par excellence;* and not less, as the Fire that shall try every man's work. Indeed, who were the "wise men" that came from the East to salute the young Prince of the manger? They were *Magi* (Gr. μάγοι); they were fire-worshipers. The "star" which they had seen in the east was the "bright and morning star;" the *Dun-pa-uddu,* "peaceful prophet of the rising sun." In Mr. Smith's recent work, "The Chaldæan Account of Genesis," the author gives three cuts, illustrative of the fragmentary inscriptions which he thinks related to the building of the tower of Babel. The cylinders from which these cuts are taken certainly represent the operations of building, of some construction in brick or tiles.[1] Three personages are seen upon each cylinder, two engaged in adjusting the tiles, the third being regarded by the author as a divinity. This leads me to doubt whether these representations relate definitely to the

[1] Pp. 158, 159. For another cut, quite similar to the third one given by Mr. Smith, and in all respects allied to the same class of representations, see *Revue Archéologique*, Paris, 1874, September, Plate XV., No. 3, being one among many Babylonian cylinders studied by M. E. Soldi.

tower of Babel; for it is obvious here that both gods and men are engaged in the work, and that, as shown in the second cut, the work is about being finished. The symbolical animals crowning the top of the columns, the folded arms of the superintendent of constructions, in an attitude of repose so different from that shown in the other examples, plainly indicate that the work hastens to completion. But more significant than all else of this idea is the fourth personage shown in this cylinder upon the extreme right. He is obviously a messenger, who goes to proclaim the happy consummation of the labors. In one hand he carries a branch. He is represented as hastening through the valleys and over the mountain heights; and in the east is seen a star, which has already mounted above the horizon.[1] He is a messenger; and it is evident that he proclaims good news. There were two classes of messengers in antiquity; those who brought good news, and they who proclaimed evil tidings. From some peculiarity in their dress, or in the symbols borne by them, it was possible to distinguish them afar off, and while they were yet on the distant mountain-tops. It is this circumstance that the prophet has improved in the highly poetic passage: —

" How beautiful upon the mountains are the feet of him that bringeth good tidings, that publisheth peace ; that bringeth good tidings of good, that publisheth salvation ; that saith unto Zion, Thy God reigneth ! " (Is. lii. 7).

Having a like reference to constructions, and to the victorious branch, symbol of new creations, are the following: —

" Behold, I will bring forth my servant the *Branch*. For behold the stone that I have laid before Joshua " (Zech. iii. 8, 9). " Behold the man whose name is *The Branch ;* and he shall grow up out of his place, and he shall build the temple of the Lord : even he shall build the temple of the Lord ; and he shall bear the glory, and shall sit and rule upon his throne ; and *he shall be a priest upon his throne* " (vi. 12, 13).

These Babylonian cylinders, then, in which, as usual, we see the substitution of brick for stone, represent the progress and final completion of the work upon which these craftsmen, divine and human, are mutually engaged and laboring together. The messenger with branch in hand, as he traverses the mountain heights, and the flaming star already risen in the east, proclaim equally the glad tidings

---

[1] For evidence that it is a *branch* held in the hand of this personage, compare it with those of the sacred tree represented on p. 91 of the same work. Still better, compare the trees shown in Rawlinson's *Five Monarchies*, i. p. 348. That this personage traverses *mountains* is proved by a comparison of this cut with the mountains shown in another plate in Rev. George Rawlinson's work just cited, Ibid., p. 466.

that the work is done, and that the glorious morning has dawned.
One loves to dwell upon these messengers of promise, who have
come down to us through the ages ; upon these great prophets and
kings, instructors and civilizers of the world, who have kindled the
beacon flames on the holy mountains of the past, and as signal lights
reflected across the centuries, mementos of humanity's struggles,
and prayers, and sacrifices, yet beautiful harbingers of final victory.
They warn us of dangers, they exhort us to labor, they encourage
our hopes, imploring us above all to keep the fires renewed, and to
transmit the signals to our own posterity.

   The investigations of the present chapter suffice to prove that the
principle of order has existed on the earth from the beginning ; that
barbarism has had its antagonists and civilization its champions
since the very first act in the great historical drama.   The human
spirit did not slumber through a long night of ignorance, all unconscious of its power and of the propitious destinies that invited it to
activity.   The first storm-cloud spreading its dark wings along the
horizon, the first thunder-bolt shot from its ragged breast, would
teach man the necessity of self-protection, even if there was no God
to direct his faltering steps, no inspiration to guide his efforts in
the midst of a strange life and a strange world.   But the genius
that could vie with the modern artist in sketching a group of animals in the reindeer period would soon penetrate into the deeper
secrets of nature, and bring to a tolerable state of perfection some
of the more useful arts.[1]   I do not admit, however, that there was

---

[1] Vid. the specimens shown in Sir J. Lubbock's *Origin of Civilization,* pp. 24–
26, and note the important admission that races of an epoch long subsequent were
far inferior in artistic skill.   But compare with these forced admissions of the
author the following frank statements of M. Paul Broca, Secretary of the Anthro-
pological Society of Paris, relating to the same specimens : —

   " It is hard to conceive how men destitute of the use of metals were able to
fabricate of bone, ivory, the antlers of the reindeer, an infinite variety of very
delicate utensils ; to carve, I had almost said to chisel, elegant forms, and to
represent by designs engraved in line on the handles of their instruments the figures
of different animals.   These figures are distinguished by an exactness and artistic
skill truly remarkable, and to find in an equal degree the sentiment of art it
would be necessary to revert, through many centuries, to the better times of
Greece.   They form a contrast so absolute with the rude delineations traced on
some Celtic monuments (*of a far later period*), that it might be asked whether
they have not been designed since the historic era by fugitives who may have sought
refuge in the caves of our ancient troglodytes.   But what other than the man of
the quarternary period could have designed in Europe on the bones or horns of
the reindeer the figure of a species of elephant which differs from all living

no Divinity shaping the course of history in those primitive ages. He whose fiery breath melts down sun and planet, ceaselessly roaring in the great furnace of existence, kindles the flames also upon the tongue of the prophet; and the Divinity who merely veils himself behind the living screen of nature, steps forth anon to lead a chosen race through unknown paths, prompting it to illustrious deeds. It is of such races that the redeemers of the world are born. It was of such men, cradled at the hearth-stones of primeval humanity, with the fire-god for their foster-father, that the first priesthoods were formed, the first mystic corporations organized, and it was through these that the sovereign-pontiffs of antiquity might trace their lineage back almost to the natal hour of humanity itself. It was these mystic fraternities, in fact, with their strong hands clasped across the dark periods, the frightful chasms of the world's history, like iron links bridging the abyss that roars below, through whom the sacred inheritances of previous epochs were transmitted, to become the germinal centres of new creations and of new eras. Finally, I believe that it was from this race of civilizers, lovers of truth, of justice, and of humanity, that a precious stone was chosen as the foundation of our own era; He who, when the sacrificial floors were to be re-sanded, and a new victim slain, placed the cup of suffering to his own lips, and ratified the eternal covenant with his own blood. I conclude the present chapter with the following extract, taken from a work already once cited : —

" After all that has been written, perhaps the symbol of Vulcan and the Cabiri may be studied with most effect in the Mosaic scriptures. Among the Harleian Manuscripts is a copy of the constitution of an ancient body of Freemasons, prefaced by a short history, commencing as follows: ' If you ask me how this science was species ? This race of men, so interesting through its civilization, led a peaceable existence. A skull found in the grotto of Bruniquel, of which M. Brun has sent us the photograph, is distinguished by the purity of its form, the softness of its outlines, the little prominence of the apophyses, the slight depth of the muscular insertions, characters incompatible with the violent habits of a savage or barbarous race. What, then, became of this indigenous civilization, so original, so different from all those which are known to us ? Was it modified by slow degrees and transformed to the extent of becoming at last wholly unrecognizable ? No ; it has disappeared in the mass without leaving any trace, and everything tends to the belief that it perished by force." (*Smithsonian Reports*, 1868, pp. 388, 389.)

The same old story from the beginning of the world — barbarism obliterating in a day what only the toils of centuries had been able to accomplish!

first invented, my answer is this: That before the general deluge, which is commonly called Noah's flood, there was a man called Lemeck, as you may read in the 4th of Genesis, whoe had twoe wives, the one called Adah, the other Zilla; by Adah hee begot twoe sones, Jabell and Juball; by Zilla hee had a sonne called Tuball and a daughter named Naahmah; these fower children found yᵉ beginning of all yᵉ craft in the world; Jabell found out geometry and hee divided flocks of sheep and lands; hee first built a house of stone and timber; Jubell found out musick; Tuball found out the smyths trade or craft alsoe of gold, silver, copper, iron and steele,' etc. This Tubal or Tubalcain we may pretty safely identify with Vulcan, the symbol of material art, or of the man understanding and working in nature. It is only in the interpretation of this symbol, and its connection in Genesis, that we can ever hope to discover the beginning of the ancient mysteries and of that system of religion and philosophy that overspread Asia and Greece. In working such a problem, the births of these 'fower children' must be looked at as so many successive manifestations of the spirit in man, producing, in fine, the Greek understanding, and the magic of Samothrace and Thessalonica. Naahmah, the last born, is the virgin Wisdom, that lies deepest in human understanding; and hence the mystic prophecy that Tubalcain, in the last days, shall find his sister Naahmah, who shall come to him in golden attire." [1]

---

[1] *Occult Science*, p. 165. The above extract, and those previously given, are from the article on the Cabiri, by E. Rich, Esq., England. We do not fully indorse the speculations of the author, but value the quotation simply for the facts, and the singular item of Masonic history given. Our investigations are in no sense related to masonry, except in so far as the facts gleaned from antiquity may be construed in this light. Our search is for the simple, naked truth, without reference to any existing organizations, political, religious, or mystical. One thing, however, is certain: the Masonic Order of to-day does not date from ancient Rome, according to the theory of an eminent French writer belonging to this fraternity. Its history evidently goes back into the night of ages. The Roman temple-craft derived their origin from the valley of the Euphrates, since the civilization of Rome is now known to have been derived, in a great measure, from this region, of all which constant accumulations of proof will be found in the chapters following.

BOOK II.

CHAPTER IV.

QUESTION OF THE ORIGIN OF THE DOCUMENTS, AND OF THE ANALOGIES EXISTING BETWEEN THEM.

SEC. 41. In the light of present knowledge, it is necessary to admit the existence of written documents at an epoch far more remote than the critics of a former period believed themselves authorized to suppose. The hieroglyphic system of Egypt seems to have been coeval with the earliest monuments, and the hieratic Accadian is supposed by some to have been brought to the Euphrates valley by the founders of the Babylonian civilization. So far as shown by any discoveries yet made, written documents existed in Babylonia and Chaldæa contemporaneously with the oldest cities and temples. Everything indicates, in fact, that the Hamites learned to write before they entered Egypt, and the Cushites also before they immigrated to the plains of Shinar. In neither country do the monuments afford the slightest hint that the system of writing had been invented after its settlement by the races creating these monuments ; on the contrary, this invention appears in both instances as a traditionary inheritance from an unknown period. But while these statements must be regarded as, in some measure at least, justified by well-known facts, it is difficult for scholars to attach much credit to those traditions, so widely prevalent in Western Asia, that certain sacred writings had existed even in antediluvian times, being preserved from destruction by the waters of the deluge and transmitted to subsequent ages. Yet the fragments of Berosus, regarded so reliable in all other matters, afford direct tes-

timony to the reality of even this supposition. In the account of
the deluge transmitted by Berosus the following passages occur : —

"God appeared to Xisuthrus (the Chaldæan Noah) in a dream,
and warned him that on the fifteenth day of the month Dæsius,
mankind would be destroyed by a deluge. He bade him bury in
Sippara, the City of the Sun, the extant writings, first and last, and
build a ship," etc. "They who had remained in the ark, and not
gone forth with Xisuthrus (after the waters retired), now left it
and searched for him, and shouted out his name ; but Xisuthrus
was not seen any more. Only his voice answered them out of the
air, saying, ' Worship God ; for because I worshiped God am I gone
to dwell with the gods ; and they who were with me have shared
the same honor.' And he bade them return to Babylon, and recover the writings buried at Sippara, and make them known among
men ; and he told them that the land in which they then were was
Armenia. So they, when they had heard all, sacrificed to the gods,
and went their way on foot to Babylon, and, having reached it,
recovered the buried writings from Sippara, and built many cities
and temples, and restored Babylon." [1]

According to Berosus, the city where Xisuthrus lived before the
deluge was the *Larsam* of the inscriptions, the modern *Senkereh*,
situated to the south of Babylon. Sippara, where the sacred books
were buried, was situated to the north of Babylon. Both cities
were noted seats of the worship of the Sun-god, to whom magnificent
temples had been raised at a very early period. We see, then, that
the existence of sacred writings even before the deluge, and their
preservation and transmission to the post-diluvian world, is expressly
affirmed by Berosus, whose remarkable fidelity on all other points
relating to antiquity is now generally admitted. But another tradition prevailed in ancient times quite similar to that just noticed,
and it tends to support it in some measure. Alluding to the children of Seth, third son of Adam, Josephus says : —

"They also were the inventors of that peculiar sort of wisdom
which is concerned with the heavenly bodies and their order. And
that their inventions might not be lost before they were sufficiently
known, upon Adam's prediction that the world was to be destroyed
at one time by the force of *fire*, and at another time by the violence
and quantity of *water*, they made two pillars ; the one of brick, the

---

[1] Vid. Rawl., *Five Monarchies*, i. pp. 145, 146. Rev. Mr. Rawlinson's version
of these " Fragments " is an improvement upon Cory's, and to be preferred for
accuracy.

other of stone; they inscribed their discoveries on them both, that
in case the pillar of brick should be destroyed by the flood, the
pillar of stone might remain and exhibit those discoveries to mankind, and also inform them that there was another pillar of brick
erected by them. Now this remains in the land of Siriad to this
day." [1]

Dr. Whiston, the translator of Josephus, appends a note to the
above, in which he remarks : —

" Although the main of this relation might be true, and Adam
might foretell a *conflagration* and a *deluge*, which all antiquity witnesses to be an ancient tradition, — nay, Seth's posterity might engrave their inventions in astronomy on two such pillars, — yet it is
no way creditable that they could survive the deluge." (Op. cit.)

M. Lenormant adopts the conjecture of M. Maury, that the pillars to which Josephus alludes, traditions concerning which are
known to have been current among the Jews, are in reality but
another form of the Babylonian tradition of the deluge, and thus
observes: " The history of the pillars in the land of Siriad is, then,
as M. Maury had conceived, only a variant of the special Chaldæan
tradition of the deluge." [2] Both M. Lenormant and Professor
Chwolsohn take note of the passage in the Egyptian Manetho,
strongly suspected to be spurious, but to the effect: That the first
Thoth, or Hermes, before the deluge inscribed upon pillars the
principles of knowledge in hieroglyphs and in the sacred language;
and that after the deluge the second Thoth translated the same
into the popular language. [3] Professor Chwolsohn, as previously
cited in these pages, shows that this ancient Thoth was identified
by various writers with the Biblical Enoch, and that he was considered the author of sacred writings. [4]

It is altogether probable, whether we admit the real existence of
these pillars or not, that the tradition concerning them should be
credited originally to those ancient " corporations " whose singular
history we attempted to trace in the last chapter, and who were
specially engaged in fabrications in brick, stone, and the metals
generally. Indeed, I think that the tradition as handed down
varied at times with respect to the materials employed for erecting

[1] *Antiq. of the Jews*, B. I. ii. 3. Whiston's version.
[2] *Frag. de Bérose*, p. 276.
[3] Vid. *Frag. de Bérose*, p. 269 ; *Ssabier*, i. p. 784.
[4] *Ssabier*, i. pp. 781–794.

these pillars, affirming that, while one was of brick, as best adapted
to resist the element of *fire*, the other was of cast brass, as admirably calculated to resist the force of *water*. Be this as it may, we
see here how extensively prevalent in antiquity were the ideas that
the art of writing existed even before the deluge, and that the
sacred science had been transmitted by some means from the antediluvian to the post-diluvian world. One has to yield very much to
the influence of incredulity not to admit that there must have been
some real basis for all these legends. But the tradition as preserved by Berosus, relating to the sacred writings buried during the
deluge, deserves, of the two, the most serious consideration, and we
proceed to show that this account of Berosus must have been fully
credited by the monarchs of Babylon.

Sec. 42. The evidences to be here adduced are drawn from the
inscription of *Nabunahid*, well known to Assyrian scholars, in which
this monarch gives an account of repeated excavations at Sippara
for the discovery of certain mysterious tables supposed to have been
buried beneath the foundation-stone of the temple Ulbar, situated
in that part of the city known under the name of Agané. The
subjoined remarks of M. Lenormant, relating to Berosus' statements concerning the sacred writings, will serve the purpose of introduction to the matter before us : —

" This history of the tables containing the principles of all knowledge, revealed by the theophanies of *Anu* (Gr. *Oannes*), which
had been buried by Xisuthrus at the time of the deluge in order
that they might be transmitted to the post-diluvian world, had
been, as we have shown, the source of the legend quite similar
relating to the columns of Thoth or Seth in the land of Siriad, to
which the Pseudo-Manetho alludes. Josephus says that these pillars existed even in his time ; and here we believe to have again a
Babylonian tradition attaching itself to a real fact, which is revealed
to us by the fragments of the Barrel (inscribed cylinder) of Nabunahid discovered at Mugheir, the ancient Ur, now preserved in the
British Museum.[1] We learn from this, in effect, that when Sagaraktiyas, a king of the first historical dynasty of the Chaldæans,
who was certainly contemporaneous with the kings of the ancient
empire in Egypt, reconstructed the pyramidal temple of the goddess
Anunit, called Ulbar, situated in that part of Sippara known as
Agané, he made certain mysterious tablets in imitation of those carried by Xisuthrus from Larsam (modern *Senkereh*), his native city,

[1] Vid. 1st Rawl. Pl. 69.

to Sippara, and buried them under the corner-stone of the temple Ulbar. These tables were probably thought to be copies of those that had been buried at the time of the deluge, and thus the king, himself really historical, thought to give to his reconstructed edifice a more august consecration in realizing a fabulous tradition. In the course of centuries these tables buried by Sagaraktiyas had become themselves famous and legendary; they had come to be regarded probably as the originals of those of Larsam, hidden for the first time by Xisuthrus. Thus, at an epoch anterior to the thirteenth century before our era, the king Kuri-galzu, who appertained to the fourth or fifth dynasty of Berosus, made excavations in the mass of the pyramid in search of these tables, but without success. Similar labors were undertaken by the kings of later periods, always for the same purpose, yet with no result. It was only at the period shortly before the ruin of the Babylonian power that Nabunahid, after protracted efforts, succeeded finally in discovering the tables buried by Sagaraktiyas." [1]

M. Lenormant then gives a translation of those portions of Nabunahid's inscription that relate to these mysterious tablets, supplying occasional expressions where the lines are broken, the sense being obvious; and these portions of the text are too important not to find a place here. Instead of M. Lenormant's version, however, I take for basis that of Dr. J. Oppert, as improved by M. J. Menant, recurring to M. Lenormant's translation in certain places, where his rendering is more clear than others: [2] —

"The tables of Larsam had been deposited under the foundation-stone (*Temin*) of the temple Ulbar, at Agané, in ancient times, by Sagaraktiyas, king of Babylon, and Naram-Sin, his son, my predecessor; they had not seen the light before the glorious days of Nabunahid, king of Babylon. Kuri-galzu, king of Babylon, who preceded me, made search for them, but he did not find the corner-stone of the temple Ulbar, and thus he made this inscription: 'I have searched for the corner-stone, and I have not found it.' *Assur-akhi-idin* (Assarhaddon), king of the country of Assur (Assyria), king of legions, made search for them." (Vid. col. ii. ll. 28–35.)

Three lines are wanting here, after which the inscription continues: —

"*Nabu-kudurri-ussur* (Nebuchadnezzar), king of Babylon, son of *Nabu-bal-usur*, my predecessor, with the aid of his army, searched for the foundation-stone of the temple Ulbar, and did not find it.

[1] *Frag. de Bérose*, pp. 291-293.
[2] For Menant's version see *Babylone et la Chaldée*, pp. 256-258.

And I, Nabunahid, king of Babylon, restorer of *Bit Saggater* and of *Bit-Zida*, in my victorious years, adoring *Istar* of Aganć, my mistress, I have caused a pit to be excavated. The gods Shamas and Bin directing me, I have searched for the corner-stone of the temple Ulbar, for my own happiness. With the constancy worthy of a king, I have directed my army in the search for this foundation-stone, where Nebuchadnezzar during three years (180 days?) had opened a trench for the excavations. They have explored to the right and to the left, before and behind; and I have searched, and I have not found it. Then they say: 'We have searched for this foundation-stone, and we have not found it. The tempest of waters has inundated everything and has ruined all.' " (Col. ii. ll. 40–57.)

There occurs now a long break in the inscription. But it appears that it gave an account of a second search, under the auspices of Nabunahid; for, where in the third column the text becomes legible again, this monarch goes on to say :—

. . . "the temple of Sin . . . and this temple . . . of the temple Ulbar . . . for the construction of this temple . . . *I have found the corner-stone of the temple Ulbar, and have read the name of Sagaraktiyas at the bottom* " (Col. iii. ll. 15–22.)

Then follows the private inscription of Sagaraktiyas himself, as copied by Nabunahid :—

"Sagaraktiyas, veritable shepherd, august lord . . . me. I say this : The god Shamas and the goddess Anunit have called me to govern the country and the people; they have filled my hand with the tributes from all nations. I say this : The temple of light, the temple of Shamas, my lord, at Sippara, and the temple Ulbar of Anunit, my mistress, at Sippara, had been overthrown even to the base, by *Zu-bu-um* in the days remote. I have cleared away the basement, I have laid bare the foundations, I have removed the heaps of earth, I have disengaged the walls, I have completed the protections of the base : I have examined its foundations, I have transported new earth, I have leveled off the foundations, I have raised above it the first stage, to the glory of Shamas and Anunit, for my happiness. May they accord to me their constant protection. May they prolong my days, may they restore to me my first life, and may they perpetuate in this palace my years of happiness ; may they protect the writing of this monument, and may they enhance the glory of my name." (Col. iii. ll. 20–40.)

There are various renderings of this inscription of Sagaraktiyas, substantially alike, but differing in some details. Following this,

Nabunahid continues, so far as relates to the tables, in this language: —

"It is thus that I have found . . . the name of Sagaraktiyas, king of Babylon, my predecessor, who had constructed the temple Ulbar in Sippara, in honor of Anunit, and who had placed there the foundation-stone. I have replaced in the foundations the *Barrel of the East, the Barrel of the West,* and the foundation-stone of the front; I have renewed the exterior of the temple Ulbar, and I have achieved its magnificence." (Col. iii. ll. 41–46.)

The monarch closes with a prayer to the "Great Divinity" for his own protection, and for the protection of his heart's hope, his son *Bel-sar-ussur* (Belshazzar); little dreaming of the sad destiny that was so soon to overtake him, and which has been so graphically related by the prophet Daniel (chap. v.).

SEC. 43. According to the "Fragments of Berosus," the abode of Xisuthrus before the deluge had been at Larancha, the *Larsam* of the inscriptions, the modern Senkereh. The writings buried at Sippara, as Berosus states, had been recovered after the deluge, and considering the fact of its being the native city of Xisuthrus, it is natural to infer that they had been restored to Larsam. This explains the phrase "Tables of Larsam," as it occurs in the inscription of Nabunahid. It was obviously a technical and current expression, designating the sacred oracles preserved from the destruction of the deluge; and it might be applied as well to officially executed copies of those writings as to the originals. All this appears perfectly rational, provided we admit that any such writings as supposed ever actually existed ; and this is the very point which the modern critic will be most likely to contest. It is evident that M. Lenormant doubts the literal existence at Larsam of any such documents; hence he feels at liberty to attribute to the king Sagaraktiyas an act of deception, in making certain "mysterious tablets in imitation of Xisuthrus," and in attempting to give a more "august consecration" to his reconstructed edifice, by realizing in this manner a "fabulous tradition." I think the author attributes here too much to the negative tendencies of modern criticism, and not enough to the intelligence and sincerity of the great men of the past. Were such monarchs as Kuri-galzu, Assarhaddon, and Nebuchadnezzar, to say nothing of Nabunahid, in their long and laborious search after the "Tables of Larsam," after all

but the ignorant dupes of a pious trick performed by Sagaraktiyas,
with a view to increase his own reputation and popularity? But
it is too late now to doubt the real existence of sacred writings at
Babylon, dating from a period so remote that it is impossible to
assign to it any definite chronology. If the verbal extracts from
those documents, transmitted to us by Berosus, were not enough,
then the "Deluge Tablets" and still later the "Creation Tablets,"
discovered and translated by Mr. George Smith, afford sufficient evidence upon this point. It is almost certain, in fact, from the circumstances connected with these new documents discovered by Mr.
Smith, that even before the time of Sagaraktiyas, whom M. Menant
would assign to the nineteenth century B. C., certain tablets of a
sacred character like those of Larsam were known to exist; so that
there is not the slightest ground for attributing any unworthy
motive or act to this ancient monarch. The documents buried by
him beneath the corner-stone of the temple Ulbar were genuine;
they were the original "Tables of Larsam," or official copies of
them. It is upon this supposition alone that we can account ration-
ally for the subsequent and protracted efforts to rediscover them,
as detailed in the inscription of Nabunahid.

At first I had seriously doubted whether Nabunahid actually
found the "Tables of Larsam." He alludes definitely to the corner-stone as being discovered, and even copies the inscription of
Sagaraktiyas relating to the reconstruction of the temple Ulbar;
but this private document in no sense answers to the Sacred Tables,
for which so long search had been made. It must be, then, the
"Barrel of the East" and the "Barrel of the West," replaced in
the foundations by Nabunahid, which are to be identified with the
"Tables of Larsam." Such being the case, they probably remain
there to the present hour; for the ruins of Sippara have never been
seriously disturbed by modern explorations. I consider it safe to
identify these *Barrels* with the Sacred Tables, first, because they
are the only documents alluded to as being found, except the private inscription of Sagaraktiyas; and, secondly, for the reason that
all the Assyriologues seem to take it for a fact that the "Tables
of Larsam" were actually discovered. Thus M. Lenormant, as
previously cited, observes: "Nabunahid, after protracted labors,
succeeded finally in discovering the tables buried by Sagaraktiyas."
M. Menant remarks: "It was alone under the last king of Baby-

lon that the tables of Sagaraktiyas were found." [1] Finally, Dr. Oppert has the following allusion to the matter : "The Babylonian king relates how he searched for and found the corner-stone of the temple Ulbar at Sippara, where the tables of Larancha (*Larsam*) had been deposited. All the monarchs before had searched for them in vain, Kuri-galzu, Assarhaddon, Nebuchadnezzar, — all had attempted to arrive at this result, which the last king of Babylon attained." [2] All depends upon the point whether the Barrels of the East and West can be properly identified with the "Tables of Larsam ; " and here I submit what appear to be the opinions of the eminent authors cited.

Thus, the tradition which Berosus records, to the effect that certain sacred writings had been preserved from destruction during the deluge and transmitted to the post-diluvian world, was in no sense a superstitious, vulgar notion. It was believed, and so fully credited by the early monarchs of Babylon that they were willing to employ their armies in undertaking immense labors of excavations, searching for supposed copies of them. They believed, moreover, in the literal existence of the " Tables of Larsam," and we trace definitely the prevalence of this idea two thousand years before our era. Still more, it is necessary to admit that such tables really existed ; for the supposition is groundless which attributes an act of pure invention and deception to Sagaraktiyas with respect to the documents deposited by him in the foundations of the temple Ulbar. That which confirms all is the recent discovery by Mr. Smith of various tablets perfectly answering to the character of these sacred writings, and whose original date evidently paralleled that of the reign of Sagaraktiyas himself. We proceed now to the investigation of the general characteristics and the comparative analogies of these ancient documents under the different forms in which we possess them, and so far as they relate to cosmogony.

SEC. 44. Allusion has been made to the theophanies or manifestations of the god *Anu*, the Greek "Oannes ; " it is from the sacred writings attributed to Oannes that Berosus takes the account relative to the creation, which is as follows, according to the version of Rev. George Rawlinson : —

" In the beginning all was darkness and water, and therein were gen-

---

[1] *Babylone et la Chaldée*, p. 97.

[2] *Hist. Chaldée et d'Assyrie*, p. 23.

erated monstrous animals of strange and peculiar forms.(*a*) There
were men with two wings, and some even with four, and with two
faces; and others with two heads, a man's and a woman's, on one
body; and there were men with the heads and the horns of goats,
and men with hoofs like horses, and some with the upper parts of
a man joined to the lower parts of a horse, like centaurs; and there
were bulls with human heads, dogs with four bodies and with fishes'
tails, men and horses with dogs' heads, creatures with the heads
and bodies of horses, but with the tails of fish, and other animals
mixing the forms of various beasts. Moreover, there were monstrous fish and reptiles and serpents, and divers other creatures,
which had borrowed something from each other's shapes, of all
which the likenesses are still preserved in the temple of Belus. A
woman ruleth them all, by name Omarka, which is, in Chaldee,
Thalatth, and in Greek, Thalassa (or 'the sea'). Then Belus
appeared, and split the woman in twain; and of the one half of her
he made the heaven, and of the other half the earth; and the beasts
that were in her he caused to perish.(*b*) And he split the darkness,
and divided the heaven and the earth asunder, and put the world
in order: and the animals that could not bear the light perished.
Belus, upon this, seeing that the earth was desolate, yet teeming
with productive power, commanded one of the gods to cut off his
head and to mix the blood which flowed forth with earth, and form
men therewith, and beasts that could bear the light. So man was
made, and was intelligent, being a partaker of the divine wisdom.
Likewise Belus made the stars, and the sun and moon, and the five
planets." [1]

There are some variations in the above translation from that of
Cory, as copied by Mr. George Smith.[2] I note these differences
according to the letter in the foregoing extract. At *a*, Cory's version adds the words, "which were produced of a twofold principle." At *b*, Cory has the sentence: "All this was an allegorical
description of nature. For, the whole universe consisting of mois-
ture, and animals being continually generated therein, the deity
above-mentioned took off his own head, upon which the other gods
mixed the blood as it gushed out, and from thence formed men.
On this account it is that they are rational, and partake of divine
knowledge." Rev. Mr. Rawlinson takes this sentence to be an in-
terpolation, and so far as relates to cutting off the head of Belus, and
the formation of man from the blood mixed with earth, it evidently

[1] Vid. Rawlinson, *Five Monarchies*, i. pp. 142, 143.
[2] *Chaldæan Account of Genesis*, pp. 40–42.

is a corruption ; for a like description occurs again in a connection more natural. The same cannot be said, however, of the other portion of the passage : " All this was an allegorical description of nature," etc., which has every appearance of being genuine. There are some phrases employed where Cory's rendering is preferable, even if it is not quite so literal, as, " cut the woman asunder," instead of " split the woman in twain ; " so, too, " divided the darkness, and separated the heavens from the earth," instead of " split the darkness, and divided the heaven and the earth asunder." The Babylonian cosmogony contemplates the work of creation as a twofold operation ; first, as *a process of division*, secondly, as *an act of generation*. The same ideas are fundamental in the Mosaic text. The art-monuments, representing the first stages of creation, are found to vary according to one or the other of these ground conceptions, and the language descriptive of the same frequently shows a corresponding variance, or the same terms may be taken in a double sense. Sometimes the creative agent is figured with sword in hand, about to cut the woman asunder ; and at others, as the mind or spirit brooding upon and impregnating the watery chaos. But I wish to present here a portion of Rev. Mr. Rawlinson's comments upon the text of Berosus, as translated by him in the foregoing extract : —

" It has been generally seen that this cosmogony bears a remarkable resemblance to the history of creation contained in the opening chapters of the book of Genesis. Some have gone so far as to argue that the Mosaic account was derived from it. Others, who reject this notion, suggest that a certain ' old Chaldee tradition ' was ' the basis of them both.' If we drop out the word ' Chaldee ' from this statement, it may be regarded as fairly expressing the truth. The Babylonian legend embodies a primeval tradition common to all mankind, of which an inspired author has given us the true ground-work in the first and second chapters of Genesis. What is especially remarkable is the fidelity, comparatively speaking, with which the Babylonian legend reports the facts. While the whole tone and spirit of the two accounts, and even the point of view from which they are taken, differ, the general outline of the narrative in each is nearly the same." [1]

The chief points in which the author considers the Mosaic and Babylonian cosmogonies different are more definitely stated in the two notes appended to the above, which I subjoin : —

[1] *Five Monarchies*, i. pp. 143, 144.

" The Chaldee narrative is extravagant and grotesque; the
Mosaical is miraculous, as a true account of creation must be ; but
it is without unnecessary marvels, and its tone is sublime and solemn."   " In Genesis the point of view is the divine : ' *In the
beginning God created* the heaven and the earth, and the *Spirit of
God* moved upon the face of the waters.'   In the Chaldee legend
the point of view is the physical and mundane, God being only
brought in after a while as taking a certain part in creation."

That which is really " grotesque " in the Babylonian account is
the long and detailed description of the preëxisting chaos, which,
in the Mosaic text, is cut down to a single verse : " And the earth
was without form and void ; and darkness was upon the face of the
deep.   And the Spirit of God moved upon the face of the waters "
(Gen. i. 2).   In a literary point of view one cannot hesitate to prefer the simple grandeur of the Mosaic description, although the fundamental idea is substantially the same in both narratives as concerns this chaos.   It might be some justification of the Babylonian
description of the preëxisting chaos to say that it is " allegorical,"
as is especially claimed for it in Cory's version of this document.
Another element of the grotesque here is the personification of
chaos under the form of a woman ; but this arose from the fundamental doctrine of the " active and passive powers of the universe,"
so widely held in antiquity, according to which *matter* was conceived as passive, as female, while *mind* or *spirit* was regarded as
active, and as male.   There is here really no difference in doctrine,
but rather in the habits of thought leading to a different mode of
treating the subject, in the apprehension of the most important facts
to be presented, whether the original chaos, or the subsequent process of creation.   The Mosaic account curtails the description of
chaos, while the Babylonian gives to this subject by far the greatest prominence.   The one describes the creative process minutely,
while the other sums up all in a few sentences.   Nor is there an
essential doctrinal difference as relates to the fundamental process
of creation.   It is conceived in both accounts as a process of *division ;* division of the primordial chaos, heaven being formed of one
portion and the earth of the other.   That which really constitutes
" the beginning," a preëxisting chaos, on one hand, or the first creative act calling forth the chaos from nonentity, on the other
hand, is the one point, if anywhere, in respect to which a wide doctrinal distinction is to be discovered.   The Babylonian system

assumes the watery chaos as this " beginning ;" but as usually interpreted the Mosaic account assumes an *absolute creation* of chaos as the beginning.

SEC. 45. The one great question here regards the assumption of a purely spiritual existence as prior to all matter on one side, or the eternity of mind *and* matter as one twofold principle, on the other. The Babylonians held the doctrine last indicated; while Biblical exegetes, for the most part, give a construction to the very first verse of Genesis which implies the opposite idea. M. F. Lenormant has correctly apprehended this whole subject in his excellent " Commentary upon Berosus," and his views merit a careful study : —

" We do not believe ourselves wanting in respect for the Holy Writings, in revealing the striking analogies in form which exist between the Biblical and Babylonian cosmogonies. These analogies are such in all their characteristic details that it is impossible for us not to consider the account of creation in Genesis, as well as that of the deluge, as ante-Mosaic documents preserved by oral tradition, and inserted in the first book of the Pentateuch; documents the existence of which was admitted and demonstrated in the seventeenth century by a great Catholic critic, Richard Simon, of whom Bossuet has well said, that hardihood in exegesis inclines to a severe orthodoxy. In my eyes, these two documents constitute an ancient traditional recital, carried from Chaldæa by the Abrahamites in their migration from Ur to Canaan."

" But if the *rédacteur* of Genesis has admitted this recital into his work, almost without any changes, he has given it a sense entirely new, and diametrically opposed to that held by the Chaldæan priesthood. An abyss, in effect, separates the two conceptions of the Biblical and Babylonian cosmogony, notwithstanding the striking resemblances in the external form. On one side we have the eternally existent material organized by a demiurge, who emanates from its own bosom ; on the other the universe is created *ex nihilo* by the infinite power of a divinity purely spiritual. To give to this recital, repeated in the sanctuaries of Chaldæa, a sense entirely new ; to transform the pantheistic conceptions the most material and gross into the light of true religion ; . . . it has sufficed for the *rédacteur* of Genesis to add at the commencement of the whole, and before the description of the chaos, with which the Chaldæan and Phœnician cosmogonies began, this simple verse: 'In the beginning God created the heaven and the earth.' Thus, the free act of a spiritual creator is placed before the existence even of chaos, which the pagan pantheism believed anterior to all ; this chaos, the

first principle for the Chaldæans, from whence even the gods themselves emanated, became *a creation* that the Eternal caused to appear
in time."

"As the lamented Abbé Le Hir, with his profound philological
science, has observed: 'God commences by drawing from nonentity
the elementary material. All other interpretation of the first verse
of Genesis is without support. It is not without intention, that
Moses has employed the verb *Bara* (ברא), opposing it to other verbs
occurring in the same account with the sense of 'to make, to fash-
ion, to form.' The first term (*Bara*) expresses more than this.
In the constant usage of the language, it denotes an operation more
divine, more radical, more creative than the others. In the present
instance, it cannot be distinguished from them in sense, if it does
not express the idea of *creation*, properly speaking."[1]

It is impossible not to admire the profound insight and critical
ability exhibited in these extracts; and I wish I could conscien-
tiously admit the author's remarks to be perfectly sound throughout.
One thing is obvious: he has seen clearly the exact point of material difference, if any, between the Mosaic and Chaldæan cosmogonies. But M. Lenormant's argument, based upon the sense of the
Hebrew verb *Bara* (ברא), is, in my view, a failure; although the
meaning attached to this term by him, that of a creation *ex nihilo*,
is usually maintained by exegetes. The reasons for doing so, however, are rather theological than etymological. All Hebrew scholars admit that the primary sense of *Bara* is "to cut, to cut out,
to carve," and thence "to form, to create, by cutting or carving,"
a process implying necessarily a preëxisting material. But that
which completely refutes the arguments of M. Lenormant and the
Abbé Le Hir is the fact that *Bara* occurs again in the twenty-first
verse, applied to the creation of whales: "And God *created* great
whales, and every living creature that moveth," etc. It would be
ridiculous to maintain *here* a creation *ex nihilo*. I shall return to
the consideration of this important word *Bara*, important for the
Babylonian as well as Hebrew cosmogony, in another connection.
For the present, and so far as relates to a comparative analysis of
the two systems, it is evident the primitive idea was one and the
same in both accounts.

No one can doubt that, as a literary production, and as an appropriate description of the work of creation, the Mosaic account is

[1] *Frag. de Bérose*, pp. 72-74.

infinitely superior to the Babylonian. The distinguishing characteristics of the Hebrew narrative are its extreme brevity on all points, and the perfect order and consecutiveness in the development of the system. The Babylonian account, on the contrary, expands, amplifies at each stage, and there is confusion in the order of events described. We see this disposition to enlarge, to go into details, in the description of chaos, as given by Berosus. In the documents translated by Mr. George Smith, the same characteristic will appear at other stages of the narrative. If we had the entire Babylonian record in a state of complete preservation, doubtless it would occupy many chapters, instead of one or two, as found in Genesis. Has the Babylonian narrative been enlarged and amplified from a briefer original? or, rather, has the Mosaic account been abridged from primitive documents much more extensive? These are matters very difficult to decide. My opinion is, that the Babylonian system has been more or less corrupted by additions, and that there were originally in the Hebrew documents some details which do not now appear.

Sec. 46.   The most valuable among Mr. Smith's important discoveries in the field of cuneiform researches is the "Chaldæan Account of Genesis." With Mr. Smith's translation, together with the chief portions of the original text, as published in Dr. Delitzsch's recent work,[1] it is possible to gain quite a clear idea of this new addition to cuneiform as well as to Biblical literature. As the tablets are quite fragmentary, affording but few continuous narratives, I select the portions more with reference to other accounts than to the order given them by the translator. The following relate to the primordial chaos, for the most part, corresponding to the second verse of Genesis, and to the main portion of the " Fragments of Berosus," already given : —

A. " (1) When above, were not raised the heavens; (2) and below on the earth a plant had not grown up: (3) the abyss also had not broken open their boundaries ; (4) the chaos (or water) Tiamat (the sea) was the producing-mother of the whole of them. (5) Those waters at the beginning were ordained ; but (6) a tree had not grown, a flower had not unfolded. (7) When the gods had not sprung up, any one of them; (8) a plant had not grown, and

---

[1] *Assyrische Lesestücke*, etc., Leipzig, 1876, pp. 40-45. Probably the text has been published in England, but I have not been able to obtain it from this source.

order did not exist; (9) were made also the great gods, (10) the
gods Lahmu and Lahamu they caused to come . . . (11) and they
grew . . . (12) the gods Sar and Kisar were made . . . (13) A
course of days, and a long time passed . . . (14) the god Anu . . .
(15) the god Sar, and"[1] . . .

The chaos personified by a woman, as in Berosus' account, is first
described; then the generation of the principal cosmical potencies
in pairs, as Lahmu and Lahamu, Sar and Kisar, after which a course
of days, a long time passes. This evidently marks the first stage,
answering to Gen. i. 1, 2, or perhaps to 1–5, including the first creation day of Moses. But I think the whole should be confined to
the first two verses of Genesis. We see here an entire absence of
the description of strange and monstrous forms inhabiting the primal chaos, such as we find in Berosus. There exists something
corresponding to this, however, in the Cutha legend relating to the
same subject, where the lines 10–23, as rendered by Mr. Smith, read
as follows : —

B. "Men with the bodies of birds of the desert, human beings
with the faces of ravens; these the great gods created, and in the
earth the gods created for them a dwelling. Tamat (Tiamat) gave
unto them strength, their life the mistress of the gods raised, in the
midst of the earth they grew up and became great, and increased
in number; seven kings brothers of the same family, six thousand
in number were their people ; Banini their father was king, their
mother the queen was Milili, their eldest brother who went before
them, Mimangab was his name, their second brother Midudu was
his name (the names of the other brothers wanting)."[2]

The foregoing passage shows that Berosus has faithfully reported
the Babylonian conceptions, relative to the monstrous inhabitants
of chaos, and moreover that they dwelt upon the earth. This corresponds to Genesis; for it was the *earth* which was "without
form, and void," not the heaven. Tiamat ruled over them on the
earth, also giving them life and strength, so that they grew up in
the midst of the earth, and became great, increasing in number.
We proceed now to that portion of the tablets pertaining to the
heavenly bodies, corresponding to the verses 14–18 of the Mosaic
narrative : —

C. "(1) It was delightful, all that was fixed by the great gods.
(2) Stars, their appearance (in figures) of animals he arranged.

---

[1] *Chaldæan Account of Genesis*, pp. 62, 63.   [2] Ibid., p. 103.

(3) To fix the year through the observation of their constellations, (4) twelve months (or signs) of stars in three rows he arranged, (5) from the day when the year commences unto the close. (6) He marked the positions of the wandering stars (planets) to shine in their courses, (7) that they may not do injury, and may not trouble any one, (8) the positions of the gods Bel and Hea he fixed with him. (9) And he opened the great gates in the darkness shrouded, (10) the fastenings were strong on the left and right. (11) In its mass (that is, the lower chaos) he made a boiling, (12) the god Uru (the moon) he caused to rise out, the night he overshadowed, (13) to fix it also for the light of the night until the shining of the day, (14) that the month might not be broken, and in its amount be regular. (15) At the beginning of the month, at the rising of the night, (16) his horns are breaking through to shine on the heaven. (17) On the seventh day to a circle he begins to swell, (18) and stretches toward the dawn further. (19) When the god Shamas (the sun) in the horizon of heaven, in the east "[1] . . .

Mr. Smith has sufficiently pointed out the resemblances existing between the foregoing passage and the Mosaic account of the fourth day of creation. This part of the Babylonian narrative was considerably longer than is shown here, yet it all appertained to the same subject, which is condensed in Genesis into five or six verses, affording another illustration of the disproportion of matter included in the two accounts. The next portion of the Babylonian text, rendered by Mr. Smith, is too fragmentary to be of much service to us. It relates to the production of "living creatures," and corresponds to the sixth day of creation as described in Genesis. There should be here an account of the creation of man; but this part of the tablet is almost wholly destroyed. The following, however, has some relation to the subject : —

D. "(5) The god Ziku (noble life) quickly called; director of purity, (6) good kinsman, master of perception and right, (7) causer to be fruitful and abundant, establisher of fertility, (8) another to us has come up, and greatly increased, (9) in thy powerful advance spread over him good. (10) may he speak, may he glorify, may he exalt his majesty. (11) The god Mirku (noble crown) in concern raised a protection, (12) lord of noble lips, savior from death, (13) of the gods imprisoned, the accomplisher of restoration, (14) his pleasure he established, he fixed upon the gods his enemies, (15) *to fear them he made man*, (16) *the breath of life was in him.* (17)

[1] Ibid., pp. 69, 70.

May he be established, and may his will not fail, (18) in the mouth of *the dark races which his hand* has *made*. (19)   The god of noble lips with his five fingers sin may he cut off; (20) who with his noble charms removes the evil curse. (21)   The god Libzu wise among the gods, who had chosen his possession, (22) the doing of evil shall not come out of him, (23) established in the company of the gods, he rejoices their heart."

We give below the address of the divinity to the newly-created human pair, before the event of their transgression.

E. "(8) Every day thy god thou shalt approach (or invoke,) (9) sacrifice, prayer of the mouth and instruments . . . (10) to thy god in reverence thou shalt carry. (11) Whatever shall be suitable for divinity, (12) supplication, humility, and bowing of the face, (13) fire thou shalt give to him, and thou shalt bring tribute, (14) and in the fear also of god thou shalt be holy.   (15) In thy knowledge and afterwards in the tablets (writing), (16) worship and goodness shall be raised, . . . (19) the fear of god thou shalt not leave, . . . (20) the fear of the angels thou shalt live in . . . (23) When thou shalt speak also he will give." [3] . . .

That which related to the fall of man is wanting ; but the curse pronounced upon man after the transgression is thus rendered : —

F. "(12) Lord of the earth his name called out, the father Elu (13) in the ranks of the angels pronounced their curse.   (14) The god Hea heard and his liver was angry. (15) because his man had corrupted his purity.   (16) He like me, also Hea, may he punish him. (17) the course of my (his ?) issue all of them may he remove, and (18) all my (his ?) seed may he destroy.   (19) In the language of the fifty great gods, (20) by his fifty names he called, and turned away in anger from him.   (21) May he be conquered and at once cut off. (22) Wisdom and knowledge hostilely may they injure him. (23) May they put at enmity also father and son and may they plunder.   (24) To king, ruler, and governor, may they bend the ear.   (25) May they cause anger also to the lord of the gods, Merodach.   (26) His land may it bring forth, but he not touch it ; (27) his desire shall be cut off, and his will be unanswered ; (28) the opening of his mouth no god shall take notice of ; (29) his back shall be broken and not healed ; (30) at his urgent trouble no god shall receive him : (31) his heart shall be poured out ; and his mind shall be troubled ; (32) to sin and wrong his face shall come." [3] . . .

SEC. 47. The primeval chaos of the Babylonian cosmogony, as

[1] *Chald. Acct. of Genesis*, pp. 82, 83.   [2] Ibid., pp. 78, 79.   [3] Ibid., pp. 84, 85.

portrayed to us in the "Fragments of Berosus," was inhabited by various species of monstrous composition, of which it is necessary to give here, if possible, some rational explanation. (*a*) "There appeared men, some of whom were furnished with two wings, others with four, and with two faces. They had one body, but two heads; the one that of a man, the other of a woman." It is perfectly easy to recognize here the representations of the original androgynous divinity, symbolizing the "twofold principle" from which everything proceeds; and which so often appear upon the art monuments. Anu and Anatu were conceived as such androgynous principle; and were assimilated respectively to heaven and earth, the upper and lower hemispheres. Alluding to Anu, Mr. Smith remarks: "He represents the universe as the upper and lower regions, and when these were divided, the upper region or heaven was called Anu, while the lower region or earth was called Anatu; Anatu being the female principle or the wife of Anu." "When Anu represents height and heaven, Anatu represents death and earth; she is also lady of darkness, the mother of the god Hea, the mother producing heaven and earth."[1] The same notion was expressed by the two halves of the sphere. When Anu represented the superior heaven, Anatu was put for the inferior hemisphere. (*b*) "Other human figures were to be seen with the legs and horns of a goat." The god Pan, and the satyrs generally, were usually figured in this form upon the monuments. Pan was put often for the cosmos, or world-all, and his flute is supposed by some to refer to the harmony of the spheres. (*c*) "Some had horses' feet, while others united the hind quarters of a horse with the body of a man, resembling in shape the hippocentaurs." The celestial horses, as found upon our modern sphere, answer precisely to this description. (*d*) "Bulls likewise were bred there with the heads of men." The human-headed bulls of Assyria are well known. These were the *Lamassu* of the inscriptions. Two classes were recognized; the Lamassu of heaven and the Lamassu of earth. Those of heaven could be no other than the celestial Taurus. (*e*) "And dogs with fourfold bodies terminated in their extremities with the tails of fishes." On the ancient spheres, two dogs were represented as guardians of the limits of the sun's northern and southern course; and it is probable there is a reference to them in this passage. They were conceived

1 Ibid., pp. 54, 55.

always as having a monstrous form. (f) " Horses also with the heads of dogs." Such figures may have existed upon the ancient sphere, and upon some art monuments, but I am not able to recall any such representations. (g) " Men, too, and other animals, with the heads and bodies of horses, and the tails of fishes." We naturally call to mind here the centaurs and the constellation Capricorn ; both appertaining to the sphere, and similar to this description. (h) " In addition to these, fishes, reptiles, serpents, with other monstrous animals." The constellations offer exact correspondences to all these forms. In the majority of instances, then, it would be easy to identify these monstrous animals with those figured even upon the modern sphere, and it is probable the ancient sphere contained many more of them.

" A woman ruleth them all, by name Omarka, which is in Chaldee Thalatth, and in Greek Thalassa, or ' the sea.' " Sir H. Rawlinson and M. F. Lenormant agree in identifying the name Omarka with the *Um-Uruk* of the inscriptions, signifying " mother of Erech." As for the other name, Thalatth, Sir H. Rawlinson holds the following: " Now the goddess thus indicated is well known to the Assyrian student under the name of *Te-li-ta*. . . . She is the goddess of the *Bar* (*Bara*), which is the first element in the name of *Bar-zip* (*Bara-zip*) or Borsippa." [1] Sir Henry's authority for assimilating *Te-li-ta* to the goddess *Bara* appears to be the cuneiform phrase which reads thus: *an nin bara = te-li-ta*, " the divine mistress *Bara* is *Te-li-ta*." [2] It is evident, then, that the Thalatth of Berosus, identified by Mr. Smith with the Tiamat, or *Ti-sal-mat*, of the " Creation Tablets," may be considered one with the " mistress Bara " of the cuneiform text just cited.

The Accadian *Bara* signifies " altar, glory ; to perfume, perfume " (Rep. 318) ; and it has a variant often confounded with it, with the values *Sara*, " to commence, commencement " (Rep. 319). As the term *Bara* is composite, it may be resolved into its two elements, the first of which is *Ba*, " to tear, to cut, to fabricate, to labor " (Rep. 108) ; and the second *Ra*, " inundation, deluge," etc. (Rep. 303). The waters of the deluge were assimilated frequently to those of the primal chaos: and the god Hea, who appears so prominently in the " Deluge Tablets," is represented also as im-

[1] Vid. Rawl. *Herod.*, i. p. 502, and note 3.
[2] Vid. Rawl. Pl. 59. Revs. l. 16.

pregnating the water of chaos.[1] The two elements, therefore, *Ba*, " to cut, to divide," and *Ra*, " watery chaos," constituting the name of the goddess *Bara*, assimilated to Thalatth or Tiamat, involve the double conception of the *division of chaos*, the separation of the primordial sea, out of the two portions of which heaven and earth were formed. But more than this, the variant of *Bara* is *Sara*, having the sense of " commencement," or of " a beginning." We might almost say that the expression *Sara + Bara* involved the express notions, " In the beginning the waters of chaos were divided."

The Babylonian zodiac was employed for marking the divisions of three distinct periods of time : 1st. The twelve hours of the day, called *kar-bu*, or " double hours ; " 2d. The twelve months of the year ; and 3d. The twelve divisions of the great cosmical year, which opened at the very dawn of creation.[2] Each of these periods commenced with the sign Aries, or the Ram, if we admit that the hours of the day were reckoned from evening, as in the first chapter of Genesis : " And the evening and the morning were the first day," etc. Hence, the zodiacal sign Aries marked the three commencements, — the beginning of the day, the beginning of the year, and the beginning of the cosmical year, of time itself, marked by the first creative act. What we have to say now is that the cuneiform character *Bara*, having its variant of *Sara*, constitutes the monogram for the first month of the Babylonian year, answering to this identical sign Aries, or the Ram. In other words, this character *Bara*, thus employed, marked the beginning of the day, of the year, and of the world.

SEC. 48. It results, from the data presented in the last section, 1st. That Anu and Anatu, as the original androgynous principle, were assimilated to the two halves of the Babylonian sphere, the upper and lower hemisphere, or the superior and inferior heavens ; and that these were put respectively for the two chief divisions of the cosmos, or heaven and earth. Ample proof of these statements will appear in future chapters. 2d. That the majority of monstrous forms inhabiting the primal chaos, as described by Berosus, may be recognized even upon the modern sphere, leading to the inference that originally all of them were to be found figured there. 3d. That the goddess *Thalatth*, *Te-li-ta*, or *Ti-sal-mat*, who

---

[1] Vid. Lenormant, *Frag. de Bérose*, p. 68.

[2] Ibid., pp. 188–191, and 234–236 ; where the existence of these three time periods is fully shown.

represents the chaos in the Babylonian system, is definitely assimilated to the zodiacal sign Aries, under the name *Bara*, or the "mistress Bara ;" that very sign which marks the primary division of the sphere from west to east into the upper and lower hemispheres, otherwise put for heaven and earth, the chief divisions of the cosmos. 4th. The name *Bara*, and its variant *Sara*, in their combined import, distinctly involve the notions: (*a*) of a *watery chaos;* (*b*) of its being *cut asunder;* (*c*) at the *beginning* of all things; (*d*) as the first act in *fabricating* the world. In view of these facts I think that it may be considered as established that the *Babylonian sphere was taken as a symbol of the cosmos ; and that its primary divisions and orderly arrangement for the purposes of a calendar and other practical uses were held to represent the process of creation and the development of order out of the original chaos.* This primal chaos was represented by the sphere in its unformed, confused state, prior to its systematic division and arrangement ; and the constellation Aries, where all the periods commenced, marked the beginning of all things, as well as the primary division of the cosmos symbolized by the two hemispheres.

I have not space in the present chapter to offer further proofs of the important hypothesis just stated ; but the course of our investigations in the chapters following will lead naturally to other facts, fully demonstrating the correctness of the position here assumed. My aim here has been to arrive, by a process as direct as possible, to the proper standpoint from which to study the Babylonian cosmogony and its relations to that of the Mosaic record. We return now to a comparison of the Hebrew and Babylonian representations of the primitive chaos, a point upon which the subjoined remarks of Mr. Smith have a direct bearing : —

"It is evident that, according to the notion of the Babylonians, the sea was the origin of all things, and this also agrees with the statement of Genesis i. 2, where the chaotic waters are called תהום, 'the deep,' the same word as the Tiamat of the Creation text and the Tauthe of Damascius." "Beside the name of the chaotic deep called תהום in Genesis, which is, as I have said, evidently the Tiamat of the Creation text, we have in Genesis the word תהו, waste, desolate, or formless, applied to this waste. This appears to be the tehuta of the Assyrians — a name of the sea-water (Hist. of Assurb. p. 59) ; this word is closely connected with the word tiamat or tamtu, the sea. The correspondence between the inscription and

Genesis is here complete, both stating that a watery chaos preceded the creation, and formed, in fact, the origin and groundwork of the universe. We have here not only an agreement in sense, but, what is rarer, the same word used in both narratives as the name of this chaos, and given also in the account of Damascius." [1]

The author goes on to identify the word Tiamat with Thalatth, also, as it occurs in Berosus' account. The important facts above stated, in relation to which no doubt seems to exist, have a bearing equally important, and deserve a careful study. The same term employed in Gen. i. 2, to designate the primal chaos, this being the Hebrew *Tehōm*, " the deep," occurs in the Babylonian account as personal name of the goddess Tiamat, or Thalatth, representing this chaos. Did the *rédacteur* of Genesis borrow this expression from the Babylonians, rejecting its mythological sense, converting a personal name into an ordinary word? Or, did the Babylonian account borrow from that of the Hebrews, adding to an ordinary designation of " the deep " a strictly personal and mythological conception? Or, again, did both systems derive this word from an account still more original, each giving to it its own local interpretation, in accordance with the spirit of the two peoples and their respective religions? These are difficult questions to answer satisfactorily, and it is probable that different views will prevail in relation to the subject. But one fact favors strongly the opinion of the priority, and even originality, of the Babylonian conception. It is, that even the Mosaic account conceives creation as in some sense a *generative process ;* and this notion is not only fundamental in the Babylonian cosmogony, but it accords perfectly with the prevailing ideas and habits of thought in antiquity. In such case, then, the idea of a passive, a female principle, associated with the primal chaos, is quite natural, and almost necessary. Nevertheless, the Hebrew verb *Rā-khaph* (רָחַף). in the expression, " And the Spirit of God *moved* upon the face of the waters," answers better to the ancient doctrine of " the mundane egg " than to that of a goddess. Whatever view is adopted, we have to admit that the Babylonian mode of conception is not without justification, even as compared with Genesis.

SEC. 49. We come again to the question as to what constitutes the " beginning " in these two cosmogonies. At first view there

---

[1] *Chaldæan Account of Genesis,* pp. 64, 65.

appears here a difference in the mode of expression, if not in fundamental idea. The Babylonian system assumes the *chaos* as the
" beginning," while the Mosaic account represents the *act*, implied
in the term *Bara*, as the " beginning." But we have seen that the
woman, personifying the chaos, is considered in the inscriptions
as the goddess *Bara*. What are the evidences that the Hebrew
and Accadian *Bara* were originally one and the same word ? 1st.
Their exact phonetical equivalence. 2d. They both involve the
notion of " cutting, carving, dividing," and thence that of "fabricating, creating," by means of cutting and dividing. 3d. Both
appear to have had reference primarily to the customs of the altar,
especially the cutting up of the victims destined for the sacrifices.
The Accadian word signifies " an altar," while the Hebrew *Bara* is
strictly cognate with *Bā-rāh* (בָּרָה), " to cut, to cut asunder," from
which comes *Be-reeth* (בְּרִית), " covenant," from the idea of cutting
up the victims offered in ratification of the covenant. 4th. The
Accadian *Ba-ra* involves the notion of " division of the waters,"
referring to the waters of the deluge assimilated to those of chaos;
while the Hebrew term signifying also " to cut, to divide," is put
in direct connection with the watery chaos. 5th. The Accadian
character constitutes the monogram for the first month of the year,
answering to the zodiacal Aries, and marks thus the " beginning "
of the day, of the year, and finally of the cosmical period, which
opened at the instant of the Mosaic " beginning," denoted by the
Hebrew *Bara*. 6th. The Accadian monogram marks also the primary division of the sphere into two equal portions, these being
assimilated to the two chief divisions of the cosmos, or heaven and
earth. If we adhere to the primary sense of the Hebrew *Bara*, it
denoted the primary division of chaos into two portions, heaven
being constituted of one, and the earth of the other. Admit, as we
do, the sense of " created " for the Hebrew term ; yet the creation
involved necessarily a process, and this process was one of " cutting," of " division." [1]

---

[1] The manuscript of the present work had been completed, so far as concerns the text, when the *Bibliotheca Sacra* (Andover, Mass.) for July, 1876,
came to hand. In the article entitled, " An Exposition of the Original Text of
Genesis I. and II.," by Rev. Samuel Hopkins, the author repudiates the notion
of " creation *ex nihilo*," as implied in the Hebrew *Bara* (pp. 512–515). In fact,
this writer's exegesis of Gen. i. 1, 2, agrees very nearly with what we have

Such remarkable coincidences, and so many of them, must be considered as demonstrative ; the two words were originally one ; and this had a technical usage in both systems, as implying the fundamental notions of creation and the creative process. It results also from these data that the idea of a creation from nonentity, or *ex nihilo*, attached to the Hebrew *Bara*, as advocated by M. Lenormant and the Abbé Le Hir, as well as by most exegetes, is perfectly foreign to the doctrine as well of the Mosaic as of the Babylonian cosmogony. The dogma of a purely Spiritual Existence separate from and prior to all material existence cannot be deduced from either system, and it was probably unknown to the world at the period when these cosmogonies were originated. Mr. Smith has stated the simple truth on this point in this passage already quoted : " The correspondence between the inscription and Genesis is here complete, *both stating that a watery chaos preceded the creation, and formed, in fact, the origin and groundwork of the universe.*" As to what constitutes the " beginning " in the two cosmogonies, if we admit a common origin of the Hebrew and Accadian *Bara*, the difference resolves itself into that between the two usages of the same word ; one as a substantive, being only another name for the chaos, yet involving the ideas of division and fabrication ; the other in a verbal sense denoting division and creation, with special reference to chaos, and the universe formed from it.

The next point of comparison between the two cosmogonies, for which the Babylonian documents afford a sufficient basis, is that which relates to the heavenly bodies. The cuneiform text appertaining to this subject, as rendered by Mr. Smith, is that which we have marked *C*. We have here about thirty lines, instead of the brief summary of Berosus, and the five or six verses of Genesis. The importance attached to this portion of the labor of creation, even in the Hebrew system, is seen in the fact that an entire day is devoted to it. The resemblances between the two documents in certain parts are very striking ; but the cuneiform text includes much in relation to which the Mosaic account is entirely silent.

explained as " the beginning," according to the Babylonian cosmogony, although he evidently conceives an abstract Spiritual Existence prior to all matter. It is obvious, I think, that the Mosaic doctrine implied a division of a previously existing chaos as the first *act* of creation ; and such seems to have been the universal traditionary idea.

This extra matter has obvious reference to the *first risings* of the sun, moon, and planets, fixing their orbits and the like. In some of the ancient mythologies, there were reckoned three distinct phases of the sun's course; there was the nocturnal sun, the winter's sun, and the sun of the primordial night of chaos; and to each of these was attached its appropriate rising. Ideas of this nature were associated with Osiris among the Egyptians. Something similar was held also with respect to the moon. The moon was thought by the Egyptians to have made its first rising at Hermopolis, and the sun at Heracleopolis.[1] These notions appear to have been quite ancient among the Hamites of Egypt, and were probably such with the Babylonians. It is by no means certain, therefore, as one might at first conclude, that they were interpolated into the Babylonian cosmical system at some late epoch. "The great gates in the darkness shrouded," mentioned in the ninth line of the text, appear to be the gates of the sun. According to classic authors, recently confirmed by the investigations of Professor Romieu, the Egyptians located the solar gates, called "Kents," at the points where the sun attained the summer and winter solstices, or in the signs Cancer and Capricorn. We shall cite Professor Romieu on this matter in a future chapter. The tenth line of the text has the words: "The fastenings were strong on the left and right." This must refer to the north and south poles; for *Su-me-la*, or *Su-mi-lu*, which in the original is put here for the "left," is a term usually designating the north, or north pole, in the inscriptions. Mr. Smith's interpretation of these two lines appears to be somewhat different, but my view conforms more, I think, to prevailing ideas. "In its mass (in the lower hemisphere) he made a boiling; the god Uru (the moon) he caused to rise out" (ll. 11, 12). This is evidently the moon's primordial rising at the beginning of time; at the commencement of the first month and the first year of the world. The notion of "a boiling" is not peculiar to the Babylonian text; it is plainly implied in the Hebrew term *Tehōm*, designating the chaos.

SEC. 50. But that portion of this text which most resembles the Mosaic account of the fourth day's labor is the following: "Stars, their appearance (in figures) of animals he arranged. To fix the year through the observation of their constellations, twelve months (or signs) of stars in three rows he arranged, from the day

[1] Vid. De Rougé, *Nomes de l'Egypte*, p. 25.

when the year commences unto the close. He marked the positions of the wandering stars (planets) to shine in their courses, that they may not do injury, and may not trouble any one" (ll. 2–7). This language certainly presupposes the existence of the sphere and of the constellations; and the labor here consists simply in arranging and methodizing them for the purposes of a calendar. The ruling expression is, "he arranged," not, as in Genesis, "God *made*," or "*let* there be." One naturally conceives here the Babylonian sphere in its primitive, confused, chaotic state. A multitude of constellations, or figures of animals having a monstrous form, had been marked out in the heavens according to fancy or some preconceived notion, and these had served the crude purpose of marking the seasons, and had been handed down from father to son. But there existed no order here, no fixed and scientific arrangement. The goddess Tiamat, or Thalatth, ruled over this incongruous mass, and thus she was the producing mother of all. There came a time when order was introduced into this chaos. The primary divisions of the sphere were definitely located, the zodiacal constellations were arranged to correspond with the months, the calendar was fixed. A consummate wisdom and an exalted symbolism presided over this work; for there was embodied here the cosmogony, the historical tradition, the religious system, and almost the entire sacred science of Babylon. We have not space in the present connection to detail the evidences in support of this last statement; but they will be constantly multiplying upon our hands throughout the present work. Suffice it to say, that the primary division of the sphere into two equal portions from east to west, marked by the month *Bara* and the sign Aries, represented the first act of creation, the division of chaos; and the two halves of the sphere thus divided were assimilated, the one to heaven ruled by Anu, the other to the earth ruled by Hea. The earth was then the abyss, corresponding to the watery signs of the zodiac. It was this region, with the exception of the labors of the fourth day, to which all the subsequent acts of creation appertained. We shall see in a future chapter that even the terrestrial paradise was mystically located in the inferior hemisphere.

As before remarked, the two accounts, in the portions referred to, are very similar; particularly as regards the fourteenth verse in the Mosaic text: "And God said, Let there be lights in the firmament

of heaven to divide the day from the night; and let them be for
signs, and for seasons, and for days, and years." The definite allusion to the "signs" and to "years," relating thus to the calendar,
can hardly be distinguished in idea from the Babylonian text,
though the language is not the same. We have seen that the account furnished by the tablets presupposes the existence of the
heavenly bodies, of the constellations, of the sphere, in fact, and
that the labor bestowed is merely in arranging them, in reducing
them to order. Does the Mosaic text imply all this? The reference to the "signs" is certainly very significant. We know now,
from the tablets, that the zodiacal signs were originally intended.
Nor do the expressions "let there be" and "God made," as usually
interpreted, differ materially in sense from the phrase "he arranged," in the Babylonian account. Professor Bush remarks:
"The original word for 'made' is not the same as that which is
rendered 'create.' It is a term frequently employed to signify
constituted, appointed, set for a particular purpose or use" (notes
in loc.). The idea, then, is almost identical with that expressed by
the terms "he arranged." This is not in either account a creation
de novo: and the reference to the "signs," understood in our present light derived from the tablets, presupposes the existence of the
constellations, if not of the zodiac itself; and the work of arranging, appointing, with a view to a calendar, seems to be implied
equally in both texts.

Briefly now, upon the disproportion of matter included at this
point of the two narratives. The question occurs here again:
which account has been derived from the other? Or have both
drawn from a common text earlier than either? It would be easy
to say that the Mosaic text has been abridged from the Babylonian;
but it would be very difficult to prove it. There are some things
extremely puerile, savoring of gross ignorance and superstition, in
the Babylonian text, suggesting the idea of later additions, as this
relating to the planets: "That they may not do injury, and may
not trouble any one" (l. 7). They were "wandering stars," and
the intention appears to be to allay any fears of collisions and
catastrophes. But, on the other hand, the reference to the first
risings, and opening of the first month, points to notions really
ancient, and strictly appertaining to the cosmical systems of antiquity. There does not exist in this any evidence of later additions.

On what theory, then, shall we account for the disproportion of matter in the two texts which in other respects exhibit such an exact correspondence? I think the difference which is here so apparent is mostly due to the fact that the Cushites of Babylon, like the Hamites of Egypt, had taken the sphere as the symbol of the cosmos, and as the embodiment of their cosmical doctrines. It is on this ground that it is possible to explain the long description of the preëxisting chaos in the account of Berosus, and which is almost wholly absent from the Mosaic text. The notions of the first risings, also, connected with those of a primordial sun and moon, located in the lower hemisphere, seem to belong to the same symbolic mode of representing the cosmos. The celestial gates, and fastenings of the left and right, belong to the same category of ideas. But in addition to all this, we must take into consideration the polytheistic conceptions involved in the Babylonian text, which hardly appertained to the primitive period. For the most part, everything relating to these habits of thought has been excluded from the Mosaic text; and for this reason it is far more brief upon nearly all the points treated. There are certain portions of the Creation Tablets in which it is easy to detect expressions of a late origin compared with the main course of the narrative. In the address to the newly created human pair, which we have marked *E*, we have allusion to "instruments," to written "tablets," to "tribute," the "bowing of the face," etc., demonstrating that the document in its present form is not primitive.

SEC. 51. In the text marked *D*, the eighteenth line, occurs the phrase, "In the mouth of the *dark races which his hand has made*." The original for "dark races" is *zalmat-qaqadi*, as read by Mr. Smith; and the translator has the following important statement in connection with this subject: —

"It appears from line 18 that the race of human beings spoken of is the *zalmat-qaqadi*, or dark race, and in various other fragments of these legends they are called *Admi* or *Adami*, which is exactly the name given to the first man in Genesis. The word Adam used in these legends for the first human being is evidently not a proper name, but is only used as a term for mankind. Adam appears as a proper name in Genesis, but certainly in some passages is only used in the same sense as the Assyrian word, and we are told on the creation of human beings (Gen. v. 1): 'In the day that God created man, in the likeness of God made he him; male and female created he them ; and blessed them, and called their name Adam, in the day when they were created.' It has already been pointed out by Sir Henry Rawlinson that the Babylonians recognized two principal races: the *Adamu*, or dark race, and the *Sarku*, or light race ; probably in the same manner that two races are mentioned in Genesis, the sons of Adam, and the sons of God. It appears incidentally from the fragments of inscriptions that it was the race of Adam, or the dark race, which was believed to have fallen, but there is at present no clue to the position of the other race in their system."[1]

Upon the term *Adam* (אדם), Dr. Gesenius has the following: (1) " *A man*, a human being, male or female, properly our *red, ruddy* (from אדם 'to be red'), as it would seem." (2) "*A man*, not a woman." (3) " *Any man*, any one." (4) " *Adam*, proper name of the first man, Gen. ii. 7, sq. At least in these passages אדם assumes the nature of a proper name in a certain degree " (*sub voc.*). Dr. Ernest Muir considers the collective sense, designating humanity in general, as fundamental in the word *Adam*, and upon this ground accounts for the fact that it takes no plural form.[2] The name of the second son of Adam, or *Abel*, is also habitually employed in the cuneiform text in a generic sense, as designating a son in general, instead of as a personal name ; thus *Ab-lu*, " a son " in the Assyrian, is the exact equivalent of the Accadian *Tur-us*, " male child, or son."[3] It is probable that the name of *Cain* had a like generic meaning, *i. e.* " offspring," considered as a " possession." It is found under the substantive form in the Himyaric inscriptions, denoting " a possession " in general.[4] All the facts, then, tend strongly to support the Babylonian view of the two races, and of the generic sense of these names. The Mosaic text cannot be said positively to contradict this theory, but on the contrary it lends more or less direct support to it. Who were these *sons of Elohim* said to have intermarried with the *daughters of Adam?* (Gen. vi. 1, 2). The opinion heretofore held by exegetes, that these " sons " were descendants in the line of Seth, and these " daughters " the posterity of Cain, although attended with difficulties, has seemed to be the most reasonable. But in the light of the " Creation Tablets " and of the facts developed by Sir H. Rawlin-

[1] *Chald. Acct. Genesis*, pp. 85, 86.   [2] *Heb. Wurzelwörterb.*, p. 359.
[3] Vid. Norris, *Assyr. Dic.*, i. p. 9. Cf. Schrader, *Keilinschriften*, etc., p. 106.
[4] Vid. Lenormant, *Doc. Math.*, notes, pp. 129, 130.

son the problem assumes quite a different aspect. It looks as though Adam and Eve ought to be interpreted as the *Primitive Church*, instead of as the *First Human Pair;* and numerous passages might be cited from the New Testament to confirm this theory. Paul to the Ephesians says (v. 31, 32) : " For this cause shall a man leave his father and mother, and shall be joined unto his wife, and they two shall be one flesh. This is a great mystery : but I speak concerning Christ and the church." Here the identical language addressed by Adam to Eve (Gen. ii. 24) is interpreted with reference to Christ and the church. We shall return to these questions in the sequel of the present work.

The actual occurrence of the name *Adami* or *Adam*, in these new documents translated by Mr. Smith, in connection with the facts before developed, seems to exclude all doubt as to the primitive and direct relationship existing between the two narratives. Only two suppositions are admissible hereafter : one account has been derived from the other, or both have been drawn from the same original, more primitive than either. Nor was the original, from which these two cosmogonies proceeded, a simple oral tradition, but rather a written document. The identity of terms, the similarity of expression, which appears here would not be found in two separate oral traditions, even if they did proceed from the same source. It is a question of some importance here : whether this original document was written in the cuneiform character, or in some other ? It would be difficult to prove the existence of the Hebrew or Phœnician character at Babylon as early as the time of Abraham ; or of any other system except the cuneiform. It is probable, then, that the document here assumed as primitive was written in the character last designated. The original of both these narratives of creation was thus a cuneiform text. It must have been preserved in the sanctuaries of the valley of the Euphrates, and under the supervision of the priests or priest-kings appertaining to them. Nevertheless, I do not believe that the " Creation Tablets " truly represent the documents assumed here as primitive. They are a later, more popular, and in some respects a corrupt version of the original cosmical narrative. The allusions to "instruments" in worship, to " bowing of the face," to the " tablets," in the address to the newly created human pair, as already suggested, are expressions inconsistent with a truly archaic and primitive account of the creation. ´ There was a more ancient, more pure, and less extended account ; one more ancient than the Assyrian copies, or the Accadian texts from which the copies were made. Whether this really primitive text will ever be discovered or not is of course very doubtful ; but if it exists to-day it is not in the ruins of Nineveh ; it must be sought, if anywhere, among the ruins of those cities nearer the mouth of the Euphrates; in Babylonia, or even in lower Chaldæa.

But the tradition of the creation did not first take its rise in the valley of the Euphrates, nor yet in the " highlands of Elam," where some Assyriologues would locate the *Karsak Kurra* of the inscriptions, or the " mountain of the world." The great Olympus of all Asia was situated far to the north and east, in relation to Babylon ; and it was from thence that this tradition was derived. But of this hereafter. I am not able to prove that the temple-craft, whose history was attempted to be traced in the last chapter, were actually those through whom the sacred writings pertaining to creation, the deluge, etc., had been transmitted to historical times ; yet I am unable to resist the conviction that such was the fact. It is possible that the future may confirm this hypothesis, or, perhaps, demonstrate that it has nothing to support it. Certainly, the existence of sacred writings at a period immensely remote, if not even before the deluge, was sufficiently indicated by the data presented at the opening of the present chapter. It was shown, also, that these writings were supposed to have been deposited under the corner-stones of temples. All this favors the supposition that the temple-craft had some active connection with such documents and their transmission to the historical period. The further progress of our investigations will develop facts showing that this conjecture is at least a consistent one.

CHAPTER V.

CREATION CONCEIVED AS A TEMPLE, AND THE TEMPLE AS AN
IMAGE OF CREATION.

SEC. 52. In the religious and philosophical conceptions of high antiquity, the process of creating the world and the method of constructing the temple were regarded as fundamentally the same, namely, that of DIVISION. It was for this reason that the notions of the cosmos and of the temple were usually assimilated to each other. Not only was the ground conception of this process one of division, but it was derived originally from the customs of the altar, the practice of dividing, of cutting up, the victims destined for the sacrifices. In the ratification of covenants the victim offered was divided into two portions placed side by side, or in case of its being cut into several pieces these were arranged in two rows, between which the contracting parties passed, both being sprinkled with the blood of the victim by a third person styled the mediator. There are three words belonging to as many different languages, and even families of languages, which are more or less directly connected with this subject, and it will facilitate these investigations to devote a special consideration to them, together with some of their cognate expressions.

1st. The Hebrew *Bara* (בָּרָא), upon which we have already made some comments. It will save frequent citations to state here once for all that my usual authorities in Hebrew lexicography have been, and will continue to be, Drs. Gesenius and Fürst, together with Dr. Bresslau, who chiefly follows Fürst.[1] The term *Bara* is thus defined : (1) " To cut, to cut out, to carve, to form by cutting or carving." (2) " To form, to create, to produce." (3) " To beget, to bring forth." (4) " To feed, to eat, to grow fat, from the

[1] Vid. Robinson's Gesenius' *Heb. Lex.*; Fürst, *Heb. u. Chald. Schulwoerterb.*; Bresslau, *Heb. and Eng. Dic.* The full titles are given in the list of authorities.

idea of cutting up food." The word is equivalent to *Barah* (בָּרָה),
which signifies : (1) " To cut, to cut asunder." (2) " To eat," like
Bara No. 4. From this last comes *Bereeth* (בְּרִית) : " Originally,
the cutting of animals of sacrifices ; transferred to covenant, treaty,
on account of the custom to pass between two rows of animals cut
for sacrifices." It will be seen at a glance that the fundamental
conception of each of these terms corresponds exactly to the first *act*
of creation as described in the " Fragments of Berosus" previously
cited : " All things being in this situation, Belus came and cut the
woman (personifying chaos) asunder, and of one half of her he
formed the earth, and of the other half the heavens." In this case
the two portions of the victim slain corresponded to the two chief
divisions of the cosmos, or to heaven and earth. The notion
obtained very anciently that creation was in some sense a sacrifice,
and we see here how closely related were the two ideas as shown in
the etymological relation of *Bara*, " to create," to *Bereeth*, " covenant," ratified by sacrifices. It was in strict accordance with conceptions of this kind that a new covenant or dispensation established between God and man, which was sealed by a sacrifice, was
often conceived as a new creation, or a new heaven and earth.

2d. The Accadian *Bara* has been already partially explained,
but it is necessary to add here a few remarks. The signification
of " cutting," and thence of " fabricating," has been shown to be
inherent in the first element *Ba*, and the signification of " altar "
attached to the term sufficiently indicates a primary reference to
the practice of cutting up the victims. We have seen that *Bara*
was connected with the zodiacal sign Aries, or the Ram, as monogram for the first month of the Babylonian year. The entire Accadian name of this month was *Bara-saggar*, " the altar of recti-
tude," literally, " the altar that makes right." Aries marked the
commencement of three distinct periods of time, that of the day,
the year, and of the great cosmical year, which was supposed to
begin with the very dawn, the first act of creation, this act being
one of division, of cutting, correspondent to the customs of the
altar. It was doubtless the practice at Babylon to sacrifice a ram
at the beginning of each day, and also at the commencement of
each year ; for this reason a ram might be conceived to have been
offered at the opening of the great cosmical year, which would give
rise naturally to the notion expressed by the Revelator, of the

" Lamb slain from the foundation of the world" (Rev. xiii. 8). The reference of this word to the goddess *Bara*, representing the primal chaos, has been shown, and also the fact that it marks the primary division of the sphere through the sign Aries, whose two halves symbolized, as will be shown, the two chief divisions of the cosmos. We shall see hereafter that the sphere was conceived as a zodiacal temple, whose chief departments corresponded to the two hemispheres. The analogy, therefore, between the temple and the cosmos, on one hand, and between the first act of creation and the practice of dividing the victim, on the other, reappears here very clearly.

3d. We come now to the Greek *Temenos*, the Latin *Templum*, " a temple." The Greek verb *Temnō*, from which *Temenos* is derived, has the primary sense of " to cut, hew, cut to pieces;" (1) of men, " to cut, wound, maim ; " (2) of animals, " to cut up, cut to pieces, to slaughter, to sacrifice ; " hence, " a covenant, truce, made with sacrifices." The substantive *Temenos* denotes " a piece of land cut or marked off, assigned as a private possession ; " " a piece of land marked off from common use and dedicated to a divinity " (Liddell and Scott, Gr. Lex. *sub voc.*). The Latin *Templum* has about the same sense and etymology ; thus, " a space marked off," particularly, in the language of augury, " an open place for observation," " the extent or circuit of the world," " a consecrated or sacred place, a sanctuary." The Latin *Tempus*, " time," is derived from the same theme ; thus, " the root *Tem*, from which the Greek *Temnō*, properly denotes a section, portion, division ; in particular of time, a period of time, daytime, and time in general " (Andrews' Lat. Dic. *sub voc.*). At the base of all these forms is the Aryan radical *Tem* or *Tam*, " to cut, to divide." There seems to be some difference of opinion, however, as to the corresponding forms of this root in the older branches of the Aryan family. The " St. Petersburg Dictionary," as stated by Professor Curtius, classes it with the Sanskrit *Tamâlas*, " sword," and *Tamas*, " darkness," the ground signification being that of " darkness " (Grundzüg, p. 221). But its original sense of " to cut, to divide," its reference to the customs of the altar, and finally its application to a cut off section of space, considered as a temple and a divided portion of duration, giving rise to the notion of time, are points upon which no doubts can be entertained. That the forms *Temenos* and *Templum* were put for the entire circuit of the world, for the cosmos itself, is also well known.

Let us see now what notions are common to these terms just noticed, either as etymologically involved in their signification, or as closely related to them by association of ideas.   1st. We have the sense of " cutting, division, separation."   2d. A reference to the customs of the altar, the practice of cutting up the victims to be offered.   The Greek *Temenos* and the Accadian *Bara* directly imply this, and the Hebrew *Bara* is nearly allied to *Bereeth*, having the same reference.   3d. The idea of a temple.   The sacrifices were naturally associated with the temple.   The Greek *Temenos*, Latin *Templum*, are ordinary words designating such structures.   The Accadian *Bara* marked the primary division of the zodiacal temple. A like reference of the Hebrew *Bara* can be traced only indirectly. 4th. The notion of time.   We have the Latin *Tempus*, "time." The Accadian *Bara* marked the commencement of three distinct time periods, one of which was the great cosmical year.   The Hebrew *Bara* is directly associated with the " beginning " of time itself, corresponding to that of the cosmical period.   5th. The cosmos.   The two Aryan terms, *Temenos* and *Templum*, were often applied to the cosmos, conceived as a temple, and the Accadian and Hebrew words were directly associated with the process of creation. If we institute a comparative analysis, then, of the conceptions attached to these three words, the resultant principle as underlying all seems to be quite apparent, namely, that the fundamental doctrine of creation and of the temple was the same, being that of *division*, and that this was derived primarily from the customs of the altar, the practice of dividing the victims destined for the sacrifices.

SEC. 53.   We proceed to offer some definite proofs relative to certain statements heretofore made, which were only partially substantiated at the time, or left entirely without support.   First, then, as regards the zodiacal temple, in the syllabaries the Accadian *Ki*, signifying " place, ground, earth," is equated to the Assyrian *As-ru* (Syl. No. 181), which has consequently the same meaning.   Now *As-ru*, like the Assyrian *Asar*, was put for the inferior hemisphere of heaven, as in the phrase *pal-asar*, " son of the lower hemisphere," and it designated also a certain class of temples conceived as images of the inferior heaven.   Referring to the two orders of temples representing the two halves of the sphere, M. F. Lenormant remarks : —

"They are designated, in effect, the first by the signs *Bit-khi-ra*, the second by the signs *Bit-mat ;* that is to say, by the same ideographic notation as the two halves of the celestial sphere ; the *As-ru* and the *Laq-qadu* or *Bit-Sadu*, the inferior and superior heaven, the nadir and the zenith. These are, then, two temples united, yet placed in opposition, as image of the two hemispheres of the universe, the inferior and superior, the infernal and celestial."[1]

It is beyond question, therefore, that the Babylonians regarded their sphere in the light of a temple, or as two temples placed in opposition, yet united in the same manner as the two hemispheres. We shall see hereafter that this mode of conception was by no means peculiar to the Babylonians, but was common to many nations surrounding them. In fact, the idea of the "celestial gates" so prevalent in antiquity only lends support to the statement here made, since the celestial gates presuppose a celestial temple.

It was stated in the last chapter, and it has been repeated in this, that the two halves of the Babylonian sphere were put for the two chief cosmical divisions, or heaven and earth. Mr. George Smith has been already cited to the effect that *Anu* "represented the universe as the upper and lower regions, and when these were divided, the upper region, or heaven, was called Anu, while the lower region, or earth, was called Anatu, Anatu being the female principle or wife of Anu." But that which the author styles here "the earth" was assimilated to the inferior heavens, or the lower hemisphere. Anu was put for the upper hemisphere, and Anatu for the lower. This is proved by various facts ; among others, that both are often equated to the Accadian *An*, "heaven ;" one as the superior heaven, the other as the inferior.[2] The goddess Tiamat, who personifies the watery abyss, or chaos, evidently appertains to the inferior heavens, and as such is the same as Anatu, producing mother of heaven and earth ; yet she is represented as on the earth,

---

[1] *Fraq. de Bérose*, p. 392. Cf. pp. 133 and 134, note.
[2] Vid. 3d Rawl. Pl. 69, No. 1, Obs. ll. 1, 2. The divine name which Mr. Smith reads *Anatu* is, I suppose, the Accadian *Tum*. In the text cited the Accadian *An* is equated first to the god *Anu*, and then to the goddess *Tum* or *Anatu ;* and as Anu is certainly put for the upper, Anatu by opposition must answer to the lower heaven ; yet, as the author correctly says, she represents the earth.

and likewise the chaos itself, as noted in the last chapter. But it will suffice for the present to add here the fact just developed, of the assimilation of *As-ru* to the Accadian *Ki*, "earth," on one side, and to the inferior heavens, represented by the temples *Bit-khi-ra*, on the other. It is perfectly apparent, then, that the two halves of the Babylonian sphere were taken as symbols respectively of heaven and earth, the two chief divisions of the cosmos. The temple and the cosmos are here shown in direct relation to each other, both being represented by the sphere or zodiac.

The same connection of ideas may be established with respect to the Hebrew tabernacle. The Holy of Holies, or interior sanctuary, as all tends to show, represented heaven. According to the Epistle to the Hebrews, the entrance of the high-priest once a year, and on the great day of the atonement, into the Holy of Holies, was a type of Christ's entrance into heaven itself. This interior sanctuary was thus put for the heavens as one of the chief divisions of the cosmos; and from this fact, it is necessary to infer that the outer tabernacle represented the other chief division, or the earth. All this was long since established by the critical researches of Dr. Bähr, who writes substantially as follows:—

"Humanly speaking, and according to prevailing ideas which were familiar to the Hebrews, the building constructed for the Deity, the house in which God dwelt, was the creation itself, including heaven and earth. In general, all the works of God were conceived as buildings; insomuch that the creation of Eve was represented as a building (Gen. ii. 22; *ba-nah*, 'to build'). But, especially, the created world is described in the technical language peculiar to architecture, as in Job xxxviii. 4–7.

"This symbolical character, which attaches generally to the sacred edifice, lends significance to the details of its description; and it affords for us, likewise, a better explanation of its order and arrangement. It falls principally into two chief divisions, of which one (the Holy of Holies) takes in a special sense the name of the whole, — the dwelling, the house, or the tent of God. With the Hebrews, accordingly, of the two divisions of creation itself, the one, the heaven, was regarded as the peculiar abode of Divinity. By analogy, therefore, it is necessary to regard the especially so-called *dwelling* as an image of the heavens, while the court corresponds to the other chief division of the world, or to the earth." [1]

[1] *Symbolik des Mosaische Cultus*, i. pp. 77–79. The language of Job, cited by Dr. Bähr is quite remarkable, and is as follows: " Where wast thou when I laid

Dr. Bähr shows that the ordinary terms employed in the Hebrew text, applied to the tabernacle as a conceived dwelling of God, such as *Beth*, "house," *Ahel*, "tent," and *Mishkan*, "dwelling," are equally used to designate the heavens as God's dwelling-place. The denial on the part of certain modern critics that the Jewish temple was conceived properly as a "dwelling of God," except perhaps as a dwelling of the *Sam*, or "name" of Jehovah, is, in my view, without a single valid reason; and it contravenes not only the fundamental principles of the Mosaic system, but its entire spirit. Jehovah was primitively the Divinity of the Hearth; and the Mosaic cultus was an expansion of this idea, as shown in our second chapter. The doctrine that God dwells, inhabits, the same as man, is both fundamental and vital to the two religions of the Bible. The two words designating the "temple" in the Old Testament, *Beth* and *Hekal*, mean "house;" and it is impossible legitimately to attach to them a sense different from that of a dwelling. The word *Beth* is an ordinary word for "house" in all the Semitic tongues. The other term, *He-kal*, is now known to have been borrowed from the Accadian language. Thus, the Accadian *E*, "house," and *gal*, or *kal*, "great," means "great house," or "palace." It frequently designates "a palace" in Hebrew; and this was its primitive sense in the sacred language of Babylon. But the first element *e*, or *he*, preceded by the characteristic of divinity, is the ordinary cuneiform expression for "heaven" put in opposition to "earth." Hence, as Dr. Bähr has correctly maintained, the especially so-called dwelling of God, that is to say, the Holy of Holies, represented the "heaven" as one of the chief divisions of the cosmos; and by analogy the outer tabernacle was put for the "earth." For the same reason, also, it inclosed a cubical space, like the superior sanctuary of the pyramidal temple, and similar to the cubical stone, both being the recognized symbolical modes of representing heaven, as proved in the third chapter.

the foundations of the earth ? declare, if thou hast understanding. Who hath laid the measures thereof, if thou knowest ? or who hath stretched the line upon it ? Whereupon are the foundations thereof fastened ? or who laid the corner-stone thereof ; when the morning stars sang together, and all the sons of God shouted for joy ?" The allusion in the closing language to the cosmical morning, when the heavenly bodies made their first risings, and the harmony of the spheres was first evoked, is quite significant ; and we shall see hereafter that it accorded perfectly with notions prevailing at a very early epoch.

Sec. 54. If the Holy of Holies represented heaven, and the outer tabernacle the earth, then the vail suspended between and separating them ought to denote the primary division of the cosmos, corresponding to the fundamental division of the Babylonian sphere from west to east, marked by the Accadian monogram *Bara;* symbolizing the first act of creation, or the division of the primal chaos into two portions, from which heaven and earth were constituted. We have quite a demonstrative proof of this in the monogram Bara, whose ordinary Assyrian equivalent is *Pa-rak-ku* (Syl. No. 255), this being one with the Hebrew *Pa-rak* (פָּרַק), "to break, to break down, to separate," from which comes *Pā-rō-keth* (פָּרֹכֶת), "a vail, curtain of separation, which separated the Holy of Holies from the outer sanctuary in the tabernacle." [1] Dr. Delitzsch, as cited below, assimilates the Assyrian *Pa-rak-ku,* equivalent of the Accadian *Bara,* directly to the Hebrew *Pā-rō-keth,* "vail" of the tabernacle. We see, then, that this vail involved the same notion as the Accadian monogram *Bara,* which marked the division of the zodiacal temple from west to east through the sign Aries, or the Ram, and that both had primary reference to the separation of chaos and the formation of heaven and earth. Thus, at every step the doctrine previously announced comes more clearly into the light and assumes fundamental importance; namely, creation and the temple proceed from one and the same ground thought, that of division; and this had its origin in the customs of the altar, the division or cutting up of the victims offered in sacrifice. The Ram, or Aries, to which *Bara* corresponds in the zodiac, was one of the animals usually selected for such purposes; as for instance, the Ram offered in ratification of the covenant between God and Abraham (Gen. xv. 9–17; xxii. 13–18).

The difference between the Assyrian and Hebrew methods of representing the cosmos by means of the temple was mainly in this: while the Assyrians employed two separate edifices, one put for

[1] Vid. Delitzsch, *Assyr. Studien,* II. i. p. 127. note. The author objects to the signification of "altar" for the Assyrian *Pa-rak-ku;* but his reasons for it are not pertinent. The idea of "division," "cutting," appertains especially to the altar, and is inherent in the Accadian *Bara.* At the same time, the notion of the vail, separating the two chief apartments of the temple, and symbolically representing the primary division of the chaos, is also perfectly in accord with the facts already developed in these pages. The author is thus correct in assimilating the *Pa-rak-ku* to the Hebrew *Pā-rō-keth,* "vail."

heaven and the superior hemisphere, the other for the earth assimilated to the lower hemisphere, the Hebrews embodied the same
notions in the two chief apartments of one and the same edifice,
these being divided by a vail. The genesis of the Hebrew tabernacle from the pyramidal temple is now perfectly plain. For the
superior sanctuary of a cubical form representing heaven, and constituting the eighth stage of the pyramid, the tabernacle substitutes
the Holy of Holies, also of a cubical form, representing the celestial region. The seven stages of the pyramid, symbols of the seven
lights of the earth, are replaced by the candlestick with seven
branches, placed in the outer sanctuary. The base of the pyramid
and the sanctuary therein answer to the outer tabernacle, both being
designed to represent the earth. The pyramidal temples, as Dr.
Bähr has shown, were primitive in all Asia. But with the Phœnicians and Hebrews the same doctrines had been given a material
expression somewhat different. The design in all was to represent
by an artificial structure, as a fit abode of the divinity, the house
built by the Deity himself, or in other words, the cosmos, such being
the archetypal abode of the Supreme Being. All the ancient cosmical theories centred in these fundamental ideas.

The statement has been made that the world, or cosmos, was
regarded as a temple by the classic nations. This is a fact too well
known to require much testimony in substantiating it. Dr. Bähr
cites a multitude of authorities to this effect, whose names even are
too numerous to appear here ; and we must refer the learned reader
to the various passages quoted by him.[1] The author himself remarks : —

" As before said, it is a conception not at all peculiar to the
Hebrews, but common to all the (ancient) nations, and inseparable
from their notion of God, to represent the world as a building or a
house of the divinity, and the heaven as his especial dwelling-place.
The universe, but in a special sense the heaven, is the real, true
temple, built by the Deity himself ; and this, as the original temple, constituted the model, the archetype, of all those constructed by
man."[2]

Beside the authorities pertaining to India, China, Egypt, and
Babylon, the author musters the writings of Philo, Plutarch, Clem-

---

[1] Vid. *Symbolik*, etc., i. pp. 91–103, text and notes.
[2] Ibid., pp. 94, 95.

ent of Alexandria, Macrobius, Varro, etc., in support of the above
statement. Herr Nissen, in a recent and critical treatise upon the
temple, holds substantially the same opinion, especially with respect
to the classic nations.[1] But these references will suffice on a point
so generally understood among scholars.

Fundamental in all these ideas is the notion that God dwells,
which was inseparable from the notion itself of divinity. The universe, or the house built by the Deity for his own habitation, is thus
the model upon which all artificial temples or dwellings of God are
constructed. From thence proceeds the fact, as already stated, that
the theories of the cosmos and of the temple are substantially the
same ; so that if we would understand the ancient cosmogonies, it
is necessary first to study the doctrine of the temple. Undoubtedly,
the divinity of the hearth was one of the chief producing causes
which determined the course of thought in antiquity, in relation to
these subjects; hence the fact that every house was in some sense a
temple. It is upon the same ground of the fundamental relation of
the temple to the cosmos that the chief divinities of the ancient
temple-craft were uniformly regarded as concerned in the fabrication of the world. But it is hardly necessary to add more testimony exclusively designed to establish the fundamental principle
that the process of creating the world and the method of constructing the temple were esteemed the same, and that this process was
believed to be one of division. The progress of these investigations
will tend continually to the support of this doctrine, which is already, I presume, quite apparent to the reader's mind. It is on
account of its superlative importance that I desire to place it
beyond all question.

SEC. 55. We proceed now to the study of the practical application, in constructing the temple, of the principle of division upon
which we have so much insisted ; upon the particular methods, also,
according to which it was attempted to model the temple after its
archetype, the cosmos. The original temple of the Etrusco-Romans
will serve an admirable initiative to this part of our subject. We
introduce here several extracts from Dr. William Smith's "Dictionary of Greek and Roman Antiquities." Upon the subject of the
Auspices, the author has the following : —

" The ordinary manner of taking the auspices was as follows. The

---

[1] *Das Templum*, pp. 2-4.

augur went out before the dawn of day, and, sitting in an open place with his head veiled, marked out with a wand *the divisions of the heavens.* Next he declared, in a solemn form of words, the *limits assigned,* making shrubs or trees, called *tesqua,* his boundary *on earth correspondent to that in the sky.* The *templum augurale,* which appears to have *included both,* was divided into four parts: those to the east and west were termed *sinistræ* (left) and *dextræ* (right) ; to the north and south, *anticæ* (before) and *posticæ* (behind)." (Art. *Auspicium.*)

In this instance, the person is supposed to face the south; in which case the left hand would be east, the right hand west, etc. In Western Asia, according to a very ancient custom, the spectator was supposed to face the east in such manner that the left or left hand was put for the north, the right hand for the south, etc. The terminology of the cuneiform texts supposes this custom, and it seems to have been common to all the Semitic nations ; thus, the Hebrew *Semol,* Aramean *Shemal,* Assyrian *Su-mi-lu,* English *Samuel,* signified " left, left hand," and ordinarily designated the north, north pole, or North Star. The Accadian terminology conforms to the same ideas. The Etrusco-Roman method was different, for reasons which Dr. Smith will explain in another extract. We see that the augural temple was laid out with reference to the cardinal points, or the primary divisions of the heavens from east to west and from north to south. This is important to bear in mind ; and not less the fact, also, that the temple thus constructed was supposed to include two elements: a cut-off space on the earth's surface correspondent to a similar space in the heavens, these being put in direct relation to each other. The system of land measuring involved a further development of these principles, and Dr. Smith thus explains it : —

" As partitioners of land, the Agrimensores were the *successors of the augurs,* and the mode of their *limitatio (divisions)* was derived *from the old augural method of forming the templum.* The word *templum,* like the Greek *temenos,* simply means a division ; its application to signify the vault of the heavens was due to the fact that the directions were always ascertained according to the true cardinal points (*a*). At the inauguration of a king or consul, the augur looked toward the east, and the person to be inaugurated toward the south. Now, in a case like this, the person to be inaugurated was considered the chief, and the direction in which he looked was the main direction. Thus we find that in the case of land surveying the augur looked to the south; for the gods were supposed to be in the north, and the augur was considered as looking in the same manner in which the gods looked upon the earth (*b*). Hence the main line in land surveying was drawn from north to south, and was called *Cardo*, as corresponding to the axis of the world; the line which cut it was termed *Decumanus*, because it made the figure of a cross, like the numeral X (*c*). These two lines were produced to the extremity of the ground which was to be laid out, and parallel to these were drawn other lines, according to the size of the quadrangle required (*d*). The limits of these divisions were indicated by *balks*, called *limites*, which were left as highroads, the ground for them being deducted from the land to be divided." (Art. *Agrimensores*.)

We add here several notes, in the order of the letters introduced into the foregoing extract, and shall be obliged in some instances to refer to the characters which the reader finds engraven in our second plate.

(*a*) Dr. Smith seems to think that the application of the terms *templum* and *temenos* to the vault of heaven, conceived as a temple, was in a measure accidental, growing out of the practice of marking out the divisions according to the directions of the cardinal points. In my view this application proceeded from fundamental ideas. That which constituted a *templum*, a *temenos*, was *division*, according to the meaning of these terms. It was the primary divisions of the celestial space that constituted it in general a temple, and the corresponding divisions of the earth's surface that made it also a temple; and, in fact, any space thus divided, limited, was a *tem*, a temple, so constituted by the simple act of division.

(*b*) The conception locating the divinities in the north was very ancient, and widely prevalent; it was doubtless of Asiatic origin, and our future investigations will tend to explain the reason of it. That the main line was considered that drawn from north to south, called *Cardo*, is a matter of some doubt, I believe, but it is not of material consequence to us at present.

(*c*) There were certain doors, or gates, turning on pivots projecting from the ends, and belonging to the temples, palaces, and even private dwellings in common use among the Romans. They were called *Cardo*, and Mr. Smith has these remarks in reference to them : —

" The form of the door above delineated makes it manifest why

PLATE II.

the principal line laid down in surveying land was called *Cardo*; and it further explains the application of the same term to the north pole, the supposed pivot on which the heavens revolved. The lower extremity of the universe was conceived to turn upon another pivot, corresponding to that at the bottom of the door; and the conception of these two principal points in geography and astronomy lead to the application of the same term to the east and west also. Hence our four points of the compass." (Art. *Cardo*.)

The author himself shows here very plainly that the conception of the entire cosmos included between the two poles, as a house, or temple, must have been fundamental and quite primitive among the Romans. Nothing more conclusive could be desired, to the effect that the cosmos and the temple were associated, and that the theory of both had for its base the notion of division. Every building, in fact, whose divisions corresponded to the cardinal regions was in some sense a temple and an image of the world. The two lines called *Cardo* and *Decumanus* cut each other in a manner to form the image of a *cross*. These were the fundamental divisions, not only of the cosmos, but of the temple and the system of land surveying. This calls to mind the symbol of the god *Anu* at Babylon, whose mystical title was *Susru*, " the founder," represented by the figure *b* shown in the first group. It had reference to the four cardinal positions of the sun in its annual course.

(*d*) With the system of parallel lines cutting each other at right angles, forming thus the plot of ground marked off into identical squares, compare the figure *h*, first group. This is the Accadian sign *U* (Rep. 238), " to measure, a measure, a cubit," etc.; and it is a determinative of names of measure in the metrical system of Babylon. Various analogies, not less striking, will be hereafter pointed out between the notions prevalent among the Romans and those expressed by the written symbols employed in the valley of the Euphrates.

SEC. 56. It appears from the statement of Dr. Smith introduced in the last section: 1st. That the augurial temple united a particular celestial, as well as terrestrial space, corresponding to each other, and put in direct relation. 2d. That the spaces thus divided off were located with special reference to the cardinal regions. 3d. That the system of territorial divisions, and of partition of the soil, was derived from the method of constructing the temple. 4th. The gate called *Cardo*, from which term our word *cardinal* is derived, was assimilated to the axis of the world, thus proving that the notions of the cosmos and temple were associated, proceeding fundamentally from that of division. The four apartments of the augurial temple are illustrated in a diagram by Herr Nissen.[1] The figure shown answers precisely to the cuneiform character marked *e*, first group. The data thus derived chiefly from Dr. Smith are very important, as proving the fact of a conceived fundamental relation between the temple and the cosmos, and of a direct analogy in respect to their theories. The cardinal divisions of the cosmos constituted it a temple, and the divisions of the temple, according to the cardinal regions, constituted it an image of the cosmos. But Herr Nissen, in the excellent treatise already cited, has several statements confirmatory of those made by Dr. Smith, and others explanatory of different points connected with our subject, which ought to appear here. Referring to the main divisions of the *templum* are the following passages : —

"The limitation (division) proceeds from the cardinal regions ; a line from east to west, cut by another at right angles, drawn from north to south, forms the basis of the entire system." "The doctrine proceeds from the first and most simple division, suggested by nature, that is, into a day-and-night side." "The midday line (meridian) is termed *Cardo*, because the heavens turn like a gate upon its hinges." "The first or common form of the limitation is the *centuriation* or division into like squares."[2]

The division into a day-and-night side is suggestive of the first day of creation : "And God divided the light from the darkness; and God called the light Day, and the darkness he called Night ; and the evening and the morning were the first day" (Gen. i. 4, 5). The Romans began the day at evening, agreeing thus with the method in Genesis. But this division into a day-and-night side was from north to south, like the *Cardo ;* while the fundamental division was that of chaos from east to west, forming the two hemispheres assimilated to heaven and earth. As the notion of the temple proceeded from that of division, so the town or city divided into squares and the entire territory of the state cut up into districts were both regarded in some sense as temples. The author explains this matter in the subjoined extract : —

"The principles according to which the city was laid out were

---

[1] *Das Templum*, p. 16.　　　　　　　[2] Ibid., pp. 11, 13, 20.

those of the system of land measuring. The *Decumanus* and *Cardo maximus* determined the direction of the two principal streets of the city, dividing it and the territory occupied by it into four regions. In the various diagrams illustrating the scheme of these divisions, the *cross* forms the basis. Our authorities represent those forms as most complete where the intersection of the Decumanus and Cardo maximus falls exactly in the centre of the city, or the forum; and it is from this point, through four gates, that the two main lines extend each way over the entire country, which is thus like the city divided into four regions."[1]

Ancient Peru, according to Mr. Prescott's history of its conquest, was laid out upon the precise plan here indicated. Not only the city, but the state itself, was thus regarded as a great temple. The ceremonies in founding a town or city, described by Herr Nissen, are quite interesting, and we shall find here an explanation of the system of "balks," to which Dr. Smith has alluded: —

"The ancient Italian town did not originate, like those of the middle ages and of modern times, by a slow process of growth from the house to the hamlet, and from the hamlet to the town. It was created at once by a single politico-religious act. The memory of the founder was perpetuated, and his veneration constituted an important part of the worship of the commonwealth. Not only the memory of the founder, but that of the year and day of its foundation was also perpetuated by annual festivals. From this cause, a regular ritual for foundations had grown up. The founder of a town, according to Cato, yoked to a plough a bull and a cow, the bull upon the right and the cow upon the left, the cow inside and the bull outside. With his head veiled, he ploughed around the space designated for the site of the town, taking care to turn all the sods inward; for the turf marked the line of the wall, and the furrow that of the ditch. Where a gate was to be located for passage in and out of the town, he lifted the plough out of the ground, and carried it over the required distance."[2]

This explains the system of "balks" to which reference has been made. The author shows that the military camp and even the vineyard, by reason of these divisions, since they were laid out with reference to the cardinal regions, were regarded in some sense as temples. The same, also, as regards the private dwellings. Everything, in fact, to which the fundamental notion of division pertained was a temple. I close the extracts for the present, from this valuable treatise, with these remarks of the author: —

[1] *Das Templum*, pp. 58, 59.   [2] Ibid., pp. 55, 56.

" All historical development proceeds from two co-related ideas, property and secure possession. Both are summed up in the notion of division, separation. As the nation separates itself from the mass of nationalities, the race from the races, so the town is an off-shoot from other towns, the class from other classes, the house from other houses. This notion of separation, of division, was incorporated by the ancients in the *Templum*, Greek *Temenos*, 'a cut-off section,' from the root *tem*, ' to cut.' " [1]

Sec. 57. Identity in the different assimilation of things divided, separated, is the fundamental law of mind and nature, as well as of all historical development. That the ancients thoroughly appreciated it is a fact so apparent that M. Lenormant and M. De Vogüe have even reproached them for this Hegelian principle. Hegelian, or not, it evinces a profound insight into the nature of things, a clear perception of laws which modern physical science demonstrates to-day to be fundamental. The reduplication of identical squares, resulting from the process of division, is a perfect geometrical expression of these ground thoughts. A single square was the basis of, and so to speak identical with, the whole. It might be cut up into smaller ones, each like the whole, or it might be reduplicated indefinitely, in such manner that the whole would be like its least portion. In general, all space, the entire sphere, might be represented by a single square. This being divided off according to fixed methods and certain fundamental ideas became thus a symbol of the cosmos. The heavens and the earth might be and actually were delineated according to this simple scheme. Every country, every city and town, and even private dwelling, was thus an image of heaven and earth; and as one was a temple, all were for the same reason temples; as one was a cosmos, a world, so all were such; thus law, unity, light, penetrated through all. But the system which prevailed among the Etrusco-Romans was derived, in its main features, from the valley of the Euphrates. This fact is becoming now more and more apparent, as the results of cuneiform research are being developed. Mr. George Smith only gives expression to the opinions justified by these researches when he says : —

" The value of the Assyrian and Babylonian mythology rests not only on its curiosity as the religious system of a great people, but on the fact that here we must look, if anywhere, for the origin and explanation of many of the obscure points in the mythology of

[1] *Das Templum*, p. i.

Greece and Rome. It is evident that in every way the classical
nations of antiquity borrowed far more from the valley of the
Euphrates than that of the Nile, and Chaldæa, rather than Egypt,
is the home even of the civilization of Europe." [1]

It is usually held by scholars that the Romans were mainly indebted to the Etruscans for their sacred science, and it is through
the Etruscans principally that we trace this science to the Babylonians, or more properly, perhaps, the Chaldæans. Augury, especially, as practiced by the Etrusco-Romans, was derived in its main
features from Chaldæa; and the augurial temple in which the auspices were taken constituted the basis of the system. M. F. Lenormant, after identifying various customs of this class prevailing
among the Romans with those described in the cuneiform texts,
very justly concludes as follows: "All these analogies are such that
they tend to make us see in the Etruscans the disciples and direct
inheritors of the auspices and divinations practiced by the doctors
of Chaldæa and Babylon." [2]

But the Asiatic origin of these notions pertaining to the temple
will be still more apparent as we see them in their more primitive
form embodied in the various symbols exhibited in our second
plate, to which we give now a more particular attention.

The figure marked a of the first group exactly represents the
basis of the Roman system of limitation or division as already set
forth in the language of Herr Nissen. It is not precisely the *hieratic* form of the Accadian *Bar*, but if the two strokes were given
the wedge shape we should have the *modern* form of *Bar;* a sign
which with the phonetic value of *mas* signifies " to cut, to separate,
sword ; " also " a measure of capacity " (Rep. 69). It is analogous
to the figure b, to which *Susru*, the mystical name of Anu, is
attached, designating him as the " founder" *par excellence.* This
is the Oannes of Berosus, who founds the Babylonian civilization.
In the bilingual phrase, which explains this figure by the term
Susru, it is preceded by the sign *Gur* repeated, which singly has
the sense of " to create." But the well-known cosmical character
of Anu and the reference of this figure to the four positions of the
sun, answering to the cardinal regions, sufficiently indicate the reference to the cosmos and temple. The couplet c is the Accadian *An*,

[1] *Assyr. Discoveries*, p. 451.
[2] *Les Sciences Occultes en Asie; La Divination*, p. 120.

" elevated, heaven, God " (Rep. 4), and is the universal characteristic of divinity preceding all divine names.   Cuneiform scholars, for the most part, believe it to have been intended for a " star," being thus a proof of the hieroglyphic theory applied to this system of writing, and of the existence of star worship at the time of its invention.   This appears to me far from correct.   The character *An*, taken alone, does not signify a " star," and there is no evidence that it ever had this meaning.   Moreover, the Accadian has quite another term for " star," namely, *Mul*.   The original reference of the figure was doubtless to the eight celestial regions, four primary and four intermediate ; and hence the meaning " elevated, heaven, god ;" but of this hereafter.   The couplet *d* exhibits the simplest element of a divided field or plot, the square.   It is the Accadian *Gil*, " to inclose, to unite, to assemble," and probably an " inclosure " (Rep. 489).   It represents the ground plan of one of the principal classes of temples in the valley of the Euphrates.   As a paleographic symbol, it forms the basis of about twenty others ; and its thoroughly symbolical character is susceptible of the plainest demonstration.   The couplet *e* is only a further development of that just considered, and of the fundamental notion of division.   It exhibits, as before observed, the form of the augurial temple of the Etrusco-Romans.   The triplet *f* appertains to the Chinese writing, showing for the first and second forms the ancient and modern character *Tsing*, " a well," denoting also " union, friendship."   It is the name of the constellation Gemini : and the third form of the character represents precisely the Spartan symbol of the Dioscuri, identified with Gemini in the western mythology.   These were the *Brothers*, *par excellence*.   The symbol was constructed with four pieces of wood, crossed in the manner shown in the figure.   It will appear hereafter that this symbol represents exactly the conceived geographical centre of the universe, according to the traditional notions of all Asia ; and it will be seen at a glance that it must be taken as the least unit of all geographical as well as political divisions, to which may be added those of the temple ; for all these were in fact fundamentally related to each other.   A further development of the triplet *f* is the figure *g*, consisting of nine identical squares, in which the nine digits are so arranged as to equal the sum fifteen, in whatever direction three consecutive numbers are added.   It was termed the " planetary seal," or " magical square," and was supposed to represent the cosmos or universe. The central division, with the numeral five, symbolized the soul of the world ; the eight squares surrounding it being put for the elements *fire, air, water,* and *earth,* male and female. The figure of a man was represented upon the seal, and this denoted that man was taken for the *microcosm,* or universe in miniature. I derive these from Dr. Bähr, and shall return to their consideration in the next section.[1] The couplet *h* in the group before us has been already explained. Its signification of " measure, to measure, cubit," etc., shows its fundamental relation to the system of land measuring, which at Rome, and probably at Babylon, had its origin in the augurial temple.

SEC. 58. A brief investigation of the second group of characters, as shown in the same plate, will conduct to the generalizing principle of nearly all the facts which have been thus far introduced into the present chapter. The three forms marked *a* are the Accadian *Mal,* " to complete, to accomplish, to fill ; to inhabit, house ; " also *E.* " house, temple " (Rep. 144, 244). In the hieratic type, we have the old Accadian or Cushite notion of the temple. It forms the basis of over a dozen composite signs, of which the others shown in the same group are examples. In the other characters, as will be noticed, the four perpendicular lines on the left are contracted to two, and the horizontal ones so far reduced as to leave an open space for the introduction of other signs, forming thus composite symbols. We introduce here a brief notice of the other characters, and then return to the one marked *a.*

The couplet *b* has the value *Kisal,* " altar, sacrifice " (Rep. 146). The altar of course appertains to the temple. But note the striking symbolism attached to this sign. The two parallel lines inclosed are the Accadian *Tab,* " to adjust, to place," etc. (Rep. 140). It is the altar that *adjusts, makes right,* and such is the notion here involved. We pass now to the couplet *d,* which has the values *Dak,* " cave, vault ; " *Dir,* " blue, deep blue " (Rep. 130). The two conceptions of " temple " and " cave " call to mind the " cave worshippers ; " but this *cave* is the *blue vault of heaven,* of which the cave temples were conceived as imitations. Even artificial caves were made for this purpose, as is proved by the etymology of the term *Dak,* evidently a contraction of *da-ak,* or *da,* " to excavate," and *ak,* " to make ; " hence, " to make an excavation " (Rep. 91, 309). The

---

[1] *Symbolik,* etc., i. p. 158.

two forms marked *c* are the Accadian *Gan*, "view, presence, inclosure;" *Gi-nu,* "presence, plain, field;" *Kar*, "summit, point"
(Rep. 155). *Gan* is identical with the Hebrew term (גן) in the
phrase *Gan-Eden*, "Garden of Eden." The value *Kar*, "summit,"
referred probably to the traditional mount of paradise. The inclosed figure, exhibiting a field cut up into squares, has not been
recognized separately in the inscriptions, so far as I know, but it is
found in the Egyptian writing as a determinative of home, settled
district, cultivated earth, answering well to the conception of *Gan-
Eden*. Finally, the couplet *e* has the values *Neu, Ekhi, Luku*, all
signifying "mother" (Rep. 148). The eight-rayed star inclosed is
the Accadian *An*, "elevated, heaven, god," as heretofore explained.
Here the notions of "temple," "heaven," "divinity," combine in
that of "mother." Was the typical idea of mother she who should
bring forth the "promised seed," the divine one? No more exalted
conception of maternity ever entered the mind of man. The
mother here is herself a goddess, or her offspring is a god! Compare this sign with the Egyptian method of representing the notion
of maternity, for an illustration of the difference between hieroglyphism and symbolism!

We return to the character marked *a*, signifying "house, temple,
to complete, to accomplish," etc. The completion of the house for
the habitation of man, of the temple for the habitation of God,
were joyous events that might well be compared to the accomplish-
ment of the work of creation itself, when God rested from his
labors. Nor were these ideas wholly unrelated; on the contrary an
inherent analogy existed between them, for the temple and the cosmos involved precisely the same theoretical principles, and were
regarded as types of each other. Compare here the nine identical
squares on the left of the Accadian *Mal*, "house, temple," with
those of the "magical square," already referred to, which was supposed to represent the cosmos. The hieratic form of *Mal*, "temple," evidently had a celestial as well as terrestrial reference. We
see here primarily a celestial field limited and divided off, with a
view to locate the position of certain asterisms, while the parallel
lines projecting toward the right served the purpose of recording
their names. It obviously represented also a corresponding system
of divisions on the earth's surface, constituting thus a regular augurial temple such as Dr. William Smith has described to us. Our

Accadian character may be taken for a definite type of this augurial temple, as recognized by the Chaldæan priesthood. Its anal gy with the "magical square," as explained, is very striking, and clearly indicates the assimilation of the temple to the cosmos, both proceeding from the one idea of division, limitation. We see here, in the sign *Mul*, as compared with the mystical square, an illustration of the practical application of the principle of division in constructing the temple and in representing the cosmos. The central division really symbolized the centre of the universe geographically, as well as philosophically the soul of the world. The eight divisions surrounding it were the eight regions of space, in relation to which we shall cite hereafter some authorities. But in a philosophic point of view, they were taken for the four elements, male and female, amounting to eight. The application of the principle of division here exhibited was undoubtedly very ancient, and it appears to have been extensively prevalent. Nevertheless, it was not the only method, nor was it apparently exclusively orthodox. With the Romans, as we have seen, only four divisions were employed, these corresponding to the cardinal regions. The two chief apartments of the Hebrew tabernacle served the same purpose, as they represented the two chief divisions of the cosmos, heaven and earth. Everywhere the fundamental idea of cutting, dividing, prevailed, but its practical application in moulding the temple into an image of the cosmos seems to have varied with different peoples. Even the numbers twelve, sixteen, thirty-six, etc., appear to have been adopted at times, but these were merely expressions of fundamental ideas, and are probably not to be regarded as in any sense typical. The numbers eight and nine, however, must have been very ancient, and had become in a measure traditional, as we shall soon show. These two modes were not essentially different; for all depended upon the manner of representing the central space, whether as a square surrounded by eight others, or as a common divergent point from which the eight regions radiated, as represented in the eight-rayed star, or Accadian *An*, to which reference has been made.

Another point of difference is to be noticed here. It is evident that the theoretical divisions of the temple, particularly of the augurial temple, were not literally represented always in material structures, at least so far as regards the ground plan. The nine

divisions of the Accadian character *Mal* must be taken as in some
sense embodying the doctrine of the temple; but no edifices of
this class in the valley of the Euphrates were thus divided off, so
far as my knowledge extends. The same probably might be said
of the Roman temples, as compared with the plot marked out by
the augur in taking the auspices. It is a fact that the axis of the
Roman temple, as actually constructed, rarely corresponded to any
one of the cardinal regions, while the augurial temple was designed
with especial reference to them. Nevertheless, both were conceived
as images of the cosmos. It would be difficult to explain, perhaps,
precisely how the doctrines of the one were materially embodied
in the other, yet originally they must have been closely related.
It is quite obvious that, in some instances, while the chief apartments represented the chief divisions of the cosmos, the others were
materially expressed by means of ascending stages, or by a certain
mystical arrangement of steps leading from one apartment to another. In some way, as we may justly infer, the fundamental conceptions were always materially expressed; but it would require a
separate treatise to explain the various methods in detail.

SEC. 59. Having sufficiently verified the general principle,
namely, that creation was conceived as a temple, and the temple
as an image of creation, the process of division being regarded as
fundamental in both; having also illustrated the practical application of the system of divisions in constructing the temple after the
model of the universe, it remains to submit some evidences tending
to show the origin of these notions, as well as their immense antiquity. In doing so, still other proofs of the correctness of the
views already put forth will be found multiplying upon our hands.
We return to the consideration of the magical square marked *g* in
the first group, especially in connection with the figure that precedes it, being the third form of the triplet *f*, the Chinese character
Tsing, " a pit or well," also " union, friendship." As before said,
the Spartan symbol of the Dioscuri, one with Gemini, or the Twins,
was constructed of four pieces of wood crossed in the manner shown
in this figure. The zodiacal *sign* Gemini is but a contraction of the
same symbol. The ancient Chinese form shows " a well " in the
centre, with the four pieces of wood so arranged as to prevent
the débris from falling back into the pit. As before observed, the
third form answers exactly to the least unit in geographical and

political divisions, as well as those in land surveying, and we proceed to show that it represented the geographical centre of the earth. M. Lenormant remarks as follows:—

" We have noticed already the system of geography, essentially symbolic and inspired by religious conceptions, which plays a fundamental part in the Book of Astrology, compiled by the orders of King Sargon, the ancient, some two thousand years before our era. He considers the country of Accad or Chaldæa as situated at the centre of the universe, and bounded by four countries that correspond exactly to the four cardinal regions: *Ilama* is east, *Martu* is west, *Gutium* is north, and *Subarti* is south." [1]

It will be seen that this symbolical scheme of geography, so singular in its nature, is perfectly represented by the figure constituting the Spartan symbol of the Dioscuri; and as Akkad is here put for the centre of the world, so the symbol itself represents this centre. But the notion had been inherited traditionally by the Accadians, in common with other ancient nations widely separated. The author just cited continues:—

" This system offers the most striking analogies with that of the division of India by the Aryans into four regions, east, south, west, and north." " Thus, we know that for the Aryans of India, this systematic division of the vast region where they had established themselves *was a reproduction of their symbolical and legendary conception of the world*, divided into four *Mahadvipa*, great islands, or continents, grouped according to the four cardinal regions around the central continent *Madhyadvipa*, in the midst of which *Meru* elevated itself, watered by the four rivers descending from the sacred mount." [2]

Mt. Meru was supposed to be situated geographically at the centre of the universe; so that we have here the same scheme as that pertaining to Accad and India. The allusion here is to the sacred mount of paradise, the abode of the first human pair, which has been identified with the plateau of Pamir, situated east of the Caspian Sea. This, too, was traditionally of a square figure, bounded by four countries like Accad, and M. Lenormant traces to it, as the divergent centre of the human races, this entire system of symbolical geography, as will appear in the subjoined extract:—

" In order to discover the origin of the analogy, so direct, or

[1] *Frag. de Bérose*, p. 321.   [2] Ibid., p. 322.

rather the identity, between the geographic system of the Chaldæan book of astrology and the conceptions of India, it is necessary to go back to the common source of primitive tradition respecting the terrestrial paradise, considered as a plateau of a square figure, having its four sides turned toward the four cardinal points, surrounded by four great countries also facing the cardinal regions, and watered by four rivers that take their rise from the central plain." [1]

We shall treat more at length upon these primitive traditions centring in Mt. Meru in another chapter. The symbolical geography of Accad, like that of the Aryans of India and various other nations of antiquity, was a sacred inheritance from the conceived first abode of humanity, to which also a similar scheme was attached. As this paradisiacal mountain was regarded as the centre of the world, so were all imitations of it. In every instance the scheme answered geographically to the geometrical figure constituting the symbol of the Dioscuri; that is, a region conceived as a square bounded by four others situated in the direction of the cardinal points. The limits of these four countries joining the central one were of course well defined; but the outer limits were not supposed to be known, and were left as indefinite. The constellation Gemini, to which the Chinese *Tsing* was applied, was regarded as the celestial " well," the source of the celestial waters; and in this respect answered to the Hindu conception of the heavenly Ganges, from which the four sacred rivers of paradise were supplied. [2] Thus, the symbol of the Dioscuri is by this means directly connected with the paradisiacal mount, conceived centre of the world, and divergent point of all geographical divisions. Again, in Asiatic mythology this asterism was associated rather with the first human pair than with the two brothers, Castor and Pollux, as in the western mythologies. The Hindu zodiac represents the sign Gemini by a man and woman, instead of by two male figures. [3] The Hindu Tama and Tami, conceived as the first human pair, derive their names from *Gem-ini*, denoting the constellation which, in China,

---

[1] *Frag. de Bérose*, p. 322.

[2] Vid. Schlegel, *Uranographie Chinois*, p. 406. Cf., for the verification of the various statements in the text relative to *Tsing*, pp. 405–411 and 673–681. On the celestial Ganges consult Obry, *Du Berceau*, etc., p. 19.

[3] *Asiatic Researches*, ii. p. 303, plate opposite. Cf. Schlegel, *op. cit.*, from whom it appears that the Egyptians and other nations represented Gemini by a man and woman.

was represented as the celestial "well." [1] Here we trace again the direct connection of the symbol in question with the sacred mount of paradise. Finally, M. Lenormant has stated the fact that the pyramidal temples of Chaldæa were considered as imitation, as artificial reproduction, of the traditional mount of paradise (Sec. 33). This notion of a terrestrial region, then, conceived in the form of a square, and bounded by four countries facing the four cardinal points, — this notion, I say, inherited from the paradisiacal mount regarded as central point of the universe, was associated definitely with the Chaldæan temple, since the latter was expressly designed as an imitation of the sacred mountain. Hence, the temple itself was regarded by analogy a centre of the universe, like Mt. Meru. It was quite common in antiquity to represent the national temple as the geographical centre, not only of the territory belonging to the state, but of the world itself. The divisions of the national domain into provinces, proceeding from the temple as the common centre, were thus allied to the geographical divisions of the world. The original type of the entire system was the sacred mount, traditionally the abode of primeval humanity. The Spartan form of the character *Tsing*, connected, as we have seen, with the traditional geography of paradise, was thus associated with the temple conceived as an artificial reproduction of the mount of paradise ; and in either case it represented the centre of the universe, as well as the unit of all geographical and territorial divisions. The association of this symbol is proved again from the fact that Gemini answers to the Accadian month called "month of the brick," or "month of constructions in brick." The symbolical scheme of divisions, then, of which Accad formed the centre, similar to that in vogue among other nations widely separated, was only part and parcel of the theory of the temple and of the cosmos which had been inherited in common from the primitive home of mankind.

SEC. 60. But the Chaldæan temple, as shown by the cuneiform character *Mal*, was connected not only with the Spartan symbol referred to, but with the system of nine identical squares, constituting the planetary seal, designed especially to represent the cosmos. This shows that the two figures were fundamentally related to each

---

[1] Whitney, *Oriental and Linguistic Studies*, p. 45. The author states that Tama was traditionally the first man, Tima, his sister, the first woman, and he derives the name etymologically from the Latin *gem-ini*.

other, and to the entire system of divisions which has been explained. In China the character *Tsing*, " a well," was directly
associated both with the partitions of the soil designed for cultivation, and with the territorial divisions of the state into provinces,
in each case connected likewise with the system of nine squares.
Dr. Schlegel thus describes the allotments of the soil constituting
the unit of Chinese society : " In antiquity *nine lots* of cultivators
formed *a well ;* four wells formed an *inclosure,* and four inclosures
formed a *community.*" [1]

In this scheme, $9 \times 4 \times 4 = 144$ families as constituting a community. The number 144 occurring in a connection so singular reminds us of some of the mystical numbers employed in the Apocalypse by the Revelator. But we see here the character *Tsing*, corresponding to the symbol of the Dioscuri applied to the zodiacal
sign Gemini, and connected with the traditional geography of Eden
as centre of the world, placed in direct relation to the other figure
consisting of nine squares, representing alike the cosmos and the
Chaldæan temple. Another fact stated by Dr. Schlegel, in the
words of an extract from the Chinese writings, is still more remarkable : —

"Since nothing is comparable to water for the purpose of level-
ing, it is for this reason that the constellation *Tsing* is the image of
rules for founding the state, for tracing out the plan of the capital,
for the demarcation of desert places, and for the divisions of the
soil." [2]

The term *Tsing*, denoting "a well," put also for the source of
celestial waters or the constellation Gemini, became the symbol of
the *level*, of *equality*, since water in a basin was the primitive instrument for determining levels, or differences in altitude. It became
also a symbol of an equal and just distribution of the soil, probably
derived from the original custom of apportioning nine lots to a well,
as set forth in the preceding extract. But it was, more than all
this, directly associated with the regulations for founding a state,
and for laying out the plan of the capital, for the territorial divisions, etc. It is impossible not to see in these notions a traditional
inheritance, on the part of the founders of Chinese civilization,
of the same doctrines concerning the temple, the method of land

[1] *Uranographie Chinois*, pp. 222, 223.   [2] Ibid., p. 408.

division, and of geographical divisions generally, which have been found to exist among the Etrusco-Romans and the populations of the Euphrates valley, — a traditional inheritance, I say, common to these various nationalities so widely separated. That which tends most strongly to prove this community of origin is the fact that the Chinese *Tsing*, corresponding to the Spartan symbol of the Dioscuri, and referring always to the constellation Gemini, appears everywhere to represent the fundamental conception.

I wish to add here some further testimony relating especially to the system of nine squares, representing alike the cosmos and the Chaldæan temple. In reference to the territorial divisions of the ancient Chinese empire, Dr. Bähr remarks : —

" The entire earth was divided into nine countries; the emperor was styled the regent of the nine earths, and the highest officers of state, the mandarins, were divided into nine orders. Under the successors of the ancient Emperor Yao, who constructed the nine canals, the kingdom was divided according to the four cardinal regions, and four mountains were taken as corresponding to these regions; over each a chief ruler was placed, and twelve mandarins, answering to the twelve zodiacal signs, ruled over the people. Somewhat later, the entire kingdom was divided into nine provinces, each of which had its ruler, but the middle province *Ki* was governed by the emperor, and the palace was situated in the centre of it." [1]

It will be seen from the foregoing that the system of nine squares was applied on a grand scale to the earth, which accords with the notion of the cosmos as represented by it ; and then to the nine provinces of the kingdom, evidently conceived for this reason as an image of the world. The middle province Ki, with the palace in the centre, symbolized the geographical centre of the universe, to be compared thus with Accad in the mystical geography of the early Chaldæans. But a still earlier territorial division of China was according to the cardinal regions, four mountains being assumed as corresponding to them.

SEC. 61. We show now the prevalence of the same method of divisions among the Aryans of India, but derived by them traditionally from Mt. Meru, reputed abode of the first men. In relation to the seven great insular continents of the world, we have the following : —

[1] *Symbolik*, i. pp. 12, 13, notes.

" Jambu-dwipa is in the centre of all these; and in the centre of this continent is the golden mountain Meru." " I have thus briefly described to you, Maitreya, the nine divisions of Jambu-dwipa." [1]

A passage previously cited from M. Lenormant has made us acquainted with the fact that the system of five regions, one central and four surrounding it, as represented by the Spartan symbol, had formed the basis of political divisions in India, and that this had been derived from traditional notions centring in Mt. Meru. In the above extracts the method of nine divisions is associated with the same locality. The two systems were thus connected in India as well as in China; and it is evident that in both countries these conceptions had been inherited from a common source, that source being the sacred mount, the primeval home of man. The four mountains assumed by the Chinese as corresponding to the four cardinal divisions of their country were connected by the Hindus with their Meru as abutting or supporting mountains, these being located also, as supposed, according to the cardinal points. This mountain system, consisting of five summits, connected with the central region of the world, and with the theory of the cosmos, in fact, must have been recognized in China, and was doubtless the basis of the notion of four mountains, to which we have alluded. Among the ancient Chinese the tortoise was a symbol of the cosmos; the upper half or back of the shell, with its thirteen divisions, representing the heavens, that is, the sun and twelve zodiacal constellations; while the lower half of the shell, having nine divisions, was put for the earth, the other chief division of the cosmos. Dr. Schlegel cites the following legend: —

" To the west of the mountain *Youen Kiao* is situated the lake of the stars, which is a thousand Chinese *li* in length. In this lake exists a divine tortoise, having eight feet and six eyes. Upon its back it carries the image of the Northern Measure (constellation of the great Dipper), of the sun, of the moon, and of the eight celestial regions. Upon its lower shell it has the image of the five summits and of the four canals." [2]

The five summits referred to are evidently the same, traditionally, as the Hindu Mt. Meru, flanked by four others according to the directions of the cardinal points. The conception in both cases is only a variation of that of the five countries, one conceived as cen-

---

[1] *Vishnu Purana,* pp. 166, 178.   [2] *Uranographie Chinois,* p. 61.

tral, the others bounding it on the four sides. The constellation of
the Dipper, to which allusion is made, fixes the original reference of
the legend to Mt. Meru, since this asterism is uniformly associated
with the sacred mount in Aryan tradition, as well as in that of the
Semitic races. The four canals are obviously to be compared with
the four sacred rivers, issuing from a common source. It is, then,
in view of the facts now placed before the reader, no longer to be
doubted that the symbol of the Dioscuri represented the geographical centre of the world, and that it was closely connected with the
magical square, as a further development of it, representing variously the notions of the cosmos or the nine earths, as in China, and
finally the augurial temple of the Chaldæans, as shown in the
hieratic sign for temple, the Accadian *Mal*. It is a fact not less
certain, that these two geometrical figures embodied certain traditional conceptions, relating alike to the temple and the cosmos,
which had been inherited in common by nations the most distantly
separated in antiquity; inherited from that central region in high
Asia which passed for being the first abode of man on earth.

The tortoise, as we have seen, was regarded by the ancient
Chinese as a symbol of the cosmos. Upon its back were supposed
to be found, among others, the images of the " Northern Bushel," or
the constellation of the great Dipper, and of the " eight celestial
regions." Here we have placed in immediate relation to each
other: 1st. The notion of the cosmos; 2d. The seven stars of the
chariot, or great Dipper ; 3d. The eight celestial regions. These
stars were definitely associated in Hindu tradition with the summit
of Mt. Meru, which was supposed to pierce the heavens precisely
at the north celestial pole, termed by them the Su-Meru. It was
upon the summit of this mountain, around which these seven stars
were thought to turn, that they located the palace of the gods,
which was thus regarded as *the heaven par excellence*. It answered
literally to the *Cardo* of the Romans, symbolized by the pivot projecting from the top of the door or gate termed *Cardo*, as explained
by Dr. Smith. This word involves the notion of " swinging," as a
gate on its hinges. Now the region of the sky thus indicated constitutes the exact point of intersection of the equinoctial and solstitial colures, corresponding to the four regions, which, with the four
intermediate ones, evidently comprise the " eight celestial regions "
referred to in the Chinese legend of the tortoise. Thus, the gen-

erative point of those regions was the *Su-Meru* of the Hindus, the *Cardo* of the Romans; and, as must be now obvious, this was the very point indicated by the eight-rayed star, Accadian symbol of "heaven, god," etc., in the most ancient inscriptions of Chaldæa. This star, I say, was intended to represent the generative point of the eight regions of heaven, and this proposed assimilation affords a perfect explanation of its three senses, "elevated, heaven, god." If the Accadian *An*, therefore, ever denoted "a star," it could have been no other than the North Star, conceived as divergent point of the principal divisions of the cosmos.

SEC. 62. The eight celestial regions were definitely associated by the Chinese, as proved by the legend of the tortoise, first, with the cosmos itself, of which the tortoise was a symbol; and, secondly, with the sacred mountain, the Meru of the Hindus, Albordj of the Persians, Gan-Eden of the Hebrews. The allusion to the seven stars of the Dipper, and to the five summits, renders it certain that these notions were attached traditionally by the Chinese to the sacred mount otherwise regarded as the birthplace of humanity. Thus, the earliest recollections, not only of the Hindus, the Persians, and Hebrews, but those also of the Chinese, centred in this one locality; the *Har-Moad* of Isaiah, of which we shall have to do hereafter. It is obvious that there were two methods of representing these eight regions geometrically, both of which were associated traditionally with Mt. Meru. One was by means of the eight squares surrounding a central one, as seen in the cosmical seal, and in the Accadian sign *Mal*, signifying "a temple." Another mode was to represent these regions by diverging lines from a common centre, as in the Accadian character *An*. Both figures symbolized the chief divisions of the cosmos. The subjoined passage from the Rig-Veda, cited by M. Carré, shows that the Hindus, as well as Chinese, received the notion of eight regions: "Savitri (the sun), the god with an eye of gold, illumes the eight regions of the earth, the beings who inhabit the three worlds."[1] The three worlds were heaven, earth, and the atmosphere. Each was divided into the same number of regions, and Mt. Meru united them all as their common centre. It was from this point that all the divisions of the cosmos diverged. Usually the four rivers were located in the direction of the four cardinal points; but the Buddhists of Thibet made

[1] *L'Ancien Orient*, i. p. 47.

them correspond with the four intermediate points.[1] But both Brahmans and Buddhists appear to have held in common the doctrine of eight regions diverging from Meru.[2] Thus, in China and in India, the system of nine squares and of eight regions had everything in common, except that in the last instance a geometrical point of intersection was substituted for the central square in the former. Both figures were associated with the primary divisions of the cosmos, and definitely with Mt. Meru as conceived centre of the world.

The number eight, which appears here in reference to the four primary and the four intermediate regions, diverging from the traditional mount of paradise, obviously has some fundamental connection with the eight Cabiriac divinities engaged in the work of creating the world. Recall here the facts developed in the third chapter, respecting the Accadian *Ak*, the Egyptian *Sesun*, and Chinese *Kuas*, paleographic symbols in which the numeral eight appears in each instance, as relating definitely to the work of creation, and to the deities engaged in it. Thus, the *Sesun* of the Egyptians is explained by M. De Rougé as referring to the eight gods who assist Thoth in superintending the labor of creation; and Sir G. Wilkinson finds an allusion in the same character to the eight regions ruled by Thoth or Mercury. The Babylonian Mercury had for his sanctuary the eighth stage of the temple of Borsippa, which represented the heaven of the fixed stars, and we have learned from M. Lenormant that this and indeed all the pyramidal temples of the Euphrates valley were but artificial reproductions of the sacred mount of paradise, whose summit, as we have shown, formed the generative point of the eight regions. Mercury was a Cabiriac divinity, the Dioscuri were also worshiped as such, and all are found associated with this mountain of the world, from whence the divisions of the cosmos take their rise. The priests who presided over the mysteries of *Shemal* in the city of Haran were called *Kabis*, that is, "Cabiri," and *Shemal* himself, as put for the North Star, was esteemed the *eighth* in relation to the seven stars of the great Dipper. He was thus the eighth Cabirus, definitely associated with the celestial region termed *Su-Meru* by the Hindus, *Cardo* by the Romans, *Su-mi-lu* in the cuneiform texts; that is to

---

[1] Vid. Obry, *Du Berceau*, etc, p. 31.

[2] On this point, see Obry, ibid., pp. 165, 166 ; cf. pp. 23, 27, 45, etc.

say, the identical point in the heavens from which the chief divisions of the cosmos were generated, and by reference to which the divisions of the temple were also located. This whole doctrine of the temple and of the cosmos is by these data definitely connected with the Cabiri; and we thus prove beyond doubt the truth of the statement put forth in our third chapter, that the doctrine of the temple, together with the cosmogony closely related to it, had been transmitted to historical times through the ancient temple-craft. The theory of the temple and of the cosmos, inherited by them from the first ages of the world, from the sacred mount, in fact, traditionally the primitive home of man, was that to which mainly the mysteries of the temple-craft appertained. The sacred mountain, of which the pyramidal temple was an imitation, was the model, the original type, of all such constructions. The cosmogony, both in theory and in tradition, was connected with the temple, and with its archetype, the mount of paradise. The fundamental conception of *division* appertained to all; and as the augurial temple included a celestial space correspondent to a terrestrial space, so the pyramid was conceived to unite the heaven and earth, just as the sacred mount united the paradise of the gods with that of the first men. Thus, in view of all the facts now before us, there can be no doubt as to the origin of the profound system of ideas which has constituted the subject of our present study; nor can there be any reasonable question as to the medium of the transmission of these doctrines from the first ages of the world to the historical period.

SEC. 63. As already stated, the image of a man was engraven upon the mystical square representing the cosmos, which denoted that man was conceived as a microcosm, or the universe in miniature. We have traced the system of divisions which this symbol of nine squares presents to the sacred mount itself, traditionally the birthplace of humanity. This accords perfectly with the history of creation, in which the formation of man, and especially his introduction into the terrestrial paradise, constituted the crowning work of the cosmos, appertaining to the last, the seventh, day of the creation week. Finally, in the New Testament, man himself appears as the temple, a building of God, a dwelling-place for the Divine Spirit; and this solves the enigma as pertains to the cosmogony, to the temple, and to man. We must advance one more step, however, in order to realize it in all its significance. The pyramidal temple had its eighth stage in the form of a cube, corresponding to the eighth division of the cosmos, the eighth celestial region, the eighth Cabirus, whose symbol was the cubical stone. The Scriptures recognize, in point of fact, an eighth day of creation, completing the octave. The Revelator thus alludes to it : —

"And he carried me away in the spirit to a *great* and *high mountain*, and shewed me that great city, the holy Jerusalem, descending out of heaven from God, having the glory of God : and her light was like unto a *stone* most precious, even like a jasper stone, clear as crystal: and had a wall great and high, and had twelve gates, and at the gates twelve angels, and names written thereon, which are the names of the twelve tribes of the children of Israel : on the east three gates ; on the north three gates; on the south three gates; and on the west three gates. And the wall of the city had twelve foundations, and in them the names of the twelve apostles of the Lamb. And he that talked with me had a golden reed to measure the city, and the gates thereof, and the wall thereof. And the city lieth foursquare, and the length is as large as the breadth : and he measured the city with the reed, twelve thousand furlongs. *The length and the breadth and the height of it are equal.* And he measured the wall thereof, a hundred and forty and four cubits, according to the *measure of a man*, that is, the angel." (Rev. xxi. 10–17.)

To show the connection of the above language with the idea of the original cosmos, it suffices merely to quote the opening verses of the same chapter : —

"And I saw a *new heaven and a new earth: for the first heaven and the first earth were passed away;* and there was no more sea. And I John saw the holy city, new Jerusalem, coming down from God out of heaven, prepared as a bride adorned for her husband. And I heard a great voice out of heaven saying, Behold, the tabernacle of God is with men, and *he will dwell* with them, and they shall be his people, and God himself shall be with them, and be their God" (vv. 1–3).

The most remarkable points to be noted in these passages are: 1st. The holy city incloses a cubical space, like the Holy of Holies of the Hebrew tabernacle ; like the sanctuary constituting the eighth stage of the pyramidal temple ; like the Arabian Chaaba, also, whose model was supposed to have been brought from heaven by the first man Adam. To these must be added the dressed stone of a cubical form, symbol of heaven, and likewise of the eighth Cabirus. 2d. The obvious reference to the zodiacal temple, corresponding to the eighth celestial region, evinced by the twelve gates, the names of the twelve tribes of Israel, and the twelve apostles. 3d. The connection with the cosmos, consisting of the two chief divisions, or heaven and earth. 4th. Its direct relation to man considered as the microcosm ; its measures are assimilated mystically to the measure of a man; calling to mind the image engraven upon the mystical square. 5th. The notion of a dwelling-place for the deity, upon which we have insisted in connection with the ancient doctrine of the temple. 6th. And finally, the great and high mountain, in which we have obviously a reference to the traditional mount of paradise, with which all these conceptions were associated.

We see in the doctrines set forth in the present chapter the vital connection between the first chapters of Genesis and the last chapters of Revelation. The Biblical *anthropology* is a perfect correspondent of the Biblical *cosmogony*. To heaven and earth answer the ordinary dualistic division of *soul* and *body*. The particular threefold division of the cosmos into heaven, earth, and the atmosphere, constituting the three worlds of the ancient Semitic and Aryan conception, correspond to the trichotomistic powers of man, the *spirit*, *soul*, and *body*. Thus, man is the true temple, the real cosmos. The fundamental relation of the cosmos and temple, and of man to both, each conceived as a dwelling of God, constitutes the ground thought of the two religions of the Bible.

CHAPTER VI.

A PARTICULAR HEAVEN AND EARTH REGARDED AS THE ARCHETYPAL TEMPLE.

SEC. 64. In the Mosaic and Babylonian accounts of creation, the formation of man is conceived as the crowning work of Divinity. In Genesis, the narrative continues with a description of the original abode of humanity, prepared by the Deity himself; and of the intimate personal relations existing at this early period between the Creator and his rational offspring. If this portion of the account is to be taken in any sense as literal, according to the usual interpretation of it by exegetes, it is necessary to give a fixed locality on the earth's surface, not only to the Garden of Eden itself, but to the personal appearance of the Divine Being in his intercourse with the first human pair. No direct communications could be imparted to man by the Deity that were not subject strictly to the limitations of time and space. Undoubtedly, therefore, a very great significance attached to this particular region during the life of Adam and Eve, and through all subsequent times within the traditional memory of their posterity. Not merely the life's experience and particular historical events connected with this primitive abode, but the image of its physical characteristics and geographical features would mingle with the saddened recollections of the past, and be engraved upon the mind of succeeding generations. Nor is it to be supposed that our first parents and their immediate descendants confined their observation and study to the configuration of the earth upon which they dwelt, or even to the natural objects with which they were surrounded. The sky, the heaven that spread its curtain over their heads, decorated with brilliant stars, and especially the sun and moon, — these objects also attracted their gaze and study. We are the more justified in such a supposition, since the narrative of the first chapter of Genesis devotes a particular account, and the fourth day of creation, to the arrangement of the heavens.

Thus, a fixed locality upon the earth's surface, and a well-defined circle of stellar observation, limited by their visual range, together with all the more prominent objects in both fields of view, would be brought naturally into immediate association, and be indelibly impressed upon the minds of the first men. It is morally certain that these various surroundings would give a deep coloring to the entire development of this primitive race, and, aside from their own acts and history, constitute the chief elements of those traditionary ideas transmitted to subsequent ages. It is equally certain, on the other hand, that those traditionary conceptions, in so far as they related to terrestrial objects, would pertain to a *particular portion* of the earth's surface, and not to the earth in general, except in a vague and indefinite manner ; that, also, in so far as they related to the stellar world, this would not be the vast expanse of the sky collectively taken, but a well-defined and limited region of the heavens, characterized by special features. A particular earth and a particular heaven, these put in direct and so to speak concrete relation to each other, would constitute exclusively the *cosmos*, or world, in which the practical development of primitive humanity had taken place. The same reasons apply in respect to personal intercourse between God and man, which is represented as existing at this period. The special relation to man here supposed, from the nature of the case, must give to the Deity an individual and local character, subject to the limitations of time and space, as all personal intercourse between rational beings on earth must be ; and this, even though the Divine Being was conceived abstractly as filling all space. In different terms, the Deity must be conceived, under such circumstances, literally to *dwell* somewhere, to Inhabit some *place*, the same as man ; and this notion of a dwelling-place, applied to the Divinity, constituted, in the minds of antiquity, one of the essential characteristics of the *temple*. Judging from the nature of the cosmos in which the intellectual and spiritual existence of the first men must have had its birth ; judging, too, from the arrangement of the temple structures, which at subsequent epochs were supposed to represent heaven and earth ; it is possible to form some idea as to the character of the originally conceived dwelling-place of Divinity, when as yet no artificial structures of this kind existed.

It consisted, in fact, of that particular and primitive heaven and earth here supposed; with which, therefore, the archetypal temple was identical. In a word, as will be shown hereafter, the heaven itself was esteemed the especial abode of the Divine Existence and the celestial paradise; while the earth itself was the abode of man, regarded as a terrestrial paradise; and the personal intercourse between God and the first human pair, as represented in Genesis, presupposes an open communication existing from one dwelling place to the other.

Sec. 65. The remarks that have been submitted thus far, being mostly of an *à priori* nature, will be thought to require an empirical basis, in order to give them in any sense a scientific value. This basis is to be sought in the concurrent traditions of the old world respecting the cradle of the human race, in so far as they are to-day available to the critic. The generalizing principles upon which it will be attempted to classify these data may be expressed in the following leading propositions, namely: —

1st. *There existed a particular, primitive, and traditionary heaven and earth, these being put in direct relation to each other; one regarded as the especial abode of the Divine Powers, the other as that of the first men.*

2d. *The conceived primary divisions of each, together with the central points of their supposed generation, coincided exactly in one case to those in the other.*

3d. *By virtue of these divisions, and upon principles heretofore established, such heaven and earth constituted at once the original cosmos and the archetypal temple.*

The first and most vital inquiry for us now is, whether the original abode of man on earth can be, with some degree of certainty, definitely ascertained? in other words, whether the geographical scheme of Genesis, pertaining to the Garden of Eden, admits of being precisely located? If this can be done, the elements of the foregoing propositions will be by this means fully supplied; since the particular earth being ascertained, its co-related stellar region becomes also known. The evidences tending to locate the terrestrial paradise must constitute for us, then, those of the first class; though no more conclusive, as we shall find, toward establishing our general theory than those pertaining to the traditional celestial paradise. It will be, by a comparative analysis of both orders of

proofs, that they will be found to afford mutual supports and confirmations of the hypothesis with which we are to be occupied.
First, then, as respects the locality of the *Gan-Eden* of Genesis.

Before the opening of the present century, the investigations of
scholars relative to this subject had been productive of no results,
except to render it hazardous for the reputation of any critic to
attempt even the solution of such a problem. The first series of
investigations, so far as my personal reading extends, that gave
promise of ultimate success in this direction were those conducted
by Colonel Wilford, and published in the " Asiatic Researches," dating from about the year 1818. Subsequently, many of the leading
Orientalists of Europe applied their learning and critical ability to
the solution of the same question, till now a well-defined theory
respecting the subject is unhesitatingly adopted by a large class of
learned critics. Although the hypothesis originally put forth by
Colonel Wilford has not been supported in many important details
by subsequent writers, owing to the unreliable character of some
sources upon which he was at that early period obliged to depend,
he was able to establish certain fundamental data that have served
the basis of more recent researches. These are substantially as
follows : —

1st. That there exists a remarkable agreement respecting the
first abode of mankind between the earliest and most authentic
traditions, preserved by the two great branches of the Indo-European race, namely, the Aryans of India and those of ancient Persia.

2d. That a like substantial agreement may be traced between
the Aryan traditions generally and those of the Semitic race ; particularly the Hebrew account of Genesis.

3d. That the common point of departure for all these races had
been outside the countries occupied by them subsequently, and since
the opening of the historical period.

It was proved, for instance, that the Sanskrit-speaking populations of India had not at first occupied this country, but had migrated from another region, entering India from the northwest.
So, too, the earliest records of the Zend or Persian tribes, although
in a measure indistinct, rendered it evident that this race had journeyed originally from a country much farther east and north. The
Book of Genesis also afforded intimations that the people primitively settled in the valley of the Euphrates had arrived upon the

plains of Shinar by migrating from the east. The question was
still a very difficult one, but it had been simplified and reduced to a
more scientific form. Everything seemed to conduct to the region
of the Hindu Caucasus as that where the earliest traditions of the
races ought to converge. The essential conditions of the problem,
as admirably stated by M. Ernest Renan many years since, were
in general these : —

1st. To find a region whose natural characteristics and geographical features corresponded to the uniform traditions respecting
the primitive home of man.

2d. A region whose situation was such, with reference to the
peoples preserving those traditions, that it was possible for each and
all to have departed from it toward the countries subsequently inhabited by them.

3d. A region to which, by the aid of their traditions and historical records, as well as by the assistance of linguistic science, tracing
the origin of names of mountains, rivers, etc., it was possible to
retrace the steps of these peoples along the routes originally taken
by them in their migrations from this central point of divergence.

It is not too much to say that where these essential conditions of
the problem before us have been clearly apprehended and strictly
adhered to by scholars in their investigations, there has been substantial uniformity and agreement in the results obtained by them.
Guided by these principles, the researches of such able critics as
Ewald, Lassen, Burnouf, Obry, Renan, Lenormant, and others have
all tended to one conclusion, namely, that the original abode of
these various races, the cradle of humanity, in fact, was the great
plateau of Pamir, situated on the high table-lands of Central Asia,
near the point where the mountains of the Belurtag unite themselves to the Himalayas. If some eminent scholars have come to
conclusions different from these, it has been generally due to the
fact, either that their criticisms were purely of a negative character,
tending to no fixed result, or that they have neglected the fundamental conditions of this problem, assuming points of departure
from which it was certain that some of the races never did depart.
There is not the slightest evidence, for instance, that the Aryans of
India ever migrated from Ararat in Armenia, from the " highlands
of Elam," or the banks of the Euphrates. But there is much proof
tending to show, on the other hand, that the Aryans entered India

from the northwest ; that the Medo-Persians migrated from beyond the Caspian ; and that the Semitic tribes, or at least the first civilizers of Babylon, came from the east. It is unnecessary, however, to dwell at greater length upon the conditions of this problem.

SEC. 66. It would lead us far astray from the main object in view to enter here upon labored investigations touching the details of this particular topic, based upon the original sources. Indeed, this work has been already and repeatedly gone over by some of the ablest critics of the present day. The principal facts pertaining to the question are well established, and the necessary conditions to be observed in relation to it have been explained. I merely serve myself, therefore, with some extracts from authors who have most thoroughly explored the entire field, and have reported the conclusion at which they have arrived. I commence with M. E. Renan as follows : —

"Thus, everything invites us to place the *Eden of the Semites* in the mountains of Belurtag, at the point where this chain unites with the Himalayas, toward the plateau of Pamir. . . . We are conducted to the same point, according to E. Burnouf, by the most ancient and authentic texts of the Zend-Avesta. The Hindu traditions, also, contained in the Mahabharata and the Puranas, converge to the same region. There is the true *Meru*, the true *Albordj*, the true river *Avanda*, from whence all rivers take their source, according to the Persian tradition. There is, according to the opinions of almost all the populations of Asia, the central point of the world, the umbilic, the gate of the universe. There is the *Uttara-kuru*, 'the country of happiness,' of which Megasthenes writes. There is, finally, the *point of common attachment* of the primitive geography, both of the Semitic and Indo-European races. This coincidence is one of the most striking results to which modern criticism has conducted ; and it is remarkable that it has been reached from two opposite directions at one and the same time, namely, through Aryan studies on one hand, and Semitic studies on the other." [1]

Prior to the investigations of M. Renan, in the article referred to here, M. Obry had published an extended review of two hundred pages, devoted exclusively to the same problem. The subjoined extract exhibits the conclusions in part, as stated summarily at the close of his work, to which this writer had been conducted : —

[1] *Histoire générale des Langues Semitiques*, etc., pp. 480, 481. Cf., by the same author, *De l'Origine du Langage*, pp. 228–230. Both treatises contain excellent criticisms upon the topic in question.

"I believe to have sufficiently established in the course of this essay: 1st. That the Semitic traditions, or better, the Semitico-Hamite, are in accord with the Aryan traditions, in placing the cradle of the human race to the north of India ; that is to say, in a country east in relation to the Semites, etc. 2d. That this region was from the first conceived as identical with that upon the mountains of which the ark of Noah rested after the deluge, as well as that of Xisuthrus and of Manu-Vaivasvata. 3d. That as Genesis affirms that the descendants of Japhet, Shem, and Ham emigrated from the east to Babylon, it is necessary to follow inversely this route, in order to find the first abode of man ; that is, to pass from the Semitic *Ararat* to that which I call the Aryan *Aryaratha*, named *Meru* by the Hindus, *Albordj* by the Persians, and *Eden* by the Hebrews. 4th. That originally Eden, Albordj, and Meru were all conceived as one and the same plateau of *a square figure, having its four sides turned toward the four cardinal points ;* and of such immense elevation that it seemed to confound itself with the heavens, the abode of the Superior Powers. 5th. That this high region, suspended, so to speak, between heaven and earth, and conceived as the cradle of the human race, passed for being watered by a single river, but which divided itself from thence into four branches, flowing toward the four great countries surrounding it, also *facing the four cardinal regions.* 6th. That the *orientation* of the four rivers (that is, their direction toward the cardinal points), and their issue from a common source, constituted in some sense two fundamental conditions of the primitive abode of humanity."[1]

A remark or two here relative to the two fundamental conditions alluded to by M. Obry in the last sentence quoted. Universal tradition does assume four rivers, the same as described in Genesis ; it assumes their issue, also, from a common source, or at least a common region of country from which all take their rise. In a certain sense, finally, tradition assumes the *orientation* of these rivers. In this respect, however, there was a variation between the *cardinal* and the *intermediate* points of the compass. The Buddhists, as cited by M. Obry himself, located these rivers in the direction of the intermediate regions ; and there is reason to believe that such notions prevailed at Babylon. In all cases, nevertheless, the idea of a certain *orientation* was fundamental, as this author insists. These were the essential characteristics of the country which, according to tradition, had constituted the first abode of humanity. A region without four rivers, these diverging from a common centre,

[1] *Du Berceau,* etc., pp. 187, 188.

and flowing in opposite directions toward four quarters of the globe, could not be identified with the geography of the Eden of Genesis. Here we have an essential condition of the problem before us that must not be neglected. It is remarkable that there exists but one such region on the earth, and it is that to which M. E. Renan and Obry allude in the extracts cited. It will be noticed that M. Obry confirms the authorities cited in the last chapter, relative to the special configuration of the traditional Eden. It was of a *square figure*, its sides *oriented*, and surrounded by four great countries also *oriented*. Uniform tradition agreed likewise in this particular.

SEC. 67. The most elaborate and recent discussion of the subject before us, so far as I am aware, is that of M. F. Lenormant in his excellent commentary upon the "Fragments of Berosus," already frequently cited in these pages. As his opinions agree so nearly with those of the authors just quoted, it is unnecessary to reproduce them here in detail. He proves the substantial agreement of the Hindu and Persian traditions respecting the locality of the sacred mount, the Eden of Genesis, and then shows that the Mosaic account points definitely to the same region. He observes: —

"That the Biblical description of Eden relates to the same country as the other (the Aryan) traditions passed in review by us is a point upon which all scholars are to-day agreed, and it is established by abundant proofs." [1]

But the most recent statement of the hypothesis, supported by the evidences adduced by the authors already referred to, is that by the distinguished Orientalist, M. G. Maspero. He does not appear to have gone over the entire ground anew, upon the basis of the original sources, but the various points referred to by him have been well established by other scholars, and his summary of the conclusions is an excellent one, which is here reproduced. Alluding to the ancient races, he remarks: —

"All have preserved, mixed with the vague legends of their infancy, the tradition of a primitive country, which had been inhabited by their ancestors before their dispersion. It was a high mountain, or better, an immense plain, of a square figure, so elevated that it seemed to be suspended, so to speak, between heaven

---

[1] *Frag. de Bérose*, p. 308. Cf. pp. 299–314, for a discussion of all the points, which is very able and critical. The author adheres strictly to the conditions of the problem, and but one result was possible.

and earth. From the interior rose a great river, which was divided
from thence into four arms or branches, which flowed toward the
four surrounding regions. There was the navel of the world, the
cradle of the human race. The peoples situated between the Mediterranean and Tigris located this legendary country in the east;
while the populations of Iran (ancient Persia) and India conceived
it to be situated in the north. The moderns have succeeded in
determining its site more definitely than the ancients were able to
do. They have placed it in the mountains of the Belurtag, near
the place where this chain unites with the Himalayas, upon the
plateau of Pamir. There in effect, and there alone, we find a
country which satisfies all the descriptions, geographically speaking,
preserved in the sacred books of Asia. From the plateau of Pamir,
or better, the mountain mass of which this region is the central
plain, four great rivers take their rise, the Indus, the Helmend, the
Oxus, and the Jaxartes, which flow in directions the most diverse,
well answering in this respect to the four rivers of sacred tradition." [1]

For the most part, the investigations of learned critics relative
to the situation of the *Gan-Eden* of Genesis have been based upon
the Semitic and Aryan traditions, with very little reference to those
preserved by the so-called Turanian races. The reason has been
that the Turanians, if we except the Chinese and those assumed to
have founded the Babylonian civilization, have preserved no such
distinct recollections pertaining to the original home of man, not
having possessed from early times a written literature. Nor has
the literature of the Chinese and that of the Babylonians been so
thoroughly investigated by scholars generally. Nevertheless, certain facts are now known, some of which have been already presented to the reader, which go to confirm our hypothesis respecting
the actual position of the sacred mount, traditionary abode of the
first men. Since China is situated to the southeast in relation to
the plateau of Pamir, we ought to be able, according to the conditions of our problem, to trace the origin of the Chinese in the northwest in relation to the region occupied by them in historical times.
On this point Professor W. D. Whitney observes: " The origin of the
Chinese people is to be sought — if it be possible ever to trace back
their movements beyond the limits of their own territory — in the
northwest." [2] This statement is abundantly supported by the facts

---

[1] *Histoire Ancienne*, etc., p. 132.
[2] *Oriental and Linguistic Studies*, 2d series, p. 63.

developed in Dr. Schlegel's recent and voluminous work upon Chinese astronomy.[1] It is necessary to recall here the cosmical legend of the tortoise, as presented in the last chapter. The matter of the five summits, of the eight regions, and the constellation of the "Northern Measure," that is, the seven stars of the great Dipper, all tend to connect the primitive Chinese tradition with the Meru of the Hindus, the Albordj of the Persians, the Eden of Genesis. With reference, now, to the Babylonians, the frequent allusion in the cuneiform texts to the " Bit-kharris of the east, the father of countries," as already mentioned, affords sufficient indications of the existence of traditions similar to those recorded in the first chapters of Genesis. But I consider the attempt of some English Assyriologues to locate the origin of these traditions, and thence the Mosaic geography of Eden, in the "highlands of Elam," or in the valley of the Euphrates itself, as based wholly upon erroneous ideas. It is quite probable that certain names of rivers and mountains had been transported from the east, and applied to corresponding objects in a country farther west. But the conditions of our problem absolutely forbid the assumption of either Elam or Chaldæa as the cradle of humanity. One of these conditions regards the peculiar geographical features, in themselves of an extraordinary character. Another is constituted of the concurrent traditions pointing to a fixed locality, preserved by peoples widely separated in historical times. Finally, we have to trace these populations, with the aid of their traditions, and by other helps, back to the original point of common divergence, and by some route physically possible. Those who are familiar with the facts pertaining to our subject will recognize at once the impossibility of assuming any other terrestrial region as the birthplace of humanity than Central and Northern Asia, if all these conditions are to be observed. But it is a most remarkable circumstance that the plateau of Pamir literally fulfills all of them.

SEC. 68. But there is another condition of the problem before us, upon which critics have never insisted, so far as I know, but it

---

[1] *Uranographie Chinois*, pp. 729-736. The extravagant theory of the author as to the antiquity of the Chinese nation may well be doubted. But it results from his investigations that the origin of this people is to be traced to the elevated plains of Central Asia, to the same region, in fact, already assumed as the primitive home of man.

seems to me very essential, and not to be neglected. I allude to the astronomical element involved in nearly all the ancient traditions relating to the sacred mount of paradise. With the particular earth inhabited by the first men, whose geographical features were so extraordinary, were directly connected certain stellar objects of a character not less notable, which constituted the particular heaven of primeval tradition, thus completing the notion of a well-defined celestial space united with a corresponding terrestrial space, previously shown to have characterized the old augurial temple of the Etrusco-Romans. We enter now upon the consideration of the astronomical element referred to, deriving our first notice from the passage in the prophecy of Isaiah, which reads as follows: —

"For thou hast said in thine heart, I will ascend into heaven, I will exalt my throne above the stars of God : I will sit also upon the mount of the congregation, in the sides of the north : I will ascend above the heights of the clouds ; I will be like the Most High " (xiv. 13, 14).

This remarkable text has become quite celebrated among Orientalists, as will be seen from the frequent allusions to it to be cited hereafter. The Hebrew name of Divinity employed in the expression " the stars of God " is *El* (אֵל), while that occurring in the phrase " Most High " is *Elyon* (עֶלְיוֹן), both being titles of the Deity common to all the Semitic nations, and thus, as Professor Max Müller has fully shown, to be referred to the period before the Semitic race had separated into distinct branches. The *Har-Moad*, or " mount of the congregation, in the sides of the north," better rendered, perhaps, as " mountain of the assembly in the extrême north," was undoubtedly, as long since suggested by Dr. Gesenius, the great Asiatic Olympus, the supposed residence of the gods. In the phrase " stars of God," that is, *the stars of El*, there is an allusion to a particular and remarkable asterism in the northern heavens, the identification of which is no longer a matter of doubt among leading Orientalists. I refer to the seven stars of the " Chariot," or "great Dipper," in the constellation of the Great Bear, whose slow, rolling motion about the north polar star has been remarked by the men of all ages. Its uniform and direct association with the traditional mount of paradise is the point to be established. Dr. Gesenius observes : —

" The place mentioned in the words of the king of Babylon (Is. xiv. 13), *the mountain of assembly* (of the gods), is probably the Persian mountain *El-Burj*, *El-Burz*, called by the Hindoos

Meru, supposed to be situated in the extreme north, and, like the Greek Olympus, regarded by the Orientals as the *seat of the gods*." [1]

The *stars of El*, then, were associated with Mt. Meru, the Albordj of the Persians, the Eden of Genesis. Relative to this point, and in allusion to Mt. Meru, Colonel Wilford remarks: —

"As it is written in the Puránas, that on Mount *Meru* there is an eternal day for the space of fourteen degrees round *Su-Meru* (the celestial pole), and of course an eternal night for the same space on the opposite side (or south pole); the Hindus have been forced to suppose that *Su-Meru* is exactly at the *apex*, or *summit* of the shadow of the earth; and that from the earth to this summit, there is an immense conical hill, solid like the rest of the globe, but invisible, impalpable, and pervious to mankind; on the sides of this mountain are various mansions, rising in eminence and preëxcellence, as you ascend, and destined for the place of residence of the blessed according to their merits. God and the principal deities are supposed to be seated, in the sides of the north, on the summit of this mountain, which is called also *Sabha*, or of the congregation. This opinion is of the greatest antiquity, as it is alluded to by Isaiah, almost in the words of the Pauranics. This prophet, describing the fall of the chief of the *Daityas* (the Babylonian king?), introduces him, saying, 'That he would exalt his throne above the stars of God, and would sit on the *mount of the congregation* in the sides of the *north*.'" [2]

The idea that an eternal day exists for "fourteen degrees round *Su-Meru*," or the north polar star, probably arose from the fact that the asterisms situated near this point never set, never leave the visible heavens, as do those constellations more distant from the pole. It will be seen that Mt. Meru was conceived as an immense conical hill, whose summit penetrated the heavens in the region of the celestial pole or *Su-Meru*; and thus, that the mountain itself literally united the heavens and earth like a vast column. Colonel Wilford also identifies the *Har-Moad* of Isaiah with Mt. Meru. But I wish to submit a passage here from the pen of Rev. A. H. Sayce, as follows: —

"The innumerable gods and goddesses, demigods and heroes, of the Accadians were adopted by the Assyrians in their popular

---

[1] Robinson's Gesenius' *Heb. Lex.*, Art. מוֹעֵד. Cf. Gesenius' *Jesaia*, notes *in loc.*; and Beylage I., especially, p. 326, b. ii.

[2] *Asiatic Researches*, vol. vi. pp. 488, 489.

mythology, in the larger proportion of cases, without any change of name. Even temples of *Kharsak-kurra*, or ' highland of the east,' ' the mountain of the world,' and cradle of the Accadian race and ritual, are founded by Assyrian monarchs. Nay, we find the same starting point of Turanian civilization mentioned in the Old Testament; Isaiah (xiv. 13) sets the king of Babylon on ' the mountain of the gods ' or ' world,' which the Jew, who had identified Accad or Urdhu, ' the highlands,' with *Ararat (Urardhu)* of the same signification, places in the north. Both Accad and Armenia are called in the inscriptions *Burbur* or ' summits.' " [1]

SEC. 69. I am not sure that Rev. A. H. Sayce would identify this "mountain of the world," the *Har-Moad* of Isaiah, with the "highlands of Elam," or of modern Susiana; but this seems to me impossible. 1st. The *stars of El* are directly associated by Isaiah with the "mount of the congregation, in the sides of the north." This sacred mountain is in the extreme north, not in the east simply, which, in reference to Akkad, was the exact direction of Elam. Besides this, no tradition connects with the "highlands of Elam " any such asterism as the *stars of El*, or such as universal tradition associated with the true Asiatic Olympus. 2d. The allusion of Isaiah, according to the general opinion of the best exegetes, was definitely to Mt. Meru, the Persian *Albordj.* But more especially, now, with reference to the seven stars of the chariot in connection with the diluvian mountain which uniform tradition regarded as one and the same with the mount of paradise. M. Obry, in allusion to the identity of the two mountains, remarks as follows : —

" But in taking this for our base, we are able to demand to-day, whether the *Ararat of Genesis* was the same as the Ararat of subsequent Biblical writings? Or, in other terms, whether this name, of doubtful origin, is not a corruption of an Aryan term, either Zend or Sanskrit, namely, *Aryaratha*, ' *chariot of the Aryas ;* ' a vague designation of a mountain of the north, situated outside Armenia, and, for instance, to the north of Media, Persia, or even Bactria, as was conjectured in the last century by the learned Abbé Millat? — this mountain being thus named, for the reason that around its summit was thought to turn the chariot of the seven Maharchis of the Brahmans, the seven Amschaspands of the Persians, or the seven *Kokabim* of the Chaldæans ; that is to say, the chariot of the seven stars of the Great Bear." " That the Chaldæans had inherited these mythical ideas is proved by the complaint of Isaiah upon the fall of the ungodly monarch of Babylon : ' this

[1] *Trans. Bib. Arch. Society,* vol. i. p. 299.

star of morning, son of Aurora, this oppressor of nations, who
boasted not to descend like other kings into the depths of *Sheol*, but
to ascend above the stars of the mighty God, and to take his place
by the side of the Most High, upon the mountain of the assembly '
(Hebrew *Har-Moad*) ; that is, assembly of the *cheba-kokabim*, or
seven stars of the Great Bear, on the sides of the north." [1]

These views respecting the transfer of the name *Aryaratha*,
"chariot of the Aryas," from *Meru* of the Hindus to *Ararat* in
Armenia, and the identification of the same stellar group to which
the name alludes, with the *stars of El*, are fully adopted by M.
Lenormant in the subjoined extract : —

"M. Obry has shown that the mountain regarded by the Aryans
as the original and sacred abode of humanity had at first received
in their traditions the name of *Aryaratha*, ' chariot of the Aryas ; '
for the reason that around its summit was thought to turn the
chariot of the seven Bramanic Maharchis, of the seven Persian
Amschaspands, and of the seven Chaldæan *Kokabim ;* that is to
say, the chariot of the seven stars of the Great Bear. This title
Aryaratha is the original of that of *Ararat ;* and it was only at a
later period that the Aryan tribes, emigrating into Armenia, transferred it to a mountain in that country, otherwise called Mt. Masis."
"The pyramidal temple of the Chaldæans was an imitation, an
artificial reproduction of the mythic ' mountain of the assembly of
the stars,' the *Har-Moad* of Isaiah, which sacred tradition placed in
the north." [2]

The author refers, in the two passages above cited, to one and
the same mountain, the Meru of the Hindus. Thus, " the chariot
of the Aryas," rolling upon the summit of Meru, designates the
same stellar group as that which Isaiah refers to as *the stars of El ;*
both associated with the traditional mount of paradise. But connected with this circle of ideas is another fact of considerable importance, and one derived from a well-known cuneiform passage.
Rev. A. H. Sayce has referred to the temples of *Kharsak-kurra*, a
phrase otherwise read *Bit-kharris*, and interpreted by this writer as
" mountain of the world." These were the pyramidal temples in
imitation of the traditional mount of paradise, to which M. Lenormant alludes. Connected with the structures called *Bit-kharris*, or
Kharsak-kurra, was another class termed *Arali ;* and a certain text
makes special mention of both as the abode of the great divinities ;

---

[1] *Du Berceau*, etc., pp. 5–7.
[2] *Frag. de Bérose*, pp. 302, 358 ; cf. pp. 317, 318.

referring, however, not to the artificial reproductions, but to that particular celestial region of which these were imitations. I quote the rendering of the text by M. Lenormant, with his comments attached : —

" ' *Nisruk, Sin, Shamas, Nabu, Bin, et Adar*, and their great spouses, who reign eternally in the interior of the great *Bit-Harris* of the east, and of the country of *Aralli*.' This is, as we find, a luminous and celestial region like the east, which serves as the residence of the great divinities, and of which the temples *Arali* are the image. Such a conception accords perfectly with that of the *Qaq-qa-du* or *Bit-Sadu*; that is to say, with the culminating portion of the superior hemisphere of heaven. As to *Bit-Harris*, ' the house well built,' which the passage cited represents as the palace of the gods, it is situated at the same time in the *Kurra* and in the *Arali*; that is to say, in the east, and in the direction of the point which serves as the pivot of rotation of the superior heavens ; and we believe that it is necessary to place at the summit of the paradisiacal mountain of the northeast, which unites the heavens and earth like a vast column, the *Har-Moad* of Isaiah, of which we have studied already the conception." [1]

SEC. 70. The concurrent opinions of so many learned critics upon the various points now before us is a fact quite unusual ; and it goes to establish the following conclusions : —

1st. That the " mountain of the assembly," to which the prophet Isaiah alludes, is none other than the traditional mount of paradise ; the Eden of Genesis, Meru of the Hindus, or Albordj of the Persians.

2d. That according to the most ancient traditions pertaining to this sacred mount, there had been uniformly associated with it a particular and notable asterism, namely, the chariot of the seven stars of the Great Bear, with which the north polar star itself was put also in direct relation.

3d. That consequently, from the singular allusion of the prophet to the " stars of *El*," in relation to which *Elyon* is considered above and " Highest," these two Hebrew names of the Deity must have been in some way connected with the same group of stellar objects, and this from the earliest periods to which the development of Semitism is to be referred.

It is probable that the statement last submitted will be received

[1] *Frag. de Bérose*, p. 393.

with some surprise on the part of Biblical critics, and not less with
hesitancy. But even the etymology of the term *El*, as generally
sanctioned by Hebrew scholars, affords a striking confirmation of
the thesis above stated. It seems to be well understood that this
name *El* (אל), "the Strong One," is derived from *ool* (אול); and
that the notion of *rolling, turning*, is fundamental in both expressions. Hence, the meaning of "Strong One, or Mighty One," has
proceeded in some way from that of "to roll, to turn." Dr. J.
Fürst explains this upon the principle that "the idea of rolling
gradually merges into that of strength" (art. אל). But it is
hardly possible that the simple conception of "rolling," without the
aid of any special circumstances, ever suggested the idea of infinite
force, leading to the selection of this term as a synonym for the
Almighty. If, however, we connect with this notion the remarkable allusion of Isaiah to the "stars of El," to the chariot of seven
stars rolling around the celestial pole, the double conception of
turning and of *strength* will at once strike the mind. As M.
Dupuis remarked long ago, but without any reference to this Hebrew divine name, the revolution of the immense mass of starry
heavens upon a single, fixed point in the northern hemisphere
would naturally attract the attention of the first men; and they
would instinctively concentrate around that point the vast assem-
blage of force sustaining this mass and causing it to revolve.
Thus, *El* was the "Strong One," who upheld the vast fabric of the
world and caused it to turn on its eternal pivots. The fact that this
title of divinity occurs in the composition of names in both lines of
genealogy from Adam before the deluge, and that neither line,
owing to an original feud between them, would be likely to adopt a
divine name from the other, tends to show that the term *El*, as
applied to the divinity, was traditionally connected with *Gan-Eden*
itself. or with that sacred summit around which turned the fiery
wheels of the celestial chariot.

The fact that the Aramaic populations of Northern Mesopotamia,
and especially the Sabæans of Haran, one of the oldest cities of the
world, in which the family of Abraham sojourned for a time after
their migration from "Ur of the Chaldees," had preserved traditions quite in accord with those already passed in review is a matter familiar to Orientalists. These recollections of the primitive
ages of humanity, constituting some of the most important ele-

ments of their original cultus, were embodied especially in the so-called " mysteries of *Shemal*," Hebrew *Semol*, a particular and critical investigation of which was published several years since by Professor D. Chwolsohn, of St. Petersburg, to which reference has been before made.[1] This ancient Semitic word *Shemal*, or *Semol*, which appears in the Assyrian texts under the various forms of *Su-mi-lu*, *Su-me-lu*, and *Sa-me-la*, signifies literally " the left, the left hand," thence employed from a very remote period to denote the north, north pole, etc., corresponding precisely to the *Su-Meru* of the Hindus. As stated by M. Obry, the Buddhist designation of *Su-Meru* is *Su-mi-lu ;* and this is exactly the Assyrian *Su-mi-lu*, applied to the same stellar region.[2] It is well known that the Haranites associated the chariot of the seven Stars of the Great Bear with their worship of *Schemal*, who was thus in relation to them, like the Hebrew *Elyon* in relation to the " stars of *El*," considered the *eighth*.[3] The reader will see at once in the character of this Haranite divinity, and in the mysteries connected with his cultus, an important confirmation of the views which have been expressed relative to the great Asiatic Olympus, to which the Hebrew prophet alludes.

Finally, we recall here the Chinese legend of the tortoise, taken as a symbol of the cosmos. The upper portion of the shell represented the heavens, and the lower portion the earth. But we see at a glance that it is not the heavens and earth in general which is intended, but a particular and limited celestial space put in direct relation to a correspondent terrestrial region. Upon the back of the tortoise were the images of the eight celestial regions, and in connection with these the constellation of the " Northern Measure," which Dr. Schlegel fully identifies with the group of seven stars, of which there is here question. Obviously, these eight regions had their common point of divergence in the pole star, the Su-Meru of the Hindus, as suggested in the last chapter. Placed in direct relation to these celestial representations were the images of the five summits and the four canals, whose reference to the traditional Eden, and exact accord with the Hindu conceptions relating to the same locality, was also pointed out in the chapter preceding.

[1] *Die Ssabier und der Ssabismus*, St. Petersburg, 1856, vol. ii. pp. 319–364.

[2] *Du Berceau*, etc., p. 83, note 4.

[3] Ibid., p. 7. Cf. Lenormant, *Frag. de Bérose*, pp. 318, 319.

Thus, we supply here another indispensable condition of the problem which relates to the precise geographical locality of the Eden of Genesis, as supposed divergent centre from whence the races departed to the various countries occupied by them since the opening of the historical epoch. It was a region with which certain peculiar geographical features, certain uniform traditions, and lastly certain stellar groups were invariably associated. It is one of the most notable facts pertaining to Oriental research that all these conditions should be literally supplied by a fixed and well-known locality on the earth's surface, from whence, above all others, it is quite easy to suppose the various races who now people the earth first departed.

SEC. 71. It is, then, a matter in relation to which it is impossible to entertain serious doubts, that the earliest traditions transmitted to us respecting the primitive home of mankind comprised the two chief elements insisted upon in the introductory remarks of the present chapter; namely, a special reference to a particular region of the starry heavens, placed in direct relation to a particular portion of the earth's surface, regarded as the cradle of humanity. That the stellar space thus associated with the terrestrial abode of the first men was conceived uniformly as the especial dwelling-place of the divine powers is also a well-established point; and it fully explains the existence of the notion in almost all antiquity that the seat of the heavenly hierarchy was in the extreme north. In fact, this highest central region of the northern heavens was deemed a celestial paradise, with which was connected the terrestrial by means of the sacred mount itself, that united the heavens and earth like a vast pyramid or column. On this point, the following observations of M. Obry have a direct bearing : —

"From all antiquity, the populations of Asia have regarded the blue vault of the firmament as a garden of delights, tapestried with brilliant stars like stones of fire." "As the radiance of the snow-capped mountains mingled with the azure of ethereal space, blending with it in the distance, the paradise of the gods appeared to confound itself with that of the first men; and to express this vague notion, they formed the Sanskrit phrase *Svarga-bhoumi*, 'celestial earth.' "[1]

We are prepared now to trace the genesis of the ancient temple,

[1] *Du Berceau*, etc., pp. 173, 174.

not merely with respect to and by means of its form, but through direct tradition derived from the primitive abode of humanity. M. Obry has shown that the diluvian mount and the mount of paradise were uniformly identified in primitive tradition. This fact was noticed and insisted upon by the older mythologists; and no doubt exists in relation to it. These two mountains identified as one, we are now able definitely to locate on the earth's surface, and to show that all the traditionary conceptions relating to it regarded a particular earth with peculiar geographical features, and a particular heaven embracing astronomical characteristics not less remarkable. The earth thus designated was the terrestrial paradise, abode of primeval humanity. The heaven put in direct relation to it was conceived as the celestial paradise, the seat of the divine hierarchy. The two were united by the sacred mountain itself, like a vast pyramidal column reaching from earth to heaven. Recall now the fact stated by Moses that the founders of Babylon had journeyed from the east to the land of Shinar, and had there undertaken to build an immense tower, whose top should reach the heaven, or, in other words, should represent the heaven; a tower which cuneiform scholars usually agree to identify with the pyramidal temple of Borsippa, in all respects similar to the brick pyramid of Sakkara in Egypt, belonging obviously to the same chronological epoch. Recall the fact also, as stated by M. Lenormant, and which we are now able to confirm, that the pyramidal temple of the Chaldæans was an imitation, an artificial reproduction, of the traditional mount of paradise, the *Har-Moad* of Isaiah, located far to the east and north, with which the stars of *El*, the seven stars of the chariot, had been uniformly associated.

In the passage from Nebuchadnezzar's inscription relating to the tower of Borsippa, a rendering of which was given in the first chapter (sec. 10), this structure is styled "the temple of the seven lights of the earth." The word "earth" here is not fully correct. The original is *An-ki*, "the astronomical, heavenly, or divine earth," distinguished from the ordinary geographical earth by the determinative *An*, "elevated, heaven, god." The allusion is obviously to the sacred mount of the east, from which the builders had recently migrated, and of which this structure was an artificial reproduction. The seven lights of this "divine earth," therefore, were the seven stars of the chariot, associated in tradition with the mount of paradise, the cradle of humanity. The seven stages of the pyramid denoted primarily these seven luminaries, the stars of El; and thus, the edifice might significantly be termed *Bab-el*, " gate of El," or *Ká-an-ra*, "gate of the god of the deluge," in reference both to the diluvian mount on one hand, and to the *Har-Moad* on the other. The direct traditional origin of the pyramidal temple from the mount of paradise is thus placed beyond question. It was an image of the cosmos, consisting of the two chief divisions, heaven and earth, as shown in the last chapter. But this cosmos was the primitive one known to man; it consisted of the particular heaven and earth of primeval tradition, that is to say, the celestial and terrestrial paradise united together by the sacred mount. The eighth stage, of a cubical form, like the Holy of Holies of the Jewish tabernacle and the cubical stone of the temple-craft, represented this particular celestial region, the *Su-Meru* of the Hindus, the *Shemal* of the Haranites, the *Su-mi-lu* of the cuneiform texts, one with the *Su-mi-lu* of the Buddhists of Central Asia. The basement of the pyramidal temple, like the outer court of the tabernacle, represented the earth; but this was a particular earth, the "divine earth," the terrestrial paradise. The seven stars were denoted by the golden candlestick, correspondent to the stages of the pyramid. It has been supposed that the seven stages of the tower of Borsippa related to the seven planets, since they were colored differently, and according to the mystical theory which appropriated a particular color to each planet. All this is quite probable, for according to the primitive conceptions everything revolved around the sacred mount as the central point of the universe; the fixed stars, the planets, but particularly the seven stars of the chariot, the divine Rishis of the Hindus. The Haranites united the cultus of the planets with that of these seven stars, in the mysteries of Samael; and it is quite probable such was the case at Babylon. In any event, there must have been a direct reference to the "stars of *El*," in the phrase to which we allude: "the seven lights of the divine earth." This supposition is confirmed by the remarks of M. Lenormant, relative to the temples *Bit-kharris* and *Arali*, representing the rotating centre of the superior hemisphere of heaven. But in addition to these facts, tracing the origin of the temple through tradition to the sacred mount of paradise, we should recall the proofs presented in the last chapter, tending to connect the system of nine squares, represented in the Accadian sign *Mal*, " temple," and in the mystical seal symbolizing the cosmos; we should recall, I say, the fact of the direct connection of this system of divisions with the traditional centre of the world, the Meru of the Hindus, the Eden of Genesis. Little doubt can exist that the primary notion of the augurial temple of a square figure exactly *oriented*, involving a particular stellar space placed in relation to a terrestrial one, had its origin likewise in the holy mount of the east, uniting a particular heaven and earth, constituting, in fact, the real cosmos known to the first men. Indeed, I believe that all the ancient cosmogonies, as well as the primitive doctrine of the temple, pertained originally to this particular heaven and earth, and not to the heaven and earth in general.

SEC. 72. The mysteries of *Shemal*, according to the researches of Professor D. Chwolsohn, and as previously stated, were celebrated annually at different periods of the year in an underground room or cave; and this writer judges, with much reason, that they had some connection with the primitive cave worship, which was held in such veneration in all antiquity. Usually, these cave mysteries comprised seven degrees of initiation, as illustrated so frequently by the ancient art monuments, where a rocky ascent, a staircase, or mystical ladder, exhibiting seven stages, is represented.[1] These degrees of initiation had reference, probably, to the seven stars of the chariot, to which were joined the seven planetary bodies in the cultus of the Haranite Sabæans; and as Shemal was unquestionably connected with the north, the celestial pole, the Su-Meru of the Hindus, these mysteries must have had reference to the sacred mount of paradise, of which the pyramidal temples with their seven stages were artificial reproductions. On the top of these pyramids was erected the sanctuary of a square or cubical form, constituting thus the *eighth stage*. This sanctuary, or eighth stage, in relation to the other seven, held the precise position that Shemal did in relation to the seven stars of the chariot; and so, of the Hebrew *Elyon*, the " Highest," with reference to the " stars of El." It is remarkable that this term *Elyon* (עֶלְיוֹן), from *ā-lāh* (עָלָה), " to go up, to mount, to ascend," strictly related, therefore, to the form *ō-lāh* (עֹלָה), " an ascent, steps, stair-way," and to *a-lia-yāh* (עֲלִיָּה), " loft, upper chamber, put for the chambers of heaven," both derived

---

[1] Vid. De Hammer, *Culte de Mithra*, Pl. V.

from the same verbal root, had reference originally to the degrees of ascent, but especially to the *highest*, the *eighth*, corresponding to the Phœnician *Eshmun*, "the eighth," all of which, in my view, points to a primitive connection with the paradisiacal mount, and to that particular stellar region associated with it, whose peculiar and striking features were well calculated to give birth, under all the circumstances, to this entire circle of primitive ideas.[1]   As expressly held by M. F. Lenormant, in which he but supports the views of his lamented father, M. C. Lenormant, the vision of Jacob's ladder, which also united the heaven and earth, in memory of which the patriarch erected a pillar, giving it the significant name of *Beth-el*, or "house of El," pertained to the same category of conceptions as the pyramidal temple itself; and an important proof of such as-similation is that the Egyptian hieroglyph of a pyramid in stages has the value *ār*, "a ladder," from the verb *ār*, "to ascend."[2]

With respect to the primitive cave worship, of which that paid to Shemal was as ancient, probably, as any, we are not to conclude, as some modern writers have done, and as Professor Chwolsohn himself is inclined to do, that it pertained to a crude, savage condition of humanity.   The archetypal cave, to which the mysteries of Shemal related, was, as heretofore affirmed, the celestial vault, the central, polar region of the heavens, to which the very name Shemal, like the Hindu Su-Meru, definitely appertained.   The cuneiform texts afford direct proof that this particular stellar space was con-ceived as a *vault*, a *cave*.   First, the Accadian character, usually employed ideographically to denote the heavens, Assyrian *Samu*, has the value *E*, signifying "vault, house" (Rep. 254).   Secondly, the Accadian *dak*, "cave, vault, blue" (Rep. 130), denotes astro-nomically the "star of the vault;" and this is equated in the texts to the god *Mak-ru*, the name of Mercury corresponding to the zodiacal sign Leo.[3]   Hence, the star of the vault, or cave, was asso-ciated in some way with the extreme north, the most elevated region of the heavens.   It was this celestial cave, therefore, to which the mysteries of Shemal, celebrated in an artificial one, had especial

---

[1] For the astro-religious conception of the god *Eshmun*, see Movers, *Phœnizier*, i. pp. 527–536.   Cf. Lenormant, *Frag. de Bérose*, pp. 382–389.

[2] *Frag. de Bérose*, pp. 358, 359 ; also De Rougé, *Chrestomathie Egyptienne*, pt. 1st, p. 73, a, 40.

[3] 2d Rawl. Pl. 49, No. 3, l. 30, and 3d Rawl. Pl. 53, No. 2, Obs. l. 6.

reference. Shemal, as being the eighth in relation to the seven stars of the chariot, must have been the eighth Cabiriac divinity, since the high-priest who presided over the mysteries was especially called *Kabir*, that is, a Cabirus.

Whether we speak of the pyramidal temple, then, which Dr. Bähr has shown to have been primitive in all Asia, or of the sacred caves, in which many have thought to find the original type of all edifices devoted to the worship of God, it is above all apparent, from the data now before us, that we have to go to the paradisiacal mountain for the absolutely primitive model of all of them. The mysteries of Shemal referred to it, and the proofs now before us are abundant to the effect that the two Semitic titles of divinity, El, the "Strong One," and Elyon, the "Highest," were originally associated with the order of ideas centring in the same traditional locality. Here was the birthplace of all the mysteries, of all the cosmogonies, of the primitive doctrine of the temple. Here was the original *Beth-el*, "house of El," the "gate of heaven," the "ladder" that reached from earth to heaven. The conical stone, even, which appertained to the mother goddess, was only a miniature mountain of paradise.

Sec. 73. We wish to recall, now, the three fundamental propositions with which we entered upon the investigations of the present chapter. 1st. There existed a particular, primitive, and traditionary heaven and earth, these being put in direct relation to each other, one regarded as the especial abode of the divine powers, the other as that of the first men. 2d. The conceived primary divisions of each, together with the central points of their supposed generation, coincided exactly in one case to those in the other. 3d. By virtue of these divisions, and upon principles heretofore established, such heaven and earth constituted at once the original cosmos and the archetypal temple. The primary division of the cosmos was coincident with the very first act of creation, the separation of chaos, out of the two portions of which heaven and earth were formed. All the ancient cosmogonies represent this first act as one of separation, and never as a creation from nonentity. It is obvious that the Mosaic text ought to be interpreted in harmony with uniform tradition derived from the same primitive source. But the cosmos to be understood here, as must be now apparent to the reader, does not regard the heaven and earth in general, but

the particular and limited heaven and earth known to the first men,
and to which exclusively all the traditions relate. In other words,
the cosmos here intended consisted of the celestial paradise on one
hand, and on the other, of the terrestrial paradise, these being
placed in direct relation, and united by the sacred mount. I think
it is safe to affirm that the cosmical doctrines of antiquity were
thus originally limited in their application. The Hindu cosmogony
was evidently associated with Mt. Meru. Such must have been the
case with the Chinese system. The images upon the upper and
lower shell of the tortoise, as we have seen, related exclusively to a
particular celestial and terrestrial space, coinciding exactly with the
Hindu tradition centring in Meru. The fact that the Mosaic and
Babylonian accounts of creation conclude with the formation of
man, placing him in a certain and definite locality on the earth's
surface, this being identical with that of Meru, tends strongly in
favor of the same conclusion. It was only in a vague sense that
the heaven and earth in general were included. The entire notion
of the cosmos had preëminently a local origin and reference. All
the temples constructed in imitation of this cosmos had a similar
limited reference. They represented the celestial and terrestrial
paradises, one as the especial dwelling of the divinity, the other as
the abode of the first men. The pyramidal temple, as M. Lenormant has correctly observed, was definitely an artificial reproduction of the sacred mount. The Hebrew tabernacle had its genesis
from the pyramid, and thus embodied a like traditional idea.

The primary division of the cosmos, as represented in the Babylonian sphere, was that from west to east through the sign Aries,
corresponding to the Accadian monogram *Bara*, which, as title of
the goddess *Bara*, related to the woman Thalatth, or Tiamat, personification of chaos. According to this, the superior hemisphere
must be assimilated to the dwelling of the divine powers, to the
celestial paradise, and the lower to the primeval abode of man, or
to the terrestrial paradise. The fact that the terrestrial paradise
had been definitely located in the inferior hemisphere will be fully
substantiated in the chapters immediately following the present.
The centres respectively of these two divisions of the cosmos,
as known to the primitive man, coincided exactly one with the
other. Meru was centre of the earth, and located precisely under
Su-Meru, central point of the heavens; while the sacred mountain

united the two. With Su-Meru, of course, are to be identified the Haranite Shemal, the *Su-mi-lu* of the cuneiform texts, and finally the Cardo of the Latins. The term *Cardo* is that from which the word *Cardinal* is derived, applied to the four principal points of the compass. It was applied, as we have seen, to a species of gate or door, having pivots projecting from each end, upon which the door turned. These pivots symbolized the two poles of the cosmos, the Su-Meru on one hand, and the Ku-meru on the other. The gate itself referred to that of the cosmical temple, of which the ordinary temple was an image. Again, the Cardo denoted the main line in land surveying; this being drawn from north to south. The gods, as Dr. William Smith has further told us, were supposed to be seated in the north, and looked upon the earth in the same direction as that faced by the augur. The seat of the gods as here assumed is now explained. The region referred to was the summit of the sacred mount, the Su-Meru of the Hindus, Har-Moad of Isaiah, Bit-kharris of the cuneiform text, Shemal of the Haranite Sabæans.

This line drawn from north to south, as will be readily perceived, united the two centres respectively of the celestial and terrestrial paradises. As represented upon the sphere, it would unite the two poles of the universe, cutting the line drawn from west to east through Aries at right angles. We should have thus the image of a cross, which Herr Nissen has stated formed the basis of the entire theory of the temple. The point of intersection, properly speaking, should be identified with Su-Meru, or the north pole, where the solstitial and equinoctial colures intersect each other. Here was the central point of divergence of the eight celestial regions, four primary and four intermediate, as was fully shown in the last chapter, and as all the circumstances compel us to assume. With these eight celestial regions, of which, as I believe, the eight-rayed star, constituting the hieratic form of the Accadian *An*, " elevated, heaven, god," was intended as a symbol, the eight terrestrial regions in all respects coincided. These were the divisions of the cosmos as known to the first men, and the two paradises, thus divided and thus united by the sacred mount, were at once the world and the temple, of which all subsequent temple structures were artificial reproductions. The plateau of Pamir, where modern criticism has definitely fixed the locality of the sacred mount, constitutes, in fact, the great water-

shed of all Asia, being for this reason regarded as the dome of the
world. It would be here naturally that the dry land would first
appear, whether at the time of the creation of man, or at the period
of the deluge. The representation of this elevated region by a
pyramid in stages doubtless proceeded from the notion of vast natural terraces rising one upon the other, from the country below to
the elevated plain of Pamir. This region is not what would be
called to-day either productive or salubrious. But there are evidences that the climate has materially changed within the period
when it was first occupied by man. In fact, the earliest portions of
the Zend-Avesta record distinct notices of this gradual modification
of climate, from one of remarkable salubrity to one of intense
severity, where there were ten months of winter to two of summer,
according to the traditional conception.

SEC. 74. It is evident that, geographically speaking, and so far
as concerns all historical tradition, we have arrived now at the primitive starting-point of humanity on earth. Additional evidences,
tending to the same conclusion, will be constantly presenting themselves during the entire progress of the investigations contained in
this treatise. But I believe these three chapters on the ancient
cosmogonies, especially the Mosaic and Babylonian, have sufficed to
settle the question as regards the original centre of the populations,
institutions, and sacred traditions of antiquity. I do not speak here
of the geological evidences now supposed to exist, thought by many
to fix the origin of man at an epoch so immensely remote as to ex-
ceed all previous conceptions, and even the most extravagant theories
of former periods. That to which I refer is the earliest *traditionary* epoch of human existence. In this regard I am confident that
antiquity does not afford a single reliable notice that conducts us
beyond Mt. Meru. This was preëminently and in all respects *the
Beginning*. In relation to this point, as the reader now fully perceives, modern criticism establishes the truth and integrity of the
Mosaic account in a manner the most complete.

Nothing results more plainly from these investigations than the
one great fact that the standpoint of the Mosaic and Babylonian
cosmogony, and indeed of all the ancient cosmical theories, was very
far different from that of modern geology. I am not prepared to
take the position that the Mosaic and Babylonian accounts of creation contain no reference, fundamentally, to physical ideas, to the

origin of the material universe in the modern sense. It is to be admitted, perhaps, that a purely physical philosophy is in some way involved. But as must be now apparent, the standpoints respectively of " Genesis and geology " are so essentially diverse that the attempt to construe one by the other, and to reconcile both upon one and the same basis, is to undertake the solution of a problem whose constituent elements are, in many respects, fundamental misconceptions. Instead of a *literal* HISTORY *of creation*, in the modern sense, I see in these ancient cosmogonies the groundwork of the *primitive* PHILOSOPHY *of the universe, of man, and of human civilization*, in the widest and most comprehensive view of these subjects. Not *history*, but *philosophy*, — this expresses precisely my theory. Physical ideas form the base of the system, sublime religious conceptions the apex. Human nature in its original condition, and in its intimate relation to the Divine Being, constitutes the point of view for the whole. The idea of the *first covenant* established between God and humanity is the central one of all; and this accounts for the fact that the notion of *division*, derived originally from the customs of the altar, appears throughout as the ground-conception. The cosmos is conceived as a temple, a house, a dwelling-place of the Almighty. The temple artificially constructed is an image of the cosmos, and as such is likewise a dwelling of God. Finally, it results that man is the true cosmos, the real temple, the actual abole of the Divine Spirit. In other words, the *church* constitutes the sum and substance of the entire theory. The primitive church was the primitive cosmos, the first heaven and earth. The creation of a new heaven and earth imparts the foundation of a new church, of a new divine dispensation; another step to regain the lost paradise.

Such were some of the grand and noble conceptions that prevailed at the starting-point of humanity, geographically and historically speaking, and so far as the earliest traditions known to antiquity afford us the slightest hint. It would be difficult indeed, as already observed, to go back of Mt. Meru. It was there that the human race was cradled. It was there that the ancient religions and civilizations had their birth. It was there, on that summit which penetrated the rotating centre of the celestial sphere, around which rolled the flaming chariot of the immortal powers, that heaven and earth, the divine and human, were first united in blissful fellowship, in happy intercourse. To retrace those mystical stages down which the fall of man precipitated the posterity of Adam has been the profound, earnest problem of all the religions of the world. Yes, and the sacred mountains up whose mystic steps the races have labored to ascend, whether Ararat or Kharsak-kurra, whether Sinai, Gerizim, or Sion, have all had their root, their traditional origin, in the mountain of the assembly of the stars, on the sides of the north; and it will be there, if ever, that the hopeful prediction of the prophet will find its spiritual fulfillment: "And he will destroy in this mountain the face of the covering cast over all people, and the vail spread over all nations" (Is. xxv. 7).

BOOK III.

THE CELESTIAL EARTH.

CHAPTER VII.

THE TERRESTRIAL PARADISE ASSIMILATED TO THE GREEK HADES.

Sec. 75. The time has been when many learned critics, not without much reason, considered it a matter of uncertainty whether the Garden of Eden ever had an actual existence on the earth's surface. In the absence of reliable data from which to proceed in attempting to fix geographically its location, the only resort was to speculation; and, as frequently occurs in such cases, the multiplicity of conflicting theories put forth tended to throw doubt upon the entire subject. But in view of the facts and considerations presented in the last chapter, together with the concurrent opinions of so many distinguished Orientalists, it is impossible to give place to any further skepticism in relation to the matter. The traditional mount of paradise had an actual historical and geographical existence, and its locality on the earth's surface has been definitely ascertained. On this point M. Lenormant makes the following observations: —

" The primitive Meru is situated to the north, in relation even to the first habitations of the Aryans upon the soil of India, in the Pendjáb and on the banks of the upper Indus. Nor is this a fabulous mountain, a stranger to terrestrial geography. The Baron D'Eckstein has demonstrated its real situation towards the Serica of the ancients; that is to say, in the southwest part of Thibet." [1]

Thus, it is necessary to give to the Mosaic description of the

---

[1] *Frag. de Bérose*, pp. 300, 301.

Garden of Eden a literal construction: 1st. So far as regards the
existence of such a region on the earth possessing the geographical
characteristics set forth in the narrative.   2d. So far as relates to
the fact that this region was the abode of the first progenitors of
mankind.   In a word, we are forced to consider the Gan-Eden of
Genesis, otherwise called the terrestrial paradise, as a literal historical and geographical fact.   Such being the case, it is a most singular and seemingly inexplicable circumstance that this traditional
abode should have been subsequently transferred in conception to
the under world, and assimilated to the state of the dead, to the
Greek Hades.   That such a transfer had been made by the Hebrews
in our Saviour's time is manifest from his language addressed to the
penitent thief on the cross : " Verily I say unto thee, To-day shalt
thou be with me in paradise " (Luke xxiii. 43).   The term " paradise," as here employed, is evidently put for the state of the dead,
for Hades.   It cannot be interpreted abstractly as "a state of happiness," according to Dr. Barnes (notes *in loc.*), for it had a definite
theological sense among the Jews, connected with the Eden of
Genesis on one hand, and with the world of disembodied spirits
on the other.   Alluding to the Jews, Dr. Campbell remarks : —

"The Greek *Hades* they found well adapted to express the Hebrew *Sheol* (state of the dead).   This they came to conceive as
including different sorts of habitations, for ghosts of different characters.   And though they did not receive the terms *Elysium*, or
Elysian fields, as suitable appellations for the regions peopled by
good spirits, they took, instead of them, as better adapted to their
own theology, *the Garden of Eden*, or *Paradise*, a name originally
Persian, by which the word answering to *garden*, especially when
applied to Eden, had commonly been rendered by the Seventy.   To
denote the same state, they sometimes used the phrase *Abraham's
bosom*, a metaphor borrowed from the manner in which they
reclined at meals."[1]

The same author observes in another place : —

"When our Saviour, therefore, said to the penitent thief on the
cross : 'To-day shalt thou be with me in paradise,' he said nothing that contradicts what is affirmed of his descent into *Hades*,
in the Psalms, in the Acts, or in the Apostles' Creed.   *Paradise* is
another name for what is, in the parable (of Lazarus and the rich
man), called Abraham's bosom."[2]

[1] Campbell's *Gospels*, Dissertation vi., sec. 19.   [2] Ibid., sec. 21.

Christ's descent into Hades was not inconsistent with the notion of his entrance into paradise at death, for the reason that the two regions had been assimilated to each other. Paradise had come to be regarded as a particular apartment of Hades. But the Hebrews were not alone in having made this transfer of paradise, whose historical and geographical character has been demonstrated by modern criticism, to be the under world, or state of the dead. It is known to scholars that a certain class of ideas prevailed in antiquity, of which the phrase "celestial earth" may be taken as expressing the underlying conception. Beside the allusions to this subject by some classic authors, it appears that the Hindu and Persian sacred books contain occasional references to it, and among these, the passage already cited from M. Obry, relating to the Sanskrit phrase *Svarga-bhoumi*, "the celestial earth," affords a direct example. This conception of a celestial or heavenly earth, as it formed itself in the minds of different peoples in antiquity, was obviously connected originally with the traditional abode of the first men, or the sacred mount of paradise, with which, as shown in our last chapter, the diluvian mount on whose summit the ark of Noah rested after the flood had been uniformly identified. But this celestial earth, also, together with various notions associated with it, and with the diluvian mountain, had been transferred to the under world, to the Greek Hades, in fact, and I proceed to offer some proofs, not only of the association of the celestial earth with Mt. Meru, but of the singular transformation of ideas to which we allude.

SEC. 76. In order to place the passage relating to the celestial earth, previously quoted from M. Obry, in immediate connection with other remarks by the author appertaining to the same circle of ideas, it will be introduced again among the subjoined extracts referring to Mt. Meru and the goddess *Ilra*, or *Ida*, personification of it: —

"We see that this name, which in the Vedas is variously written as *Ila, Ida, Ilra*, or *Ira*, designated primitively 'the earth,' as has been observed already by M. Wilson and M. Lassen. We may conclude with M. A. Kuhn and M. Alfred Maury that the Greek name *Era*, and that of Ireland *Ire*, 'earth,' came from the Sanskrit *Ira*, and afterwards that the title of the goddess *Rhea* was formed from it by metathesis. But as *Ida* of the Vedas was a veritable *Parvati*, or 'mountain goddess,' it is allowable, perhaps, to find here the origin of the name *Ida* applied to certain mountains in Phrygia, Crete, etc."

"It is understood, also, that Meru with its four great supports, or abutting mountains, rises in the middle of the central continent *Madhya-dvipa*, itself very elevated, to which they give these various titles: *Svarga-bhoumi*, 'celestial earth;' *Souvarna-bhoumi*, 'earth of gold;' *Akrida-bhoumi*, 'earth of pleasures;' *Touchita-bhoumi*, 'earth of joy,' but more generally those of *Ila-varcha, Ila-vrita, Ila-varta*, 'section, province, or region of *Ila*,' daughter and wife of *Manu*, and considered the mother of the human race."

"As the light of the snow-capped mountains mingled with the azure of ethereal space, blending with it in distant perspective, the paradise of the gods appeared to confound itself with that of the first men ; and to express this vague notion they formed the Sanskrit phrase *Svarga-bhoumi*, 'celestial earth.' "[1]

These extracts go to establish the various points : 1st. That the celestial earth was associated directly with Mt. Meru. 2d. That the goddess *Ira*, or *Ida*, primitively put for the "earth," was connected likewise with this sacred mountain. 3d. That *Ida* was the reputed mother of the human race ; in which character she is to be identified, of course, with the mother of Eden. There can be no doubt, therefore, of the reference of this circle of conceptions to the traditional paradise. Their subsequent transfer to the under world, termed *Hades* by the Greeks, is the next point to be investigated ; and here I serve myself with some lengthy extracts from the great

---

[1] *Du Berceau*, etc., pp. 22, 23. note, 174, note. We shall see that this "celestial earth" = *Svarga-bhoumi*, otherwise termed *Soucarna-bhoumi*, or "earth of gold," was one with the *Aralla* of the cuneiform texts, situated, like Mt. Meru, in the extreme north. According to Dr. Oppert, in a recent treatise, this *Aralla*, or *Arallu*, was also situated in the north, and was conceived "the earth of gold," like the *Soucarna-bhoumi* of the Hindus. Dr. Oppert's language is as follows :

"The Babylonian Noah, Xisuthrus, was translated to an eternal life without passing through death : other indications of the same species are found in the texts distinguishing the sojourn of the living from that of the dead ; this last is called the country of *Arallu*. In the conception of the Assyrians, this locality was found in the country of the north, the region of the disappearance of the sun. This country of *Arallu* is at the same time the *earth of gold* (according to the ancients the hyperborean country was rich in gold), and it is very probable that the name of the lake of *Aral* is but a relic of this ancient Assyrian name." (*De l'immortalité de l'âme chez les Chaldéens*, p. 4.)

The *Aralla* was one with the *Mat Nudea*, or Hades, of the cuneiform texts, into which the goddess Ishtar descends, a notice of which will be given in the sequel of the present chapter. The conception of the north as the region of the disappearance of the sun is hardly rational. The true explanation of it is that the *Aralla*, as we shall show, was identified astronomically with the lower hemisphere of heaven, where the sun does disappear.

work of Dr. G. S. Faber, which not only go to establish the fact of the transformation in question, but will afford other valuable materials for future use. The author proceeds: —

"Such being the universal intercommunion between the moon and the earth, the great mother being alike deemed a personification of each, both these planets bore the common name of Olympias, or Olympia; by which was meant the world; for Mount Olympus, as we have already seen, was no other than the Indian Mount Ilapu, or Meru, which is fabled to be crowned with the mundane circle of *Ila* or *Ida*. Accordingly the moon was deemed a sort of *celestial earth*, bearing a close affinity to this our nether world."

"This will lead us to understand the import of some very curious particulars which Plutarch mentions as being presented to the imagination of Timarchus, in his vision of the infernal regions. The friendly spirit, who acts the part of an hierophant (for the pretended vision seems evidently to describe the process of an initiation), informs him that Proserpine is in the moon, and that the infernal Mercury or Pluto is her companion. This moon is wholly distinct from the celestial (astronomical) moon; being what some call a *terrestrial heaven* or *paradise*, and others a *heavenly earth*. It belongs to the genii or deified mortals, who tenant the earth; and it is described as wearing the semblance of a floating island. It is surrounded with other islands, which similarly float on the bosom of the great Stygian abyss; but it is loftier than them all, and therefore not equally exposed to the destructive fury of the infernal river. In this navicular moon or lunar island there are three principal caverns. The largest is called the Sanctuary of Hecate; and here the wicked suffer the punishment due to their crimes. The other two are rather doors or outlets than caverns; the first looking towards heaven, the second towards the earth. These serve for the ingress and egress of souls; for the moon is the universal receptacle of them; into her they enter by one door, and from her they issue by the other door. She receives and gives, compounds and decompounds; and on her depend all the conversions of generation. While the moon thus floats on the waters of the Styx, the infernal river strives to invade and overwhelm it. Then the souls through fear break forth in loud lamentations; for Pluto seizes upon many who happen to fall off. Some, however, who are plunged in the raging flood contrive, by dint of great exertion and good swimming, to reach the shores of the moon; but the Styx, thundering and bellowing in a most dreadful manner, does not allow them to land. Lamenting their fate, they are thrust headlong into the abyss, and are hurried away to partake of another regeneration."

" Here, therefore, we may perceive the origin of that singular intercommunion between the earth, the moon, a ship, and a floating island, which may be traced throughout the whole system of paganism in every quarter of the globe. The earth was a greater world ; the ark, a smaller world ; the earth, a greater ship or floating island ; the ark, a smaller ship or floating island. But the lunette was the astronomical symbol of the ark. Therefore the moon became at once a ship, a floating island, and *a celestial earth.* Hence, what was predicated of the one was also predicated of the others ; and as the ark was a floating moon, as the earth was a ship, and as the moon was a boat and *a heavenly earth* and a floating island, one and the same goddess was deemed an equal personification of them all ; one and the same set of symbols was employed equally to typify them all. Accordingly, the great mother is declared to be at once the earth, the moon, and a ship ; nor is this singular intermixture of ideas to be found only in a single country ; it pervades the whole pagan world, and thus affords an illustrious proof that all the various systems of Gentile idolatry must have originated from some common source. That source was the primeval Babylonian apostasy." [1]

SEC. 77. That Babylon was one of the great centres of religious corruption in early times is not to be doubted ; yet in the light of present knowledge, it would be impossible to trace to this source exclusively all the idolatrous conceptions and customs of antiquity. But Dr. Faber's analysis of some of these ideas, as contained in the foregoing extracts, is mostly correct, and it is very able. To those not familiar with Asiatic mythology, however, the above statements will appear somewhat intricate and confused ; and it will be necessary to offer some comments in explanation : —

1st. It is a fact that the great goddess was ordinarily identified with both the earth and moon ; and it was owing to the resemblance of the lunette to a ship floating on the bosom of the celestial sea that the same goddess was assimilated to a ship, especially to the diluvian ark.

2d. The notion of a floating island appears to have reference to the summit of the mountain on which the ark rested, as it rose above the immense ocean of devastating waters, seeming like an island floating upon its dark abyss. Thus the ark and the floating island were assimilated to each other, and both these to the earth and moon.

[1] *Origin of Pagan Idolatry,* vol. iii. pp. 5, 13, 14. 21.

3d. The ark was conceived as a world, since during the deluge it contained all that remained of the human race ; and the earth was regarded as a ship, a floating island, for the reason that the summit of the diluvian mountain projecting above the watery abyss was for the time the entire earth then visible.

These remarks, it is believed, will sufficiently explain the origin and nature of the singular conceptions to which Dr. Faber alludes. I think his statements are substantially correct, and the elucidation of them just offered is probably the true one. It is this moon, assimilated to the earth, to the ark, to the great goddess, and to the diluvian mount, the latter identified with the mount of paradise, which is variously termed a *celestial* or *heavenly earth*, a *terrestrial heaven*, etc., correspondent to the Sanskrit phrase *Svarga-bhoumi*, "a celestial earth," applied to Mt. Meru, traditional abode of the first men. The most extraordinary circumstance connected with this circle of conceptions is that the whole has been transferred to the infernal regions, to the Greek Hades. The sacred mount of paradise, regarded as the cradle of humanity, being the Gan-Eden of Genesis, has been demonstrated to have had an actual historical and geographical existence. The diluvian mountain was identified with it, and was thus not less historical. To the same locality the celestial earth had been assimilated. Yet everything has been transferred to the under world, to the state of the dead, and the sacred mount, conceived as a ship, floats on the bosom of the Stygian abyss. I doubt whether the entire field of antiquarian researches affords a fact so remarkable and seemingly so unaccountable. It exhibits a widespread and complete misconception, at a later epoch, of traditionary ideas relating to the first ages of the world. In point of fact, a *fourth* and wholly factitious division or region, of which the primitive man had not the slightest conception, has been added to the cosmos; a circumstance that demands a brief yet careful study in the present connection.

According to the traditions most primitive among the ancients, the cosmos or universe consisted of two principal divisions, the heaven and the earth. But equally primitive, if not more so, and extensively prevalent, especially among the Aryans, was the notion of three regions or divisions, usually called the "three worlds," these being regarded as heaven, earth, and the intervening space, or the atmosphere. The idea of a *fourth* world or region, as included

in the cosmos, never entered the mind of man during the first ages.
It is only at a later epoch that the notices of another division of
the cosmos, of an under world, a subterranean region, becomes more
and more frequent in the sacred writings of the ancients. We submit here some quotations illustrating the primitive conception. M.
Carré gives the following passages from the Rig-Veda: —

"O Agni! The three worlds, the earth, the heaven, and the
atmosphere, are thy work. By thy light thou hast illuminated the
heaven and the earth." "Of the three worlds, two appertain to
the domain of Savitri (the sun); the third (the earth) is the abode
of Yama (god of the dead) and the sojourn of the dead." [1]

M. Carré makes the perfectly correct statement in a note as follows: "The Aryans recognized three worlds, the heaven, the
earth, and the intermediate space." [2] The second passage cited
above from the Rig-Veda contains a very important statement in
relation to the point before us. Yama, to whom the third world or
the earth appertains, was the reputed first man, and his sister Yami
the first woman, the two corresponding to Adam and Eve of Genesis. The Persian Yima, the same personage as the Hindu Yama,
was traditionally the founder of paradise. [3] At the same time Yama
was regarded by the Hindus as god of the dead, and as such the
earth or the third world appertained to him, and the earth is expressly represented in the foregoing passage as "the sojourn of the
dead." Notwithstanding these ideas the Aryans held distinctly the
doctrine of a future life. Something quite similar to the character
of Yama as god of the dead is that of the Babylonian divinity Hea,
in allusion to whose wife, Gula, Rev. A. H. Sayce remarks: —

"Gula, 'lady of the house of death,' was the wife of Hea, the
earth, and so originally the same as Nin-ki-gal, 'lady of the great
earth,' the queen of Hades. Nin-ki-gal was a form of Allat or
Istar; and the name Bahu (one with Gula) is merely the Bohu
(בֹּהוּ) of Genesis, the primeval 'wasteness' or chaos of night and
the under world." [4]

Some confusion of statement appears here, growing out of the
anomalous circumstance of which there is question. The term

[1] L'Ancien Orient, t. ii. pp. 46, 55.
[2] Ibid., p. 46, note 3.
[3] Professor Whitney, Oriental and Linguistic Studies, p. 45.
[4] Jour. Bib. Arch. Society, London, vol. iii. p. 173, note 3.

Nin-ki-gal means, literally, "mistress of the great earth;" from which it is evident that her peculiar domain was the earth, the third world or division of the cosmos according to the Aryans, and not the *fourth* region or world understood as Hades. Again, Hea is often qualified as "lord of mankind," at other times as "dwelling in the great deep;" while Rev. A. H. Sayce in the above note assimilates him to the "earth;" yet he answers in Babylonian mythology to the infernal Mercury or Pluto, god of the dead. Finally, Bahu, the Bohu, or chaos of Genesis, assimilated to Gula by Rev. A. H. Sayce, although "lady of the house of death," appertains strictly to the earth, for, according to the Mosaic text, it was the earth which was "without form and void," etc. Tiamat also, in the Babylonian account of creation, who represents the chaos, has her domain upon the earth. In all the cosmogonies the earth constitutes the very lowest region known and recognized. No fourth world is ever mentioned. Thus, what are termed the "infernal regions" by Dr. Faber, denoted by the Greek word *Hades*, must be regarded as a later conception, to which primeval tradition makes no allusion. This was literally a fourth and wholly imaginary division, that had been added to the cosmos of primitive times.

SEC. 78. Our investigations thus far have sufficiently established the fact that the traditional mount of paradise had been conceived as a terrestrial heaven, or a celestial earth ; and that this, together with the sacred mount itself, under the form of a floating island, ship, or ark, had been transferred at a later period to an under world, answering to the Greek Hades, constituting thus a fourth division of the cosmos entirely unknown to primitive tradition, and wholly inconsistent with the actual historical and geographical character of the Gan-Eden of Genesis, as verified in the last chapter. This anomalous circumstance in the historical development of religious ideas is without parallel, and it challenges an attempt on our part at some rational explanation of it, which will be made in the sequel of this chapter. But as preparatory to this, it is necessary for us to study more thoroughly the conception of the so-called celestial earth, or terrestrial heaven ; and such is the labor upon which we now enter. In this study, the cuneiform texts will be found, I think, to afford us the most reliable and adequate information.

There are two cuneiform phrases very frequently put in the relation of opposition to each other in the texts, although their employment separately is perfectly legitimate and even more frequent.
One of these is the Accadian *An-e*, the Assyrian reading of the
same characters being *Il-same*. This is the ordinary expression for
" heaven," considered as a divinity. The literal sense of the Accadian phrase is heaven + vault, or house, hence the " heavenly
house." The other locution to which I refer is the Accadian *An-ki*,
Assyrian *Il-irziti*. The sense attached is " the earth " considered
as a divinity or goddess. Assyriologues are probably correct in
referring the phrase to the earth goddess. But *An-ki* means literally heaven + earth ; that is, "celestial or heavenly earth." Allusion has been made to this expression once before, but it is necessary to bestow upon it now a more particular attention, since it
evidently answers in the cuneiform texts to the Sanskrit phrase,
Svarga-bhoumi, or " celestial earth," being the precise topic of our
present research. When not preceded by the determinative *An*,
the element *Ki* has the sense of " place, ground, earth," ordinarily
put for the geographical earth. It constitutes also the determinative of place, district, city, etc., as in *Bar-sip-ki* for Borsippa, *Bab-
el-ki* for Babylon. Obviously, then, there is a marked difference in
sense between *Ki* and *An-ki*, although cuneiform scholars do not
usually make any distinction in practice. Mr. Norris has very happily expressed the notion to be attached to *An-ki* as follows : " The
sphere ; the astronomical earth, distinguished from the geographical
earth by the determinative (*An*)." [1] This appears to me exactly
the fundamental idea, but it does not by any means complete the
conception involved. An astronomical earth could be no other
than a terrestrial field divided off according to the cardinal regions,
like the augurial temple for example, and put in direct relation to
a correspondent celestial space. But there are two cuneiform passages, one of which has been already noticed, that go to fix precisely the *traditionary import* of the Accadian *An-ki*. The tower
of Borsippa is termed by Nebuchadnezzar, " the temple of the
seven lights of *An-ki*." [2] The tower of Babylon is styled by the
same monarch, "the temple of the foundation of *An-ki*." [3] These

---

[1] *Assyr. Dictionary*, iii. p. 939.
[2] *Bit urme 7 il-irziti (An-ki)*, ibid.
[3] *Bit temin il-irziti (An-ki)*, ibid.

temple structures were regarded as primitive in the valley of the Euphrates, being erected soon after the Cushite emigration from the east to the land of Shinar. They were, in fact, as we have seen, designed expressly as imitations of the diluvian mount, identified with that of paradise. The phrase *An-ki*, therefore, has a direct reference to this sacred mountain, conceived as the abode of primeval humanity. Such being the case, there can be no doubt that we ought to connect with the Accadian *An-ki* precisely the same notion as involved in the Sanskrit expression *Svarga-bhoumi*, "celestial earth," applied to the same traditional locality. The particular heaven and earth united by Mt. Meru are thus expressly denoted by the two elements of the Accadian, as well as by those of the Sanskrit expression, namely, heaven + earth, or "heavenly earth."

Again, the Vedic goddess *Ilra*, or *Ida*, a name denoting primarily the "earth," was associated by the Hindus with their Meru, and this affords another point of connection with the Babylonian tradition. As before stated, the cuneiform expression *An-ki* is often interpreted by Assyriologues in special reference to the earth goddess, with which *Is-tir*, the Babylonian Venus, is frequently assimilated. M. Obry conjectures that the Vedic *Ida* may have some connection with the Mt. Ida, "mountain of the hand," of Phrygia, Crete, etc. Venus was doubtless under some form associated with the Mt. Idas of antiquity, and there is much reason for the supposition that the primitive Mt. Ida was Mt. Meru itself. The following distich, as rendered by Mr. Fox Talbot, shows that *Is-tar* was associated with such sacred localities : —

> "But Ishtar smiles upon him with a placid smile,
> And comes down from her mountain, unvisited of men." [1]

Having traced the connection of *An-ki*, considered as a celestial earth, with the sacred mountain of the north and east, it is probable that the same phrase as denoting the earth goddess, assimilated to *Is-tar*, had a like reference; and we may thus identify the mountain of *Is-tar* with that to which the Vedic *Ida* appertained. In either case, this is probably the great mother, the mother of Eden, in fact, whose immediate relation to the celestial earth will be abundantly established as we proceed, and may indeed be presupposed from the facts already before us.

[1] Vid. *Trans. Bib. Arch. So.*, ii. p. 31.

Finally, the seven stars of the chariot, traditionally associated
with Meru of the Hindus, the Eden of Genesis, and the *Har-Moad*
of Isaiah, have been identified with the "stars of El," to which the
prophet alludes. The Hebrew *El* is the same as the cuneiform *Il*
or *Ilu*, and Mr. Norris cites a certain text which connects this
divine personage with the celestial earth, thus: "The god Il, the
ornament of the celestial earth." [1] We should compare this qualification of the god Il with that applied to the tower of Babylon
or "gate of Il," "the temple of the foundation of the celestial
earth;" also to the tower of Borsippa, "temple of the seven
lights of the celestial earth;" evidently referring to the seven stars
of El, or Il, associated with the sacred mountain. Thus, it is obvious from the data now before us that the Accadian *An-ki*,
whether as a celestial earth, as the mother goddess, or as associated
with the god Il, Hebrew El, is to be referred primarily and traditionally to the *Kharsak-kurra*, "mountain of the world," identified
with the Gan-Eden of Genesis. The pyramidal temples were in
a certain sense material expressions of this heavenly or celestial
earth, at the same time that they were artificial reproductions of
the mount of paradise, to which these various conceptions originally appertained. The antiquity of these notions also must have
been very great, for the earliest temple structures in the Euphrates
valley had a direct reference to them as their material embodiment.

SEC. 79. Mr. Norris, as already cited, connects with the Accadian
An-ki the notion of the sphere, and of an astronomical earth. Both
are involved strictly in the idea of a celestial earth, which is literally an astronomical earth. Nevertheless, the special relation of
this phrase to the Babylonian sphere, which, as has been shown,
was taken as a symbol of the cosmos, needs some further elucidation. A certain bilingual text affords us some very valuable equa-
tions for our present purpose, among which I select the following:
1st. *An = the god Anu*. 2d. *An = the goddess Tum, or Anatu*.
3d. *An-ki = the god Anu and the goddess Tum*.[2] I suppose it is
the Accadian *Tum* which Mr. George Smith reads *Anatu*, wife of
Anu.[3] But formerly he read the same character *Anunit*, I believe,

<hr>

[1] *Assyr. Dic.*, iii. p. 940: *il Il Supar il-irziti (An-ki)*.
[2] 3d Rawl. Pl. 69, No. 1, Obs. ll. 1-3. Cf. 2d Rawl. Pl. 54, No. 3, Obs. ll. 2, 3.
[3] *Chald. Acct. of Genesis*, pp. 54, 55.

which is possibly the same as *Anatu*, though this appears not to have been so understood heretofore.[1] The Accadian *Tum* is the old Hamite or Cushite name of a deity assimilated to the lower hemisphere of heaven, a personage which appears as a goddess in the cuneiform texts, but as a male divinity in the hieroglyphic inscriptions of Egypt. They are the same astro-mythological character, as will appear hereafter, and since they are such I prefer the reading *Tum* to that of any other; but that the goddess is the wife of *Anu* is not to be overlooked. Mr. Smith's remarks relative to these two divinities, a portion of which have been before cited, are quite important to us here, and will be reproduced : —

" He (*Anu*) represents the universe as the upper and lower regions, and when these were divided, the upper region, or heaven, was called Anu, while the lower region, or earth, was called Anatu." " Anatu, the wife or consort of Anu, is generally only a female form of Anu, but is sometimes contrasted with him ; thus, when Anu represents height and heaven, Anatu represents depth and earth ; she is also lady of darkness, the mother of the god Hea, the mother producing heaven and earth."[2]

But both Anu and Tum are equated to *An-ki*, that is, to heaven + earth, which proves Mr. Smith's remarks upon their respective characters to be correct. They must equally represent the upper and lower hemispheres of heaven, since each is equated to the Accadian *An*, "heaven, god," etc. The term *An* in this case must be taken in the sense of heaven. Thus, Anu is heaven, and Anatu, or Tum, is heaven ; but these are the two halves of the sphere placed in opposition to each other, like the *Qaq-qa-du* and *As-ru*, symbolized by the two orders of temples, the *Bit-mat* and the *Bit-khi-ra*, heretofore explained to us by M. Lenormant. The apparent contradiction in conceiving one and the same goddess to represent the earth at one time, and the inferior hemisphere of heaven at another, is due to the fact already insisted upon by us that the Babylonians assimilated the lower half of the sphere to the earth. I propose soon to offer still further proof of the fact of such assimilation. But that which is to be particularly noticed here is the exact equivalence of the two cuneiform phrases, namely, *An-ki*, heaven + earth, considered as a "celestial earth," and *Anu* and

[1] Vid. *Assyr. Discoveries*, p. 173, l. 39.
[2] *Chald. Acct. of Genesis*, pp. 54, 55.

Tum, heaven + earth, regarded as mythological or astro-mythological characters, personifying heaven and earth. I have shown that *An-ki* related especially to the sacred mount, uniting a particular celestial and terrestrial space, of which the pyramidal temple was an imitation, and to which was applied the Sanskrit phrase *Svarga-bhoumi*, " celestial earth." It will be easy to show now that the primary reference of *Anu* and *Tum*, as astro-religious representations of heaven and earth, was also to the traditional " mountain of the world." We note, first, the following passage from Mr. George Smith : —

" The heaven or region of the blessed was called *Samu*, and was divided into various sub-regions bearing different names, the highest being the ' Heaven of Anu,' the supreme celestial god." [1]

Thus, the heaven which *Anu* especially represents is not the entire expanse of the sky, but the highest central region ; that is to say, the region of the polar star, one with the Su-Meru of the Hindus. This and no other, as is well known to Assyriologues, is the " heaven of Anu," usually so designated in the texts. But another proof comes readily to our hand here. The seven stars called *Sabi*, or " Seven," in the inscriptions, evidently those of the chariot, and one with the " stars of El," were directly associated with *Anu*, as will appear from the words of M. Lenormant following : —

" Perhaps it is necessary here to note the facts furnished by a tablet of the British Museum not yet edited, which seems, in effect, to identify *Anu* with a god named *Sabi*. This name, which signifies ' seven,' is written with the numeral 7, followed by the phonetic compliment *bi*. The god *Sabi* is then the 'god seven,' or perhaps more exactly the 'god of the seven,' the god who presides over the group of seven other divinities, and unites them. This is only a different mode of expressing the notion involved in the name *Ashmunu* or *Eshmun* (*the eighth*). Thus, *Anu* identified with *Sabi* is precisely *Anu* in relation to the group of seven planets, united in the cultus of the tower of Borsippa." " In all cases, we comprehend now how the myth of the Cabiri born from the *hand of Anu* is applied to the pyramid in stages at Borsippa, which is the 'temple of the seven lights,' and at the same time the 'temple of the divine hand.' " [2]

We connect and confirm here so many points at once that it is necessary to particularize them. 1st. The tower of Borsippa is

---

[1] *Assyr. Discoveries*, p. 221. [2] *Frag. de Bérose*, pp. 389, 390.

called *Bit-Zida*, "temple of the right hand ; " and the myth of the
Cabiri born from the hand, among whom the Phœnician *Eshmun*
was reckoned " the eighth," was shown to have been connected with
this very tower, in our third chapter. 2d. Under the name *Sabi*,
" seven," M. Lenormant would identify *Anu*, who had his sanctuary
in the basement of the tower of Borsippa, with the Cabirus called
Eshmun, or " the eighth," in relation to the other seven. 3d. But
M. Lenormant sees here only a reference to the seven planets ;
while it is obvious to me that the primary allusion is to the seven
stars of the chariot, associated with Mt. Meru. The reasons are
manifold in proof of it. (*a*) The tower of Borsippa was an express imitation of the sacred mount. (*b*) *Eshmun*, *Anu*, like the
Haranite *Shemal*, represented the highest celestial region, which
was " the eighth," and was identical with the *Su-Meru* of the Hindus. (*c*) The stars of El, in relation to which Elyon represented
the Highest, were connected traditionally with the same terrestrial
and celestial localities. 4th. We have another proof here that the
primitive Mt. Ida was Mt. Meru, to which the Vedic Ida appertained, since it has been shown that the pyramid of Borsippa, imitation of the sacred mount, was otherwise regarded as a Mt. Ida, or
mountain of the hand. 5th. The tower of Borsippa, besides the
title of "temple of the right hand," was especially called the " temple of the seven lights of *An-ki*," or " the celestial earth," which
was certainly connected with Mt. Meru.

SEC. 80. We return now to the consideration of the identity of
An-ki with *Anu* and *Tum*, according to the bilingual texts already
cited. The first element *An* answers here to *Anu*, and the second
element *Ki* to *Tum*, or *Anatu*. We see here that the reference is to
a particular heaven, the so-called "heaven of Anu," and not to
celestial space in general ; another proof that the eight-rayed star,
constituting the hieratic symbol of *An*, related to the rotating centre of the superior heavens, and was no mere hieroglyph of a star.
This was the " heaven " *par excellence*, associated with the mount
of paradise. *Anatu*, or *Tum*, consequently, was not put for the
earth in general, but for the particular earth related definitely to
the " heaven of Anu," that is to say, the terrestrial paradise. But
the traditional paradise had been transferred to the infernal regions ; and Mr. Smith's language will afford evidence that such had
been the case even among the Babylonians. After alluding to the

divisions of heaven, of which the highest was the "heaven of Anu," in the passage last cited from this author, the heaven thus designated being regarded as the abode of the blest, he adds the following: "Hell, on the other hand, was generally called *Mat-nude*, or *Aralli*, but has various other titles" (*op. cit.*). Mr. Smith gives a long and graphic description of these regions, which I introduce. First, however, I wish to show that *Mat-nude*, or *Aralli*, termed "hell," by the author, and obviously one with the Greek Hades, designates precisely the region put in opposition to the "heaven of Anu," represented by the goddess *Tum*, wife of *Anu*. We have three bilingual phrases which go to establish the identity here claimed, and that in the clearest manner. They are as follows:[1]—

> A-ra-li   = A-ral-li.
> Bit-mat-bat   = A-ral-li.
> (U-ru-gal) tum = A-ral-li.

In the first equation, the Accadian *A-ra-li* is explained by the Assyrian form of the same word, or *A-ral-li*. In the second, the Accadian *Bit-mat-bat*, or properly *E-kur-bat*, is equated to the same Assyrian expression. Finally, the Accadian *Tum*, with the special Assyrian reading of *U-ru-gal* in this instance, is also explained by *Aralli;* which Mr. Smith assumes as only another name for *Mat-nude*, the infernal regions, Hades of the Greeks. The equation of *Tum*, in the bilingual phrase just quoted, to *Aralli*, only another name for *Mat-nude*, or *Hades*, proves the assimilation of *Tum*, or *Anatu*, wife of Anu, and otherwise put for the "earth," to the lower hemisphere, or under world, to which Ishtar descends. I reproduce now in full Mr. Smith's description of the two regions respectively of *Anu* and *Tum:—*

"The abodes of the dead were supposed to consist of two regions, one in the sky, presided over by Anu the god of heaven and Bel the god of the earth, and the other beneath the world, presided over by Hea the god of the ocean and infernal regions. In the upper regions, or heaven, were the abodes of the blessed; there the departed wore crowns, they drank beautiful waters, and consorted with the gods; but the notions of glory and honor at that day come out in the description of the inhabitants of this happy region: they are the kings and conquerors of the earth, the diviners and priests and great men, in fact, the strong and successful among mankind. On the other hand, the description of the infernal regions is most vivid

---

[1] Vid. 2d Rawl., Pl. 30, Revs. ll. 11-13.

and powerful, and is almost the same as that in the splendid inscription of the descent of Ishtar into Hades, where we read : —

" (1) To Hades the land of my knowledge ;
(2) Ishtar, daughter of Sin, her ear inclined ;
(3) Inclined the daughter of Sin her ear ;
(4) To the house of the departed, the seat of the god Iskalla ;
(5) To the house from within which is no exit ;
(6) To the road the course of which never returns ;
(7) To the place within which they long for light ;
(8) The place where dust is their nourishment, and their food mud ;
(9) Light is never seen, in darkness they dwell ;
(10) Its chiefs also, like birds, are clothed with wings ;
(11) Over the door and its bolts is scattered dust.

" This dark region, where the inhabitants in their hunger devour filth, and thirst for light, is guarded by seven gates, and surrounded by the waters of death ; it is the home of the weak and conquered ones, of wives who stray from their husbands, and men who abandon their wives, and disobedient children. These are represented as weeping in misery and corruption in their dark and eternal prison-house, ' the place from which there is no return.'

" By the power of Hea, who here corresponds to Pluto, the lord of Hades, the ghost of Heabani was delivered from this hell, and, rising out of the earth, soars up to heaven. These religious ideas are remarkable on account of their close similarity to those of later religions and subsequent races, and their importance is increased by their antiquity, as at the latest they date more than two thousand years before the Christian era." [1]

Then follows the reference to the "heaven of Anu" on one hand, and to *Mat-nude* or *Aralli*, regarded as Hades, on the other, in language already quoted from the author. There can be no doubt that the foregoing exposition of religious ideas, prevailing in the valley of the Euphrates two thousand years B. C., is correct and faithful. Mr. Fox Talbot, in several learned papers published in the " Transactions of the Society of Biblical Archæology," London, already cited in these pages, has fully confirmed Mr. Smith's statements above by evidences drawn from the legend of Ishtar's descent into Hades, and from other sources. But the religion of Babylon was predominantly *astral*, as all Orientalists are aware, and an astronomical element is undoubtedly fundamental in all these conceptions as exposed by Mr. Smith. The descent of Ishtar into Hades is

[1] *Assyr. Dic.* pp. 220, 221.

really a descent into the inferior heaven, to meet her beloved *Adonis*,
or the sun-god, who has met with a violent death, and is now in the
lower hemisphere. Ishtar is delivered from Hades, or *Mat-nude*, by
Hea himself, just as Heabani was, only in a different manner. But
while the astronomical element referred to was at the base of, and
really primitive in, these religious conceptions, it had become wholly
misconceived and perverted. The absolutely primary reference
had been to the mount of paradise, of which, as shown in a previous chapter, the two halves of the celestial sphere were taken as
symbols, the upper being put for the celestial paradise, the abode
of the great divinities, of whom Anu was chief, and the lower half
for the terrestrial paradise, to which *Tum* or *Anatu*, as wife of
Anu, ought to appertain as great mother. But the terrestrial paradise, symbolized by the lower half of the sphere, had been transferred to a fourth world, to the infernal regions, to the Greek
Hades, as previously shown in the extracts from Dr. Faber, and as
evinced still more plainly in the foregoing quotations from Mr.
Smith. Here is evidently a terrible confusion of ideas, a manifest
misinterpretation of primitive doctrines, a network of inconsisten-
cies, from which it is necessary to work ourselves out, and explain
the causes which have led to it.

SEC. 81. To this end, it is advisable to recapitulate briefly here,
as an aid to the memory, the principal points that have been established in the few last sections.

1st. The primary reference of *An-ki*, conceived as a heavenly or
celestial earth, was to the traditional mount of paradise, uniting a
particular heaven and earth, to which primeval tradition especially
pertained. The proof of this reference is: (*a*) That the two
pyramidal temples, the most ancient and typical representations of
the paradisiacal mountain, were styled, one, " the temple of the
foundation of (*An-ki*) the celestial earth;" the other, "the temple of the seven lights of (*An-ki*) the celestial earth." (*b*) The
obvious connection of these seven lights with the seven stars of the
chariot, or great Dipper, with which also the seven planets had been
united in the cultus of different peoples. (*c*) The exact agreement
in the meaning of *An-ki* with the Sanskrit *Svarga-'houmi*, or "celestial earth," definitely applied to the sacred mount, and proving that
such notions were ordinarily attached to it.

2d. The exact equivalence of *An-ki* in such original reference to the mythological titles *Anu* and *Tum*. This is shown: (*a*) By their equation to each other in the bilingual text cited. (*b*) By the fact that Anu represents the highest and central region of the heavens, correspondent to the Su-Meru of the Hindus, and viewed as the original seat of the gods, as in fact the celestial paradise. The fact that *Tum* or *Anatu*, as wife of *Anu*, is put in opposition to him, representing the lower region, the earth, proves her primitive assimilation to the terrestrial paradise, united to the "heaven of Anu" by means of the sacred mountain itself. (*c*) By the other title of Anu, or *Sabi*, by which he is assimilated to *Eshmun*, "the eighth," also put for the highest heaven, and obviously termed "the eighth" in relation to the seven stars of the chariot.

3d. The identity of the region denoted by the title *Tum* with the *Aralli*, the *Mat-nude*, conceived as the infernal abode, like the Hades of the Greeks. This is proved: (*a*) By the cuneiform phrase that explains the Accadian sign *Tum* by the Assyrian *Aralli*. (*b*) By the fact that the term *Aralli*, as proved by a certain text cited and explained by M. Lenormant, whose language has been previously introduced, related to a traditional region, "the Aralli of the east," put in direct relation to the *Bit-kharris*, or palace of the gods, identified with the Su-Meru of the Hindus, consequently with the "heaven of Anu;" that is to say, with the rotating centre of the superior heavens.

4th. All goes to show that at Babylon, as well as by the authorities referred to by Dr. Faber, the terrestrial paradise as a celestial earth had been transferred to the Greek Hades. Some confusion will arise here in the absence of an explanation. Anu represents originally the upper and lower regions, as already stated by Mr. Smith. So *An-ki* primarily includes the celestial and terrestrial space, united by the sacred mount; that is, the heaven and earth of the original cosmos. But when the two regions were separated, Anu was put for the upper, and Anatu or Tum for the lower. Thus, heaven is always male, and earth uniformly female. On the same principle, *An-ki*, as female, came to represent exclusively the divine earth, put in opposition to *An-e*, Assyrian *Il-same*, or divinized heaven. The texts often place the two in contrast, just as Anu and Tum are so conceived. Thus, *An-ki* came to denote exclusively the terrestrial paradise, regarded as a "celestial earth," though the notion of a particular heaven united with it was always involved, and was indeed primary. It is, then, this particular celestial earth as distinguished from the heaven of Anu, from the paradise of the gods, that was transferred to the infernal regions represented by the Greek Hades.

It will be better here to offer a concise statement of my theory as to the causes which led to this transfer, this singular perversion of primitive doctrines, so that the reader will be able to occupy the same standpoint as myself in relation to the evidences which have been and are to be introduced. Such a statement can be made now to advantage, whereas before it would have loaded the memory with too many important ideas, all of them somewhat complicated, at the starting-point.

1st. It has been shown in a previous chapter that the Babylonian sphere had been taken as a symbol of the cosmos. The two divisions, the upper and lower hemispheres, represented respectively the two chief divisions of the cosmos, namely, heaven and earth, the upper portion being put for heaven, and the lower for the earth. In this way, the zodiacal temple, with its main line of separation drawn through the sign Aries, denoted by the Accadian *Bara*, corresponded precisely to the Hebrew tabernacle, with its two apartments separated by a vail. The term *Bara* equals the Assyrian *Pa-rak-ku*, which Dr. Delitzsch, heretofore cited, compares with the Hebrew *Pa-ro-keth*, denoting the vail of the tabernacle.

2d. But the original cosmos, that to which in point of fact the earliest traditions referred, consisted of the particular heaven and earth known to the first men, and which had been conceived, the one as the celestial paradise, especial seat of the divine powers, the other as a terrestrial paradise, happy abode of primeval humanity. These two abodes, therefore, constituting the original cosmos, were symbolized by the two divisions of the Babylonian sphere, since it had been taken to represent creation in its chief apartments. The upper portion, the superior heaven, was put for the celestial, and the other, the inferior heaven, for the terrestrial paradise. Hence, this lower hemisphere had a double assimilation *to the earth* and *to the terrestrial paradise*, just as the upper hemisphere represented *heaven* and the *celestial paradise*.

3d. The assimilation of the inferior portion of the sphere to the earth, to a particular earth, identical with the paradise of man, had been neglected, had been in fact forgotten. A sufficient proof of this neglect is the fact that it is necessary to-day to enter into these labored investigations in order to prove the reality of such original assimilation. The fact that a terminology, also, primitively applicable only to a particular and limited celestial and terrestrial space, was afterwards interpreted of the entire heavens and earth, and is so generally understood by scholars at the present day, is another proof that a great and primitive doctrine appertaining to the cosmos and to the zodiacal representation of it had been at an early epoch almost entirely forgotten. Yet not entirely, for I think it had been preserved in the mysteries, and a careful comparative study of the cuneiform texts relating to these matters shows that the priest-kings of Babylon had perpetuated the ancient dogma on which so much really depended.

4th. The consequence of the neglect referred to was that the inferior hemisphere, corresponding to our *nadir*, instead of being understood as symbolically the earth, according to the dogmas of the zodiacal temple, became the "under world," the "region of darkness," the "infernal abodes," the Hades of the Greeks; and the terrestrial paradise, the celestial earth, originally identified with the lower hemisphere, was consequently in the same way referred to this *fourth* world, of which the first men had no conception. I speak here of a *fourth* world in relation to the "three worlds" of the early Aryan tradition and sacred books. It is obvious that the superior and cosmical triad of Babylon, consisting of Anu, Bel, and Hea, had reference to a similar division of the cosmos into three regions. But Hea, as we see from Mr. Smith, was at a later epoch assigned, as god of the dead, to this *fourth* region wholly factitious.

SEC. 82. A fact that goes far to demonstrate the general theory just set forth is the definite location, on the part of the ancients, of the Greek Hades, and particularly of the infernal river, the Styx, in the inferior heaven or lower hemisphere. Its astronomical locality was well understood and exactly described. It will be sufficient for us upon this point to cite a passage from Professor A. Romieu's critical treatise upon the "Egyptian Décans," being a series of letters addressed to Dr. Lepsius. But before introducing this passage, it is necessary to produce another one relating to the gates of souls, located one in the upper, the other in the lower hemisphere, and which is as follows: —

" We recognize on divers sides the general conception of two celestial gates, situated upon the milky way, giving passage to human souls; by one of these gates the souls descend from the most elevated regions of space, which they were supposed to inhabit (before birth), and by the other they return towards their first abode, after having accomplished the period of sojourn on the earth that had been allotted to them. The precise position of these gates of the firmament varies somewhat with different authors; but all agree, nevertheless, that the gate by which the descent is made was situated in the northern heavens, and the other in the southern." " Permit me, monsieur, to recall here the passage of Macrobius which develops the point in question : ' Behold the path by which the souls descend from heaven upon the earth (at birth). The milky way extends so far in the heavens that it cuts the zodiac at two points, in Cancer and in Capricorn, two signs that give their names to the tropics. The physicists call these two signs the gates of the sun, because they mark the limit of the sun's northern and southern course, which never passes beyond them. It is, they say, by these gates that souls descend from heaven upon the earth, and remount from the earth toward heaven. It is by that of men, or by Cancer, that the souls take their route toward the earth, and it is by Capricorn, or gate of the gods, that they return to the seat of their own immortality, where they are numbered with the gods; this is that which Homer has figured in the cave of Ithaca.' " [1]

The author shows that these gates were called *Kents* by the Egyptians, and he attempts to locate them definitely upon the Egyptian sphere. The *Kents* do not correspond exactly with the signs Cancer and Capricorn, yet they appear to have been situated at opposite points in the two hemispheres. It is in treating upon the position of these gates of the souls, as they pass from heaven into the body at birth, and from the body at death toward heaven, that Professor Romieu fixes the locality of the infernal river, or Styx, in the inferior heavens : —

" The astrologue Firmicus informs us that the Styx takes its rise in the eighth degree of Libra; we know that the positions of the Greek constellations have submitted to variations in process of time, of which we have to-day no precise knowledge, and particularly in respect to Libra, which did not exist primitively upon the Greek sphere. Nevertheless, according to the position of the curve, which fixes the first décan of the (southern) Kent, we see that this curve ought to pass near that part of the celestial sphere where the astrologues place the eighth degree of Libra. This leads us to con-

[1] *Décan du ciel Egyptien*, pp. 16, 17.

clude that the Styx was simply the Egyptian idea of the Kent, transported into the Greek religion, and the coincidence of the two risings affords a confirmation of all the preceding. The identity of the two myths being admitted, we ought to note this passage of the Georgics where, in speaking of the antarctic pole, Virgil says: 'But the dark Styx and the infernal shades behold that (pole) under their feet.' This situation of the Styx and of the ghost, enabling them to behold the south pole, and to approach it, seems to indicate that the Styx, and consequently the Kent, extended itself far to the south, from whence it results that the four décans of this Kent had been placed upon the milky way."[1]

That which results absolutely from the facts developed in the two extracts from Professor Romieu is: 1st. The primitive astronomical character of the infernal river, the Styx, the "Stygian abyss," as termed by Dr. Faber; consequently of the Greek Hades, also, since the Stygian sea is definitely located in Hades, or the under world. 2d. The location astronomically of this river, and thus of Hades, in the inferior hemisphere, the river itself taking its rise in the sign Libra, and stretching off to the south pole. 3d. The astronomical character of the gates of souls, the Egyptian Kents, and their connection in some sense with this infernal abode and river. What we have now to do is to connect with these facts Dr. Faber's description of the *celestial earth*, otherwise termed a *terrestrial heaven* or *paradise*, under the form of a ship, or floating island, conceived as floating upon this Stygian abyss, whose terrific waves and tempests engulf the unfortunate souls who are unable to secure a foothold upon the island. It seems to me that these facts ought to be regarded as perfectly conclusive to the effect that the terrestrial paradise, or celestial earth, had been primitively as-

---

[1] Ibid., p. 22. The entire distich cited by the author from the Georgics is as follows : —

"Hic vertex nobis semper sublimis; at illum
Sub pedibus Styx atra videt, Manesque profundi."

(i. 242, 243.)

"The north pole is always elevated above us; but that (the south pole) the dark Styx and the infernal shades behold under their feet." With the ancients the north celestial pole was the centre of the heavens, like our zenith, and the south pole was thus like our nadir. A recent annotator of Virgil observes: "*Videt*, of course, does not mean that the south pole is actually visible from the shades." But this is precisely what Virgil *does mean*, and the fact stated by Firmicus, that the Styx takes its rise in the eighth degree of Libra, proves the definite location astronomically of all these ideas in the inferior heavens.

similated astronomically to the inferior hemisphere, and that at a
later period, by some singular misconception, this traditional abode
of the first men had been converted into the Greek Hades, astronomically located also in the lower hemisphere. We cannot assume
here, in accordance with the principles of the school of M. Dupuis,
that all these conceptions were purely astronomical from the beginning, and that no really historical and geographical element attaches to them. The historical and geographical facts have been
demonstrated as absolutely primitive, and it has been shown that
the sphere had been taken as a symbolical representation of them.
The superior and inferior hemispheres were regarded as symbols
respectively of the two principal divisions of the cosmos, that is to
say, of the particular heaven and earth known to the first men, conceived as a celestial and a terrestrial paradise. But the assimilation
of the lower hemisphere to the earth had been forgotten, and as a
consequence this hemisphere had assumed the character of a distinct cosmical region lower than the earth, becoming thus the under
world, the abode of darkness and of the dead. Still, the tradition
of paradise attached to it, and consequently the primeval abode of
humanity was transferred also to the Greek Hades, whose astronomical location in the inferior heavens has been fully established.

SEC. 83. While upon the subject of the Egyptian sphere, it will
be well to introduce some other facts connected with it which relate
to our general topic. In his treatise upon the " Nomes of Egypt,"
M. J. De Rougé has the following relative to Heliopolis : —

"The god of Heliopolis was the sun under its two principal
forms: *Tum*, that is to say, ·the hidden sun,' the sun of the primordial night before its manifestation to the world; and *Ra*, the
sun after its birth." [1]

M. Eugène Grébaut translated an Egyptian hymn to Ammon
Ra, in which occurs this sentence: " Sole form who produces all
things ; the sole One who produces all beings ; men are issued from
his eyes, and his words become gods." To which the translator
adds the following comments in a note: " That is to say, according
to my view, men are produced from the luminous manifestation of
Tum, the god ' who exists alone in the abyss of waters.' As diurnal
sun he rises upon the world vivified by the light of his eyes, when
all beings take their birth. His word becomes the gods ; the texts

1 *Nomes de l'Egypte*, p. 38.

say habitually, 'the gods issue from his mouth,' that is to say, he manifests himself by his word, the truth." [1] Another fact in relation to the Egyptian Tum is to be noted. An ancient scribe employs the following language in praise of his master, the king: "Thou dost enter thy palace, as *Tum* into the solar mountain." [2] According to Champollion and Dr. H. Brugsch, the "solar mountain" was located in the zodiacal sign Libra, and connected with it were represented the mother goddess and her child Harpocrates. [3] I have already expressed the opinion that the Accadian and Egyptian *Tum* should be identified as a primitive Hamite divinity connected with the lower hemisphere. The different local developments of one and the same original character, so frequent in mythology, are not at all surprising. Their primitive sameness may be inferred: 1st. From the phonetical identity of the two names. 2d. From the fact that both appertain to the inferior heavens. 3d. From their connection with a sacred mountain. The Egyptian Tum was associated with the solar mountain, located in the sign Libra, consequently in the east. A bilingual phrase seems to put the Accadian Tum in relation to a sacred mountain, thus, "The sublime mountain of Tum is Istar." [4] The phrase *Tul-kû*, which I render "sublime mountain," is the Accadian name of the seventh month, which corresponds to the sign Libra, in which the solar mountain of the Egyptians was represented, together with the mother goddess and her child. The mountain of the Accadian *Tum* is identified with *Is-tar* in the text cited, who was considered the goddess mother. The coincidences here are very striking, and if we call to mind the facts heretofore presented tending to connect *Is-tar* with Mt. Meru, under the name of *Ida*, it will be necessary to admit that the Egyptians had preserved a tradition of the paradisiacal mountain, and of the mother of Eden, recording them, so to speak, in their sphere. With the fact of *Tum's* association with a sacred mountain, should be connected also the data previously

---

[1] *Revue Archéologique*, June, 1873, p. 387, text, and note 3.

[2] Vid. M. G. Maspero, *Bibliothèque des Hautes Etudes, Genre epistolaire chez les Egyptiens*, etc., p. 96.

[3] Brugsch, *Nouvelles Recherches*, etc., pp. 55, 56.

[4] 3d Rawl. Pl. 68, Col. 2, Obs. l. 27. The text has the reading: *An Tum tul-kû-ga = An Is-tar*. The element *Tul* has the sense of "mound, column, rampart," and is sometimes applied to mountains.

established in this chapter which tend to identify Anu and Tum with the particular heaven and earth known to the first men, that is to say, the mount of paradise.

Having determined the locality, astronomically, of the Greek Hades and of the Stygian abyss, it will be possible with greater certainty to ascertain the precise domain of Hea, whom Mr. Smith very properly represents as god of the dead, ruler of the infernal regions, like the Pluto of classic mythology, to whom in fact Hea has been frequently compared. A cuneiform text well known to Assyriologues contains a list of the twelve names of Mercury, one to each month of the year, assumed by this planet successively as it attends the sun in its annual course. In the Assyrian month Adar, corresponding to the sign Pisces, or the fishes, Mercury takes the name of *Kha An Hea*, " fish of the god Hea." [1] This seems to connect Hea definitely with the inferior hemisphere, and especially with the zodiacal division Pisces. Moreover, in such connection he seems to be a form of Mercury ; and this explains the phraseology of Dr. Faber in depicting the vision of Timarchus. as already cited, namely, " the infernal Mercury or Pluto." We know now that this infernal Mercury appertained to the inferior heavens, and thus we have a double proof that Hea had his domain in the same astronomical locality. Not only his character, so similar to that of Pluto, but his connection with the sign Pisces, go to establish this point.

Again. a tablet giving the thirty-six titles of Hea, who rules the infernal world, has the two following as first and second : " Lord of the earth (Ki)," and " Lord of the celestial earth (An-ki)." [2] This is an important text for our present purpose. Hea appears as lord of the earth, of the celestial earth, and of the infernal regions. He is termed in other places lord of mankind ; and the name of his wife is *Nin-ki-gal*, " mistress of the great earth." All this is perfectly consistent with our theory. The primitive earth known to mankind was the celestial earth, the terrestrial paradise. It had been assimilated to the lower hemisphere ; and it is this region that becomes the especial domain of Hea as ruler of the under world, converted by misapprehension into a world by itself, distinct from the earth, and identical with the Greek Hades. It is to this abode, and floating on the Stygian river taking its rise in the sign Libra,

---

[1] 3d Rawl. Pl. 53, 2 Obs. l. 13.
[2] 2d Rawl. Pl. 55, Col. 2, ll. 17, 18.

that the terrestrial paradise has been transferred under the form of a floating island.

Sec. 84.   I wish to introduce here some other leading facts tending to establish our general theory, drawn from the legend of *Ishtar's* descent into Hades, some passages from which, as rendered by Mr. Smith, were presented before (Sec. 80).   This legend, being indeed a fine literary production, has been a favorite study among Assyriologues.   M. Lenormant has published two or three versions of it, the later improvements on the first; Mr. Fox Talbot has given to the public as many more, and Mr. George Smith one or two.   Dr. Schrader devoted a special study to the subject some time since, and recently Dr. Oppert has done the same.   So far as relates to a true rendering of the text, there is now a substantial agreement among cuneiform scholars.   But its real intent and meaning are probably not yet fully understood.   That an astronomical element is fundamental in it is apparent not only from internal evidence, but is naturally to be inferred from the fact that the religion of Babylon was preëminently astral.   But we seem to be gradually accumulating overwhelming proofs that the astronomical element in the ancient religions, especially the Chaldæo-Assyrian, was at the first thoroughly symbolical, relating to traditions and doctrines more primitive, but whose real import had been at later epochs sadly misconceived and perverted.

A remark or two here in relation to some of the principal characters which appear in this legend, and to some circumstances connected with it.   Ish-tar, as we know, was the Babylonian Venus, answering to the goddess of love of Greek mythology.   Upon the real object of her descent into Hades, Mr. Talbot observes: " I conjecture that she was in search of her beloved Thammuz-Adonis, who was detained in Hades by Persephone or Proserpine." [1]   Thammuz or Adonis, as is well known, was the sun of the lower hemisphere, the youthful sun-god, who had suffered an untimely and violent death.   He is detained in Hades by Proserpine, companion of the infernal Mercury or Pluto, according to the vision of Timarchus, already related to us by Dr. Faber.   In the opinion of Mr. Talbot, Proserpine is to be identified with *Nin-ki-gal*, "mistress of the great earth," wife of Hea, whom Mr. Smith has compared to Pluto.[2]

---

[1] *Records of the Past*, vol. i. p. 142.

[2] Ibid., p. 143, note 2.

Both Hea and *Nin-ki-gal* play important parts in the legend to which we refer. Respecting the object of Ishtar's descent, namely, to meet her beloved Adonis, the *Dú-zu* or *Tur-zi* of the cuneiform texts, M. Lenormant holds substantially the same views as those expressed above by Mr. Talbot.[1] Hence, as Adonis represents the sun in the inferior heavens, like the Egyptian Osiris, it is necessary to locate in this astronomical region the particular Hades into which Ishtar makes this descent. The notions respecting this abode of the dead, as set forth in the legend, may be gathered from the portion of it already cited from Mr. Smith's version. This local- ity into which Ishtar descends is also definitely fixed by the proofs heretofore introduced relating to the domain of Hea, and conse- quently of his wife *Nin-ki-gal*.

We wish to study now the import of some of the expressions employed in designating the region to which Ishtar goes to meet her beloved. I shall refer to the lines already presented from Mr. Smith's version on one hand, and on the other to the original text as published by Mr. Talbot, some of whose notes are to be cited below.[2] In the first line: "To Hades," etc., the cuneiform for " Hades" is *Mat-nudea*, literally "land of no return." There seems to be no question as to the meaning of the phrase. Its Ac- cadian etymology would be *mat*, "land," *nu*, "no, not," and *de*, "to return." The second name for the same region occurs in the fourth line: "To the house of the departed," etc. The text has *Bit-E-di-e;* and the rendering "house of assembly," suggested, but not adopted by Mr. Talbot, would be much better. Of this, however, hereafter. In the fifth line we have the cuneiform *Bit sha E-ri-bu*, rendered: "To the house from within which," etc., by Mr. Smith. This version is far from being literal, although on the whole it is not objectionable. *E-ri-bu*, as Mr. Talbot observes, must be allied to the Greek *Erebos*, " region of darkness," " under world," having original reference to the place where the sun sets at night, and like the Egyptian *Amenti*, " west," put for the abode of the dead, where the sun descends at close of day. But *E-ri-bu* in

---

[1] *Premières Civilisations*, t. ii. pp. 94–96. Dr. J. Oppert gives countenance to the same view, *L'Immortalité de l'âme chez les Chaldéens*, p. 4.

[2] For this text, see *Trans. Bib. Arch. So.*, vol. ii. pp. 179–212. The most of it is now published likewise in the fourth volume of *Cuneiform Inscriptions*, often cited in these pages as 4th Rawl.

our text has a verbal sense, that of "entering" in opposition to departing. "The house of entering, from which is no departing," conveys the idea with sufficient exactness. The terms *Erib* and *Atzu*, placed here in relation, as Mr. Talbot states, mean his *setting* and his *rising*, when applied to the sun. It is evident that they have nearly these senses in the present instance. But in the sixth line Mr. Smith has the rendering: "To the road the course of which never returns." For "road," the text employs *Khar-ra-ni*, "road, path," which is an ordinary term designating the "zodiac," and there seems to be a reference, a sort of double allusion to the zodiac, the same as to the west, in the use of *E-ri-bu*.

We return now to the phrase in the fourth line: *Bit-E-di-e*, "house of the departed," as interpreted by Mr. Smith. The following valuable note relative to this expression is by Mr. Talbot:

"Hades is here called *Bit Edi* or *Bit Hedi*, בית עדה, 'the house of Assembly,' because the spirits of all past generations are assembled there, Heb. עדה, coetus, conventus, turba. In the Syriac New Testament, עדתא is continually used for 'Ecclesia,' the assembly. Similarly in Job xxx. 23, Hades is called בית מועד, 'the house of assembly,' to which is added לבל חי, 'of all living.'" "Considering this eastern usage of the word *Hedi*, עדה, I think it probable that the Greek 'Hades' is derived from it."[1]

The Semitic *Hedi* (עֵדָה) is thus defined by Dr. Fürst: "An assembly, a congregation, especially of Israel, but also the members of a family, or house; also, host or mob that join for a certain purpose" (*sub voc.*). Its application to the dead in the Hebrew is quite exceptional, to say the least. It is one of the ordinary expressions for "the congregation" of Israel, assembled before the tabernacle or in the temple. Under the Syriac form, and in the New Testament, as remarked by Mr. Talbot, it is put for the "Ecclesia," the congregation or church. As for the Hebrew *Moad* (מוֹעֵד), its application to the dead in the passage, Job. xxx. 23, cited by Mr. Talbot, is the only instance, I believe, of its employment in this sense. Its primary meaning is "appointed time, fixed time, festival;" thence put for appointed assembly, usually designating the "congregation, assembly" of the house of Israel, like the term just explained. But notwithstanding these facts, Mr. Talbot's statements in the foregoing extract appear to be correct.

[1] *Trans. Bib. Arch. So.*, vol. iii. p. 125.

His assimilation of the cuneiform *Bit Hedi* (*Bit-E-di-e*) to the
Hebrew word first referred to admits of no doubt, and his conjecture that the Greek *Hades* was derived from it may be accepted,
at least for the present.

SEC. 85. We have in the data comprehended in the last section
another series of striking confirmations of our position that the
infernal abodes, understood as the Hades of the Greeks, were definitely located astronomically in the inferior heavens, the region into
which the sun descends as it sets in the west, and which answers in
some sense to our nadir. There can be no doubt that the Egyptians located their *Amenti* or Hades astronomically in precisely the
same locality. Various other facts revealed by a comparative study
of different cuneiform expressions might be introduced as a still
further support of our theory. Among these is the bilingual equation of the Accadian *An-ki* to the Assyrian *Il-sar*, in which *Sar*,
the Accadian *Khi*, " to be good, to be happy," is a monogram for
Assur, or supreme divinity of the Assyrians.[1] The later form of
Sar is *Asar*, and its reference to the inferior heavens is proved by
the phrase *pal-asar*, " son of the lower hemisphere." For the present, however, we will consider the fact of the astronomical location of both the celestial earth and the Greek Hades in the inferior
heavens as sufficiently established. Indeed, no point in the whole
domain of antiquities was ever better supported than the proposition referred to by the evidences already presented. Again, the very
name for Hades, the cuneiform *Bit-E-di-e*, Semitic *Hedi*, to which
Mr. Talbot would trace the origin of this Greek term, as we find
it employed in the fourth line of the legend in question, affords
another proof of the conversion of the terrestrial paradise, regarded
as a celestial earth, into the infernal regions as conceived by the
Greeks. It was shown in our chapters on the temple and cosmos,
especially the fifth and sixth, that the Hebrew tabernacle and temple were architectural imitations of the traditional mount of paradise, the same as the pyramidal temples of the Euphrates valley.
This mountain is termed the *Har-Moad*, or " mount of assembly,"
by Isaiah, evidently in strict analogy with the sense of *Beth-Moad*
and *Beth-Hedi* or *Hedah*, designating the " congregation " of Israel,
assembled before the tabernacle or in the temple. Bishop Lowth

---

[1] Vid. 2d Rawl. Pl. 54, 3 Obs. l. 6. This identification of the two expressions, *An-ki* and *An-khi*, the Assyrian *Il-sar*, admits of no doubt.

puts the *Har-Moad* and *Beth-Moad*, equal to the *Beth-Hedi*, in direct relation (note Is. xiv. 13) ; and we show that the latter, as a place of "assembly," was an imitation architecturally, an inheritance traditionally, as regards the former. But the *Beth-Hedi*, imitation of the paradisiacal mount, central object of the Israelitish congregation and church, is only another name for the infernal regions in the legend of Ishtar, regarded also by Mr. Talbot as having furnished to the Greeks their name *Hades*. A similar transfer of ideas is seen in reference to *Beth-Moad*, as it occurs in the passage cited from Job, where this location denotes the region of the dead. In a word, we have here the most striking proof that Hades and the church, whether that of the Hebrews, or the "Ecclesia" of the New Testament, originally meant the same thing ; that their traditional origin was the terrestrial paradise ; the connection with which of the Accadian *An-ki*, the Sanskrit *Svarga-bhoumi*, and the *celestial earth* of the vision of Timarchus reported by Dr. Faber, may be regarded as already fully established. As previously stated, the primary cause of the singular perversion of original doctrines, which we behold in the transformation of the terrestrial paradise into the Greek Hades, arose from the neglect or forgetfulness of the fact that the lower hemisphere, taken as a symbol of the first abode of humanity, was at the same time assimilated in conception to the earth itself ; originally the particular earth identical with paradise. But that which had a powerful tendency in the same direction, as we can now fully appreciate, was the fact that the sun's course had been taken as a symbol of human existence. The following statements of M. Mariette-Bey will serve us here as illustration and proofs : —

"Originally Osiris is the nocturnal sun ; he is the primordial night (like the Egyptian *Tum*) ; he precedes the light ; he is consequently anterior to *Ra*, the diurnal sun. From this principal character flows a multitude of allegories, which are grouped around Osiris, making of this personage a type of divinity the most curious to study. The life of man had been assimilated by the Egyptians to the course of the sun above our heads ; the sun that sets and disappears below the horizon is the image of the death of man. Hardly has this supreme moment arrived, when Osiris takes possession of the soul, which he is charged to conduct to the eternal light. Osiris, they say, had formerly descended upon the earth. The *good being par excellence*, he had softened the manners of men

by persuasion and beneficence. But he had fallen into the snares of
Typhon, his brother, the genius of evil, and while his two sisters,
Isis and Nephthys, collected the fragments of his body which had
been thrown into the river, the god was raised from the dead,
appearing to his son Horus," etc. " The image of death had been
derived from the sun, as it disappears at the horizon of evening ;
but the resplendent sun of morning was taken as a symbol of the
second birth into a life which this time knows no death." [1]

Another instance of the assimilation of the sun's course to human
existence, quite similar to the one just noticed, appears in the double
character of the Hindu Yama, reputed first man, yet unquestionably assimilated to the sun. Conceptions of a like nature prevailed
at an early epoch among the Chinese. It will be easily perceived,
then, since the descent of the sun below the western horizon was a
type of man's death, and its reappearance in the east a type of the
resurrection to a new life, how powerfully such conceptions would
tend to locate the Egyptian Amenti, the Greek Hades, in precisely
that astronomical region where we find situated the abode of the
dead, according to the conceptions of many peoples of antiquity.
But so long as this portion of the sphere was assimilated to the

_____

[1] *Musée à Boulaq*, etc., pp. 100, 101. It is remarkable that the transcription of
the name of Osiris, when written with vague vowels is *Asar* (vid. E. De Rougé,
Chrestomathie Egyptienne, pt. 1, p. 73), a reading identical with the Assyrian
Asar, signifying the lower hemisphere. The Assyrian term is the name of the
chief divinity, whose monogram is the Accadian *Khi*, " to be happy, to make
happy," designating this deity as the " good ; " thus *An Khi* is the " god good."
The hieratic form of *Khi* is identical with that of *Zid*, " the sun." The two
phrases *An-khi* and *An-ki* are often equated in the texts. These facts seem to
warrant the following conclusions : —

1st. That the Assyrian God *Asar* was originally one with the Egyptian *Asar*,
or *Osiris*. They were both solar deities, both being forms of the sun in the lower
hemisphere. They were both esteemed the " good " *par excellence*, and their
names were phonetically identical.

2d. We have thus another evidence of a community of ideas at an extremely
remote epoch, inherited by the populations of the Nile valley and those settled in
the country of the Euphrates and Tigris.

3d. Both the Assyrian and Egyptian *Asar*, as the " good being," the " beneficent," seem to have been connected or identified closely with the " celestial
earth." The cuneiform *An-khi* and *An-ki* are equated to each other ; and in the
" Litany of the Sun," translated by M. Naville, that which he renders the *Empyrean* is the peculiar abode of Osiris, and it answers very nearly to the " celestial
earth " transferred to the inferior heavens (vid. note 1, chap. xvi. upon this last
point, p. 425).

earth as one chief division of the cosmos, in strict harmony with the primitive doctrines of the zodiacal temple, it would be impossible to suppose a factitious world of which the first men had no ideas. So long, too, as the earth, to which all the cosmogonies originally related, was recognized as the particular earth identical with the traditional paradise, there was no danger that it would develop itself into the later Hades of Greek mythology. In reality, the Hebrews of Moses' time had preserved the primeval doctrine in its purity, or better, Moses had restored it. He maintained the strict relation of the *Beth-Moad*, "house of the assembly," to the *Har-Moad*, "mount of the assembly," from which the former had been traditionally derived, being architecturally an imitation of it. This fact, finally, reveals the fundamental connection of the Mosaic cosmogony, as previously interpreted in these pages and represented in the *Har-Moad*, with the Mosaic theocracy represented in the *Beth-Moad*, or *Beth-Hedah*.

SEC. 86. As observed in the last section, Bishop Lowth assumes a direct relation of the *Har-Moad*, in a theological or religious sense, to the *Beth-Moad*. Our investigations have served to demonstrate this relation in a traditional sense. The *Beth-Moad*, or Hebrew tabernacle, was expressly designed, like the pyramidal temples of the Euphrates valley, as an imitation, an architectural reproduction of the *Har-Moad*, or the "mountain of assembly" in the sides of the north, traditional abode of primeval humanity. In other terms, the two apartments of the tabernacle were intended to represent the celestial and terrestrial paradise, united by the sacred mountain, of which Mt. Sion and all the sacred mountains of antiquity were but reflections. Bishop Lowth's language is as follows : —

" ' *The mount of the divine presence*' (*Har-Moad*). It appears plainly from Exod. xxv. 22, and xxix. 42, 43, where God appoints the place of meeting with Moses, and promises to meet with him before the ark, to commune with him and to speak unto him ; and to meet the children of Israel at the door of the tabernacle ; that the tabernacle, and afterward the temple, and Mount Sion whereon it stood, was called the tabernacle, and the mount, of convention, or of appointment; not from the people's assembling there to perform the services of their religion (which is what our translation expresses by calling it the tabernacle of the congregation), but because God appointed that for the place where He himself would meet with Moses and commune with him, and would meet with the

people.   Therefore, *Har-Moad* (הר־מועד), or *Ohel-Moad* (אהל־מועד),
means the place appointed by God where he would present himself,
agreeably to which I have rendered it, in this place, ' the mount of
the divine presence.' "   (Notes, Is. xiv. 13.)

Undoubtedly, the idea of an immediate divine presence is implied
in both expressions, the *Har-Moad* and *Beth-Moad.* The Accadian
character *Gan*, a phonetic value assimilated to the Hebrew *Gan* in
the phrase *Gan-Eden* by most Assyriologues, whose hieratic form
has been exhibited and explained, has the especial sense of *presence*,
as if in direct allusion to the idea upon which Dr. Lowth insists.
Nevertheless, the notion of the divine presence on one hand does
not exclude, but rather presupposes that of assembly, of congregation, on the other.   In either case, the relation of the *Beth-Moad*
to the *Har-Moad*, in a religious, and we may add, in a politico-religious sense, is clearly recognized.   It is with the *Har-Moad*, or
" mountain of assembly " in the extreme north, that the prophet
directly associates *El* and *Elyon*, and likewise the " stars of El,"
evidently the seven stars of the chariot, represented, as I have supposed, by the golden candlestick with seven lights.   This particular region of heaven, the celestial paradise, was the especial
dwelling-place of divinity, of which the Holy of Holies was designed
as a symbol.   We see here a confirmation of the doctrine maintained by Dr. Bähr, that the Hebrew tabernacle was properly conceived as a dwelling of God, a doctrine upon which we have also so
much insisted.   The Holy of Holies was religiously and traditionally
a representation of that particular heaven regarded as the seat of
the divine hierarchy in the primitive traditions of all Asia.   This
was the abode of *El*, or of *El-elyon*, " the Most High God," whom
Abraham expressly identifies with the *Yahveh* or Jehovah of the
Old Testament (Gen. xiv. 22).   We have shown the original reference of *An-ki*, " the celestial earth," to the same sacred locality
conceived as the dwelling of *El* in connection with that of primeval
humanity.   It is through this Accadian expression that we are able
to prove the existence of similar ideas at Babylon to those just
noticed among the Hebrews.   As heretofore cited, we have the
phrase *Il-su-par Il-irziti (An-ki)*, " El, the ornament of the celestial earth."   But a variant of the name of Babylon affords us a still
more direct connection, namely *Bab-il il-irziti*, " gate of the God
of the celestial earth." [1]   The expression *Bab-il*, " gate of El," an-

1 Vid. Norris, *Assyr. Dic.*, i. p. 70.

swers very nearly to the Hebrew *Beth-Moad*, with the idea of "celestial earth" attached, all referring primarily to the traditional mount of paradise. These doctrines were primitive, no doubt, to the entire Semitic race, and were directly associated with El. Moses restored them to their pristine purity. At Babylon they had been perverted in a measure by the transfer of *Beth-Hedi*, equal to *Beth-Moad*, to the under world conceived like the Hades of Greek mythology. The Mosaic books contain nothing of this nature. Among the Haranites, also, a corruption had taken place. The divine names *Elyon* and *Shemal*, Hebrew *Semol*, appear to have referred equally to the rotating centre of the northern heavens, with which the "stars of El," the "Strong One," were directly associated. But with Moses, *Elyon* was not the "Highest" in relation to *El*, that is, was not superior to El, since such a conception was polytheistic. The two names were held to denote the one God, *El-elyon*, "El the Highest." With the Haranites *Shemal*, answering to *Elyon*, was put for the pole star, considered thus as higher, as superior to the stars of El. In other words, and according to Jewish tradition, *Shemal*, English *Samael*, was primitively the highest archangel, but he aspired to the rank of supreme divinity, in consequence of which he fell from his first estate, and became the chief of rebellious angels, identified with *Satan*.[1] Although as a divine name, *Shemal* seems to have pertained to the early epochs of Semitic development, a false notion had been associated with it by the Haranites, and it was thenceforth rejected by the Jews. It is the same blasphemous idea that the prophet attributes to the king of Babylon, supposing him to say in substance : "I will ascend *above* the *stars of El*, and seat myself by the side of *Elyon*, the very highest in relation to them" (Is. xiv. 13, 14). This was to put *Elyon* higher than *El*, and also to exalt himself to a higher estate, that of the Haranite *Shemal*, whom the Hebrews supposed to have fallen into the depths of *Sheol*.

As before observed, the Hebrews had preserved the original and true sense of the *Beth-Hedi*, as relating to the mount of paradise, and thence also the "celestial earth" identified with it.[2] It was

[1] Vid. Chwolsohn, *Ssabier*, ii. p. 221 ; whose various extracts from rabbinical writings are given, relating to *Samael*.

[2] It is not to be doubted, however, that the Jews of our Saviour's time had conceived the notions of different paradises, the heavenly, the earthly, and the

embodied likewise in the earliest temple structures of the valley of
the Euphrates, and even temples called *Aralli* were constructed by
the Assyrian monarchs,' which shows that the term *Aralli* put for
the under world, or Greek Hades, in one instance, was in such case
a corruption of the original notion, placing it in connection with
the sacred mount of the east, of which the temples *Aralli* were
imitations. Our further investigations relating to the "celestial
earth" will have regard to the primitive notions attached to this
phrase, attempting to trace the expansion of the idea into the
politico-religious institutions of the old world.[1]

infernal. Dr. Dillmann shows that such was in fact the case, and that some of
the rabbis identified the infernal paradise with *Gehenna* itself (Schenkel's *Bibel-
Lexikon*, vol. i. pp. 377–379. Cf. vol. ii. pp. 42–50).

[1] In the 4th chapter of *La Litanie du Soleil*, edited and translated by M. Ed.
Naville, there are several direct allusions to the pyramid in stages in connection
with the sun-god, under the form of Osiris, or the nocturnal sun. The region is
that of the dead, the *Ament*, astronomically the inferior heavens, which, as we
have shown in a previous chapter, symbolized the earth, that is to say, the particular earth known to the first men, or the terrestrial paradise. This " Litany
of the Sun," inscribed upon the walls of the royal tombs at Thebes, dates from
the period of Seti I., and embodies the cosmical doctrines of the period ; but
many passages contain traditional notices and ideas appertaining to the earliest
times, preserved in a greater state of purity than even the corresponding passages
in the *Book of the Dead*. The gods of the pyramid in stages, to which allusion
is made in the Litany, bear the same name as those mentioned in the 144th and
147th chapters of the *Book of the Dead*, in connection with the *Seven Ari or Aris*
(vid. Lepsius, *Todtenbuch. Vorwart.*, p. 16. Cf. De Rougé, *Rituel Funéraire*, Introduction, pp. 24, 25). This mention of the pyramid in stages in the Litany is
quite important, and M. Naville offers the following comments upon it : —

 " It is impossible not to recognize in this word (*Ar*) an edifice having the form
of a pyramid in stages, like that of Sakkara. This class of constructions, rare
in Egypt, was common in the country of the Euphrates, where the tower of Borsippa was the most famous example. This tower was consecrated to the ' seven
lights of the earth,' that is to say, to the seven principal stars. The gods of the
pyramid are those whom we find called the *Ari*. Thus, the 144th chapter of the
Book of the Dead shows us that there were seven gods having this name ; this is,
then, in accord with the Babylonian cultus, and we are able to recognize in these
seven divinities the sun, moon, and five planets. If at a later period a distinct
sanctuary was assigned to each of these gods, as seen in the *Book of the Dead*,
this is no evidence that they were not originally the gods of the degrees of the
pyramid. Here, at least, the determinatives recall nothing similar to the representations in the 144th chapter, but rather the ancient tradition analogous to that
of Babylon." (*La Litanie*, p. 93.)

M. Naville follows M. Lenormant in interpreting the "seven lights of the
earth," associated with the tower of Borsippa, as the "sun, moon, and planets."

I have shown in the present chapter that, while these seven luminaries were doubtless joined in the cultus, the primary reference must have been to the seven stars of the chariot, traditionally associated with the mount of paradise, of which the pyramid in stages was an artificial reproduction. One of the proofs to this effect, and quite conclusive, is that the "earth" illumed by these seven lights is not the *Ki*, or "geographical earth," but the *An-ki*, the "celestial earth," proved to have been originally one with the sacred mount, and that with which the seven stars of the chariot were directly connected. But M. Naville is undoubtedly correct in assimilating the seven gods of the *Ar* to the seven lights of the *An-ki*.

I have already insisted upon the point in the first chapter (Sec. 10) that the pyramids of Borsippa and Babylon, the most ancient in the valley of the Euphrates, embodied the same traditional ideas, and appertained to the same epoch, as the brick pyramid at Sakkara, the oldest in Egypt. It will be seen that the facts developed by M. Naville go far to confirm this position. If, in the Nile valley, and at an early period, the geometrical pyramid had taken the place of that in stages, the traditionary conceptions, primitively associated with the latter, had been preserved in the *Book of the Dead*, though partially obscured ; but in the "Litany of the Sun" these conceptions were far better preserved, where we find them connected with the cosmogony and the mystical doctrines concerning the future life. In both instances, the scene is laid astronomically in the inferior heavens, or the Egyptian *Ament*, answering to the Greek *Hades*. Here, too, we find what corresponds perfectly to the "celestial earth ;" that which, in his version of the Litany, M. Naville calls the *Empyrean*, analogous to the Elysium of the Greeks, the paradise of the later Jews. It must be admitted, then, that the Egyptians had preserved distinct recollections of the sacred mount of the east, and of the notions which we have shown were primitively connected with it. As among other peoples, the Egyptians had transferred all to the inferior heavens.

I venture to offer here one or two suggestions of an etymological character, somewhat hazardous no doubt, but pertaining to the Egyptian *Ar*, phonetic value of the hieroglyph of a pyramid in stages. According to M. E. De Rougé, both the phonetic value and the meaning of "ladder," a " pyramid in degrees," are derived from the verb *Ar*, " to ascend" (Chrest. Egypt. pt. 1, p. 73). Compare this radical with the same phonetic element in the Semitic *Ar-arat*, name of a mountain in Armenia, on which Noah's ark is supposed to have rested. That the diluvian mount, identified with that of paradise, was conceived as a mountain of degrees, or stages, is proved by the fact of its being artificially represented by the pyramid in stages, as heretofore verified. Another proof is, that the Accadian monogram for *Akkad*, when applied to the Armenian mountain, has the reading of *Til-la* (Lenormant, Rep. 341). The first element *Til*, or *Tul*, signifies "mount, mountain," and *La* means "ladder, degree ;" hence *Til-la* denotes a mountain in stages, or degrees. The *Ar-arat* in Armenia was traditionally called the " mount of the descent of Noah," just as the Hindu *Meru* was termed the " mount of the descent of Manu ;" and it is obvious that some mystical sense is involved in these expressions. On the other hand, this *descent* implies an *ascent* also mystical in character. Again, the ordinary Assyrian term applied to the divisions or stages of the pyramidal temple is *Par-su*, from the Semitic radical *Par*, softened form *Bar*, signifying the act of " cutting or breaking in, separating, dividing, as with a sword or plough," etc.; hence the *divisions* or stages of the temple. It will be seen that the element *Ar* is involved in these two radicals as *P-ar* and *B-ar*. This, however, only by way of suggestion. It is a fact, I believe, that the sacred mountain of degrees was sometimes conceived as a *ploughed mountain*, from the idea of a vineyard planted upon the side of a mountain, which had been cut up into terraces or stages by means of the plough, etc. This reminds us that Noah planted a vineyard after leaving the ark. We come now to the element *Ar* in *Ar-ya-rata*, "chariot of the Aryas," a name primitively applied to Mt. Meru, and from which the Semitic *Ar-arat* was derived according to MM. Obry and Lenormant. The element *Ar-ya*, in the foregoing title of Meru, like *Ar-yas*, is that from which comes the designation of the *Ar-yan* races, etc. I believe that the Hindus sometimes gave the name *Ar-yas* to the seven stars of the chariot, traditionally associated with Meru. The root *Ar*, from which the various forms *Ar-ya*, *Ar-yas*, etc., are derived, signifies "to plough." The English term *earth* is from the same radical, meaning primitively a *ploughed land*, afterwards *earth* in general. But the original application of this term was probably to Meru itself, the primitive home of the Aryans: and, if so, to the *ploughed mountain* of which the pyramid in stages was an imitation. The mountain masses, rising one above the other in the distance, could not be better represented than by a pyramid in stages. Such seems to be the conception involved in the Hebrew term *H-ar*, "mountain," in which the element *Ar* appears again. But to return now to the Egyptian *Ar*, denoting a pyramid in stages, which in Babylon was regarded as an imitation of the mount of paradise. I cannot but entertain the suspicion that this old Hamite expression was a technical one, inherited from the same locality as that designated by the Aryan *Ar-ya-rata*, afterwards corrupted into the Semitic *Ar-arat*. If such may be considered as having been the case, there can be no more question whether the Egyptians had preserved the traditions common to so many nations relative to the sacred mount of paradise. I make these suggestions, however, only in the hope that some one better qualified to deal with such matters may decide respecting their value. Aside from the Semitic radicals, such as *H-ar*, "mountain," in the phrase *Har-Moad*, applied to the Asiatic Olympus, *P-ar* and *B-ar*, "to break through," "to cut in," etc., I should include the Accadian *K-ar*, "summit," in the phrases *As-kar* and Scandinavian *As-g-ar-d* and cuneiform *K-ar-Sak-Kurra*, all evidently referring to the same locality

CHAPTER VIII.

Sec. 87. In his description of the destruction of Babylon contained in the 13th chapter, or the one immediately preceding that, from which has been cited the passage relating to the fall of the Babylonian monarch, in connection with the *Har-Moad*, or "mountain of the assembly," the prophet Isaiah employs language of a most remarkable character, of which the following verses will afford an example: —

"Behold, the day of the Lord cometh, cruel both with wrath and fierce anger, to lay the land desolate: and he shall destroy the sinners thereof out of it. For the stars of heaven and the constellations thereof shall not give their light : the sun shall be darkened in his going forth, and the moon shall not cause her light to shine." "Therefore I will shake the heavens, and the earth shall remove out of her place, in the wrath of the Lord of hosts, and in the day of his fierce anger" (vv. 9, 10, 13).

There is little difference of opinion among all judicious critics as to the general import of this class of figures so frequently occurring in Scripture, and the subjoined notes by Bishop Lowth, appended to the 10th verse, may be taken as representing the views of the majority of exegetes: —

"The Hebrew poets, to express happiness, prosperity, the instauration and advancement of states, kingdoms, and potentates, make use of images taken from the most striking parts of nature, — from the heavenly bodies, from the sun, moon, and stars, which they describe as shining with increased splendor and never setting ; the moon becomes like the meridian sun, and the sun's light is augmented sevenfold (see Isa. xxx. 26) ; new heavens and a new earth are created, and a brighter age commences. On the contrary, the overthrow and destruction of kingdoms is represented by opposite images ; the stars are obscured, the moon withdraws her light, and the sun shines no more ; the earth quakes and the heavens tremble, and all things seem tending to their original chaos (see Joel ii. 10 ; iii. 15, 16 ; Amos viii. 9 ; Matt. xxiv. 29)."

Although really and highly poetical, the language to which Dr.
Lowth alludes in the above extract is not wholly the product of
imagination, is not entirely figurative. In one sense it is almost
technical, for it is based upon ideas and customs of a character
quite ordinary. I refer to the fundamental conceptions constituting the theory of the ancient civilizations. For the most part the
kingdoms of antiquity were regarded as *astronomical* or *celestial
earths*, and this fact explains the origin and nature of the language quoted from Isaiah, and the various texts cited by Dr.
Lowth at the close of his note. With the ancients, the state was
an expansion of the idea of the temple, which involved, as we have
seen, the double conception of a celestial space systematically
marked off into divisions, — put in direct relation to a terrestrial
space similarly cut off and divided. Not only this, the state was
in point of fact, though the traditionary origin had been in some
instances lost or forgotten, an expansion of the idea of the particular heaven and earth previously shown to have constituted the
actual cosmos or world as known to primeval humanity. Each
kingdom was thus a world, a heaven and earth, within itself. At
a later period the entire heavens and earth were symbolically
represented in the organization of the kingdom and the territorial
divisions. But in the very earliest epochs I am inclined to think
the practice was more frequent simply to represent the heaven and
earth constituting the traditional paradise, or " mountain of the
world," according to Rev. A. H. Sayce's interpretation of the phrase
Kharsak-kurra heretofore cited. The division of the territory,
however, in a manner to represent the twelve signs of the zodiac
was a practice certainly very ancient.

The extensive prevalence in antiquity of the custom of representing the celestial world in the organization of the kingdom or
empire, is a fact which was long since fully established by Dr. Bähr,
in his critical treatise on the Mosaic worship already referred to in
these pages. At the period when he wrote, the facts were not so
well known as at present, but later researches have tended only to
confirm, and to place beyond doubt, the general hypothesis so ably
supported by him. Such being the case, the origin of the Scripture
phraseology to which reference has been made is quite apparent,
together with the rules of exegesis that should apply to it.

The question of the origin of the ancient civilizations is one of very great importance, and one upon which quite contradictory theories have been and are to-day entertained. Perhaps the most popular hypothesis among writers at the present day is, that the theory and organization of the states of antiquity were a gradual growth and improvement from a very crude and even savage condition of humanity. I believe myself able to produce an array of facts in the present chapter, sufficient not only to overthrow completely this hypothesis, but to show that the foundations of the ancient kingdoms proceeded from grand and noble ideas, whose actual genesis may be traced back to the first ages of humanity. Indeed, the data already included in the previous chapters must be regarded, I think, as affording a very strong presumption in favor of the view just expressed.

SEC. 88. The first order of facts to which I wish to call the reader's attention appertains, properly speaking, to the symbolical geography of the ancients, some notices relative to which have been heretofore presented. M. Lenormant was cited upon the singular system of geography as revealed in the great book of astrology, compiled by the orders of Sargon the ancient, about the year 2000 B. C. (Sec. 59). He had made allusion previously in the same work to this subject in nearly the same terms, yet so far different that I desire to reproduce them here : —

" It would be curious to devote a study to the geographical system of this book (of astrology), which indicates in itself an epoch extremely remote, and when the knowledge of the Chaldæans outside their own country was quite limited. The *rédacteurs*, in effect, conceive the country of *Akkad* as situated in a central position between four stranger countries, which correspond to the four cardinal points, to the ' four regions of heaven,' of which the ancient Chaldæan monarchs styled themselves kings (*sar kibrativ arbaiv*), corresponding to the primitive Cushite tetrarchy (Gen. x. 10)." [1]

It is to be noticed, first, that the author attributes this curious geographical scheme apparently to the ignorance of the Chaldæans. He corrects himself in his second reference to the same subject as heretofore quoted, styling it a "symbolical system" and "inspired by religious conceptions," in connection with which he refers to similar systems prevailing in other countries, schemes so exactly similar, in fact, that he finds himself compelled to refer the origin

[1] *Frag. de Bérose*, p. 27.

of all of them to primitive tradition respecting the sacred mount
of paradise, with which a like scheme was associated. The author's
fine archæological tact and immense antiquarian lore rarely permit
him to be deceived in the end. He has stated, in fact, in the passages before cited, the real nature and origin of this geographical
system. But in the extract here produced he yields unguardedly
to the spirit, so prevalent among certain writers of our day, which
attributes to the ignorance of the ancients that whose real import
and significance they are unwilling to take the trouble to ascertain.
It was customary with the nations of antiquity to consider their
country as the centre of the world, and especially the national temple as the central point of the universe. Neither vanity nor ignorance suggested this notion. It appertained to a symbolico-religious
system of geography that had been inherited from the great Olympus of all Asia, of which the national temples were designed as
artificial reproductions. The sacred books of the ancients sometimes allude to this mountain as the root of all others, and to the
sacred river that watered it as the source of all the rivers of the
world. It was truly the root of all the *holy* mountains, and the
source of all the *sacred* rivers, and such in point of fact is the idea
intended. Such notions were by no means the offspring of ignorance, but of a widespread symbolism, whose origin is to be traced
to the primeval abode of humanity.

M. Lenormant refers to the primitive Cushite tetrarchy, of which
Moses has the following notice, in connection also with that of
Asshur: "Whence it is said, Even as Nimrod the mighty hunter
before the Lord, and the beginning of his kingdom was Babel, and
Erech, and Accad, and Calneh, in the land of Shinar. Out of that
land went forth Asshur, and builded Nineveh, and the city Rehoboth, and Calah, and Resen, between Nineveh and Calah" (Gen.
x. 9–12). It is generally understood, I believe, among Assyrian
scholars that these four cities, forming the basis of Nimrod's kingdom, were conceived as a sort of mystical square, or tetrarchy, the
notion of which had been traditionally inherited, as M. Lenormant
supposes, from the paradisiacal mountain. Four other cities seem
to constitute the foundation of Assyria, or the kingdom of Asshur,
and it is probable that a similar mystical idea attached to them.
The ancient kings of Chaldæa, whose inscriptions have been preserved to us, appear to have inherited the same conception, this

being the interpretation attached by M. Lenormant to the frequently occurring phrase, *sar kibrativ arbaiv*, "king of the four regions." Precisely analogous ideas had been handed down in the line of Shem, as well as that of Ham, according to the opinions of Drs. Tuch and Herzog, relative to the import of the genealogy of Aram. We note the following observations in an article attributed to Dr. Herzog : —

"Finally, in the catalogue of nations (Gen. x. 22, 24), Aram appears after Elam, Assur, Arphaxad, Lud, as Shem's fifth son, and his sons are : Uz, Hul, Gether, Mash. Now Dr. Tuch interprets these four names as the 'termini of the Arameans.' Uz, the southern, against the Edomites and Arabs; Hul, the western, against the Canaanites; the unknown Gether (probably Gutium of the cuneiform texts), perhaps the eastern, against Elam and Assur; lastly Mash, undoubtedly the northern, against the Japhetic Armenians."[1]

Thus, Aram answers in this scheme to *Akkad* in the system of Sargon the ancient. We come now to still other proofs of the existence of the same arrangement. Rev. George Rawlinson, after having described the natural division of the country of Chaldæa, proceeds to remark : —

"We have no evidence that the natural division of Chaldæa here indicated was ever employed in ancient times for political purposes. The division which appears to have been so employed was one into northern and southern Chaldæa, the first extending from Hit to a little below Babylon, the second from Niffer to the shores of the Persian Gulf. In each of these districts we have a sort of tetrarchy, or special preëminence of four cities, such as appears to be indicated by the words: 'The beginning of his kingdom was Babel, and Erech, and Accad, and Calneh, in the land of Shinar.'"[2]

The division into upper and lower Chaldæa, having no correspondence with the natural characteristics of the country, taken especially in connection with a tetrarchy in each district, has every appearance of a symbolical reference to the upper and lower portions of the sphere, separated by an equatorial line. This will appear the more probable if we call to mind the fact that the Hamites of Egypt are known to have symbolized the two worlds, or two chief divisions of the cosmos, by upper and lower Egypt.

[1] *Protest. Eccl. Encyc.*, i. p. 227, art. "Aram." Cf. Tuch, *Commentar u. d. Genesis*, p. 204.

[2] *Five Monarchies*, i. p. 15.

Passing now to India, we find a system of political divisions the same as that pertaining to *Akkad* and Aram, a notice of which is thus given by M. Obry: —

"It is necessary to remark that after their installation in Hindustan, taken in the widest sense, the Aryans of India divided this country into four regions, east, south, west, and north, . . . and that they placed between them a middle country . . . all in imitation of the four *Maha-dvipas* and of the (central) *Madhya-dvipa* of the entire earth. I add in proof of this imitation that, after the dismemberment of *Indra-prastha* or Delhi, the four chiefs, or *Radjas*, who partitioned Hindustan among themselves replaced the great king who turns the golden wheel, assuming titles similar to those which a Buddhist tradition (of uncertain date) attributes to the kings of the four *Maha-dvipas*, China, India, Persia, and the Chinese Turkestan, being an extension of the revolving circle or wheel of the four regions." [1]

M. Lenormant cites the foregoing passage, mentions a like political division of ancient Iran or Persia, comparing all with the primitive Cushite tetrarchy, deriving them equally from a primeval tradition centring in Meru of the Hindus, the Albordj of the Persians, both identical with the Gan-Eden of Genesis.[2] The existence, then, of a symbolical system of geography, inherited from common tradition, inspired by religious conceptions, and prevailing among nations widely separated in antiquity, must be regarded as an established fact. The division of the world into nine earths by the Chinese, and of their own country into nine provinces, evidently in imitation of the cosmos, was shown from the statements of Dr. Bähr in a previous chapter.

SEC. 89. In the examples of political divisions thus far presented, although their symbolico-religious, as well as traditional character admits of no question, and notwithstanding a design to represent the cosmos is quite apparent in many of them, there is no direct, positive proof inherent that a celestial space was placed in relation to the terrestrial space thus marked off and divided. Of this, however, there does not exist much reason for doubt; first, since the scheme of four countries surrounding a central one, the latter conceived as centre of the universe, was uniformly arranged with particular reference to the cardinal points, and was derived from the Asiatic Olympus, with which a celestial or astronomical

---

[1] *Du Berceau*, etc., p. 47.   [2] *Frag. de Bérose*, pp. 321–323.

element was always associated. Secondly, we know that in the case of *Akkad* and the four countries surrounding it twelve stars were connected with each of the five regions named. In his second allusion to the geography of *Akkad*, as cited by us (Sec. 59), M. Lenormant refers to a well-known cuneiform text where, as he observes, "twelve stars preside over the destinies of each of these regions, and the influences exerted upon them during each month are described."[1] The text cited is quite fragmentary, but we have one list of asterisms styled: 12 *mul-nus mat akkad ki*, "the twelve stars of Akkad;" and another entitled: 12 *mul-nus mat martu ki*, "the twelve stars of the west or of Phœnicia." In its perfect state, the tablet appears to have assigned a definite series of twelve asterisms to each of the five regions, of which Akkad formed the centre. Thus, we have the best of reasons for the supposition that it was customary to connect with the geographical scheme of five terrestrial regions a definite reference to the stellar world placed in immediate relation to it. These mystical divisions, therefore, were in the strict sense celestial or heavenly earths, intended to represent or reproduce the original cosmos to which universal tradition pertained. Each of these countries was, for itself, a particular heaven and earth, modeled after the sacred mount, reputed abode of the first men. The antiquity to which, without any doubt, we must assign the geographic scheme of the first Sargon, and the primitive character of that of which Aram formed the central region, as already noticed, must be regarded as sufficient proof that these symbolical ideas were fundamental in the theory of the ancient civilizations.

But there are numerous instances in antiquity where the territorial divisions had an especial reference to the number *twelve ;* and in all such cases we may be quite sure, considering the conformity of the notion to that involved in Sargon's system, that there was a direct intended relation to the twelve zodiacal constellations. Although Dr. Bähr's reference to the four castes in the passage cited below appears to me doubtful, the numerous instances cited involving the number twelve must be held to establish the principle, namely, that it was customary in antiquity to arrange the terrestrial

[1] *Frag. de Bérose*, p. 321. The tablet referred to will constitute the subject of a special study in the sequel of the present treatise. It is published in 2d Rawl. Pl. 49, No. 1, Obs.

240   HAR-MOAD.

kingdom in a manner to represent the celestial world, divided according to the zodiacal constellations or signs. The author is correct, also, in assuming a close relation between the numbers four and twelve. The four-sided plot, in the form of a square, was most frequently chosen to represent the zodiacal divisions, three signs, like the three gates of the holy city of the Revelator, being assigned to each side of the square. The passage in which the writer just alluded to musters so many examples in proof of his theory is as follows: —

"Since the stars were considered, not only as animated, but actually as divinities, so the starry heaven was conceived as a celestial city, or as a heavenly kingdom; and since this kingdom, upon which all terrestrial life and existence depended, was arranged in twelve divisions, it constituted really the type for the order and arrangement of the cities and kingdoms of the earth, or the lower cosmos. Hence the division of the people and of the territory according to the number twelve. The king or chief ruler of the state represented the highest divinity or the sun, and was at the head of the twelve tribes or races, over each of which a subordinate ruler was placed, answering to each division of the zodiacal land. This arrangement proceeds on the same principle as that of the division of the population into four castes, namely, the representation of the order of the universe, although the partition according to the number twelve appertains rather to the heavens as its type than to the entire cosmos. Some examples may be cited here, and we begin with the east. The successors of the Chinese Emperor Yao report that he installed twelve mandarins as rulers over the empire, after having divided it into four parts, according to the directions of the cardinal points. We find from the Mosaic narrative (Gen. xvii. 20; xxv. 16) that the Arabians were divided into twelve tribes proceeding from as many ancestral heads; and in the time of Mahomet, the Saracens and Nabataeans formed twelve tribes, to each of which a division of the zodiac was dedicated. A like division existed in Persia. The imperial palace inclosed a public court divided into four parts for the young men, the overseers, the men, and the ancients, over each of which divisions twelve archons ruled. Egypt also in ancient times was partitioned into twelve chief provinces, connected with the division of the population into four castes. Even the division into thirty-six nomes, already mentioned, was only an expansion of the earlier system, being an imitation of the thirty-six heavenly decans, three to each of the twelve zodiacal signs. According to an ancient tradition the populations of interior Africa, the Ashantees with others, constituted twelve races. With the Greeks this system was very common. Twelve tribes formed the confederation of the original populations of Delphi; there were the twelve cities of the Ionians of Asia; twelve cities of the Achæans of the Peloponnesus; the twelve communities of Attica, of the Cecrops; also the twelve phylarchs or phratries of Athens, proceeding from the four original races. The highest tribunal, that of the Areopagus, consisted primitively of twelve members, and twelve chiefs of cities were originally associated with the king by the Phæacians. So, too. among the Etruscans, and their neighbors the Romans, this arrangement according to twelve prevailed. There were the twelve Etruscan cities," etc.[1]

SEC. 90. The author then proceeds to the twelve tribes of Israel, and to the frequent occurrence of this number in the Mosaic ritual generally. In the majority of the instances cited in the foregoing extract, the writer is perfectly correct according to our best information at the present day; and the examples, being drawn from the most widely separated quarters of the old world, render it impossible to entertain a doubt that the principle involved was fundamental in the theory of the ancient civilizations. The ancient kingdoms were regarded generally as celestial or astronomical earths; as imitations, in fact, of the order and arrangements of the heavens. But it is necessary to introduce here another passage, in order to complete Dr. Bähr's view of the underlying ideas involved in the politico-religious institutions of antiquity : —

"It is a fact upon which no doubt can be entertained that the ancient systems of worship did not grow out of the state regulations, as if the state and the political ordinances had a prior existence to the religious institutions; on the contrary, the political organizations were modeled entirely according to religious ideas. Nor was it a fact, after a ruler had been established and the civil policy had been formed, that these notions were transferred to a divinity, the conception of an earthly kingdom giving rise to that of a divine government; but rather was the notion of a divine ruler the origin of that of a temporal one. What God is for the entire world, such was the earthly monarch conceived for his individual kingdom, namely, a miniature divinity who dwelt and ruled on earth, just as God dwells and rules in the heavens. Indeed, the order and arrangement of the heavens, the notion of a heavenly and divine government, constituted in the minds of the people of the east the originals of which the institutions of this lower world were but imitations, and it was for this reason that they were thought to possess the stamp of divinity and the sanctions of religion. It was

[1] *Symbolik*, i. pp. 203, 204.

for these reasons, likewise, that monarchs were frequently adored under the same forms as the deity himself ; but the divine worship was never borrowed from that of kings." [1]

The foregoing statements by Dr. Bähr are extremely important, and if they are to be admitted as correct, go far to establish the actual theory of the civilizations of the old world. The assumed priority of religious institutions, or those of a politico-religious nature, to those of a purely civil character, seems to be fully justified by known facts. No class of official characters were more primitive than the ancient priest-kings, whose general history and functions were traced in our third chapter. The well-known fact that sacerdotal and civil functions were combined in these personages implies that religious ideas exercised a potent influence in the establishment of the primitive institutions on earth. But we have a proof even more conclusive than this, in the fact that the form of government in Egypt, prior to the most ancient civil dynasties headed by Menes, was properly speaking *a theocracy*. In Dr. Henri Brugsch's " History of Egypt," based upon the results of modern investigations, he makes the following observations as cited by M. Léon Carré : —

" We have no historical tradition relative to the form of the first government of the Egyptians, but everything tends to the belief that it was a theocracy, that is, a form of the state where the priests govern and administer the laws of the country." [2]

That which confirms the opinion here expressed by Dr. Brugsch is the fact, which is now perfectly verified, that the ancient nomes or provinces, into which the territory of Egypt was divided under the monarchy, were primitively of a strictly religious character. This was affirmed some years since by Dr. Uhlemann, and recently M. Mariette-Bey has confirmed the statement as follows : —

" The division of Egypt into nomes, or provinces, had for its basis an anterior division into religious districts." [3]

The civilization of Egypt, if we leave aside the question of that of Babylon, was the most ancient of any at present known. Yet the beginning of this was a theocracy, and the nomes were a reli-

---

[1] *Symbolik*, i. p. 11.

[2] *L'Ancien Orient*, etc., t. i. p. 24.

[3] *Musée à Boulaq*, p. 101. Cf. Uhlemann, *Handb. d. Ægyptischen Alterthumskunde*, b. iii. pp. 66, 67.

gious inheritance from an epoch anterior to the first dynasty. In the valley of the Euphrates, it is certain that religious ideas exercised a powerful influence from the first ages. The pyramidal temples of Babylon and Borsippa, whose construction dated from the earliest traditionary period, sufficiently evince by their character, as already partly explained by us, the order of ideas which were current with the founders. Dr. Bähr's statements, then, in the last extract cited from him, must be regarded as substantially correct, and those of the first quotation are mostly so, according to our present knowledge.

SEC. 91. It was stated in a previous section that upper and lower Egypt were taken as symbolical representations of the two principal divisions of the cosmos, or heaven and earth. This seems to be proved by the subjoined remarks of M. Chabas, relative to the mystical region called Sutensinen : —

"Sutensinen was otherwise the theatre of the triumph of Osiris (over his enemy Set or Typhon) ; and this god receives there the double crown, which symbolizes the royalty of upper and lower Egypt. He dies there and is raised again under the form of the *beneficent soul*. At this moment the *organization of the two worlds*, that is to say, *the two Egypts*, and their union under one sceptre is definitely accomplished. This important event appertains thus to the divine dynasties, or to the heroic times of Egypt; and Menes, the first human king whose name has been transmitted to us, receives properly the title of 'king of upper and lower Egypt.' " [1]

Osiris is the nocturnal sun, the sun of the primordial night of chaos, and his renewal in the region of Sutensinen, or the lower hemisphere, was a type of the organization of the two worlds, heaven and earth, to which upper and lower Egypt were assimilated. The hieroglyphic expression denoting these divisions is *Ta-ui*, explained by Dr. Brugsch, "the two worlds, upper and lower Egypt." [2] These conceptions, as will be seen, do not appertain to comparatively late epochs; they were primitive in the Nile valley, and belonged to a system absolutely anterior to the first dynasty. We see here the notion of a terrestrial kingdom modeled

[1] *Les Papyrus hieratique de Berlin*, p. 19.

[2] *Grammaire hieroglyphique*, p. 5. For additional proof that the double crown symbolized the sovereignty of the two worlds, understood cosmically as well as of the two Egypts, consult M. E. Grébaut, "Hymne à Ammon-Ra" (*Revue Archéologique*, Paris, 1873, p. 386, note 2.

expressly upon the principles of cosmogony, organized in a manner
to represent the sphere. Upper and lower Egypt, according to this
scheme, with the thirty-six nomes corresponding to the thirty-six
heavenly decans, was literally "a celestial earth," judged by the
meaning obviously involved in this phrase. The classic authors
repeatedly assert that these nomes were intended to represent the
decans, and the statement has been made to this effect by modern
investigators. I am not aware of any list of the decans recovered
from the monuments which can be assigned to a period much earlier
than the twelfth dynasty. That the Egyptian sphere dates from
an era far more ancient than this admits of no doubt. Its particular division into thirty-six decans, instead of the ordinary twelve
signs, was possibly not primitive. Nevertheless, it is obvious that
the Egyptian empire was modeled after the sphere from the first.
This is proved by the remarks of M. Chabas just cited. Some
authors affirm that this country was at one time divided into twelve
provinces in imitation of the twelve zodiacal constellations. As
regards the nomes, they were doubtless primitive. M. Mariette-
Bey reads the names of six of them in a tomb belonging to the
sixth dynasty, and his theory of their original religious character
tends to the conclusion that they appertained to the period of the
theocracy. Their number varied at different times ; it is probable,
however, that the typical number corresponded to the decans, or
perhaps to the forty-eight primitive constellations. In matters
appertaining to such an extreme antiquity it is usually very difficult
to settle beyond question all the details, and such is the case relative
to the matter of the nomes and decans. It is enough if we prove
the existence of the ground conception, the assimilation of the two
Egypts to the upper and lower hemisphere, to have been actually
primitive.

It is hardly necessary to offer more evidence in verification of the
fact that the ancient kingdoms were regarded as astronomical or
celestial earths, and that this conception was fundamental in the
theory of the ancient civilizations. We have dwelt longer upon the
archaic ideas prevailing in the Nile valley, since the Egyptian civilization, so far as existing monuments afford definite proof, was the
most primitive of any known to us. It cannot be objected, therefore, that the notion of a celestial earth as a fundamental doctrine
of the old world appertained only to comparatively late epochs. It

was ancient, and the most ancient of any dogma known pertaining to the institutions of antiquity. More than this, it had been inherited from tradition even by the Egyptians. Similar ideas prevailed in the valley of the Euphrates at the founding of Babel, as proved by the mystical titles applied to the pyramidal temples, such as, "foundation of the celestial earth," and "seven lights of the celestial earth." These conceptions, at an epoch so very remote, had not been imported from Egypt. On the contrary, the proofs are abundant that they had been brought from the far east; that for the Cushites of Babylon and the Hamites of Egypt, in fact, the doctrines of which there is here question were a traditional inheritance from a civilization which the world has forgotten, that of which the great Asiatic Olympus formed the original centre. We see now, in the light of the facts which have been presented, the original import of those singular locutions employed by those of high antiquity. The monarch was accustomed to style himself "son of the sun." The Chinese regarded their kingdom as the "celestial empire." The Chinese emperor was the "regent of the nine earths," in the same sense that the Egyptian monarch was "king of the two worlds," and the ancient Chaldæan ruler "king of the four regions." It was not ignorance, nor was it vanity, that prompted these expressions at the first; they had their origin in the profound symbolism which was inherent in the political and religious institutions of high antiquity. Every kingdom was a celestial earth, a cosmos, a world within itself. The process of founding it was assimilated to that of creating the world, and was thus but the creation of a "new heaven and earth," after the model of the "first heaven and earth," to which all the traditions pertained. The internal organization of the state and the divisions of the national domain were arranged with special reference to the same leading idea. Nature, the heavens, the universe, — these were the models according to which the foundations of the ancient empires were laid. Nevertheless, the original idea was a traditional one, derived from the particular heaven and earth known to the first men; and this fact is one which should never be forgotten. It is the great fact upon which Dr. Bähr never insisted as he should have done, and probably would have done, if he had possessed the advantages afforded by more modern researches. It was impossible in his day to prove that the civilizations of the old world had a genealogy, and could be traced to the birthplace of humanity for the origin of their fundamental ideas. But all the evidences now before us plainly indicate that such was the case, and there is much more to come having a like tendency.

SEC. 92. According to the Hindu conceptions, the celestial Ganges, regarded as original source of the river of paradise, poured its waters upon the summit of Mount Meru, which descended from thence through the three worlds, in each of which the waters were gathered into a single source, from whence they were divided again into four branches, flowing toward the four cardinal regions; the Buddhists, however, as heretofore stated, conceiving them to flow toward the intermediate points of the compass.[1] The Aryans of Persia had a similar notion connected with their river Avanda, which must have been conceived as having a celestial source, according to the following text of the Zend-Avesta: "I invoke, I celebrate the height, the divine summit, source of waters, and the water bestowed by Mazda;" to which M. Carré adds the note: "This source is the *Arduissur* (or *Avanda*), at the summit of the sacred mountain, the *Bordj* (*Albordj*), from whence issue all the waters that flow upon the earth."[2] Among the Chinese, the constellation *Tsing*, "a well," identical with Gemini of our sphere, was regarded as the source of heavenly waters; hence the name often applied to this asterism *Tien-Tsing*, "the celestial wells."[3] This constellation was associated with the sacred mountain, as was previously shown. I proceed to show now the existence of similar ideas at Babylon.

The inscriptions usually cited by cuneiform scholars, as 1st, 2d, and 3d Mich., are deeds or grants of land, donated by royal personages to certain distinguished individuals, and are evidently of a typical character for this class of documents.[4] They were inscribed upon conical stones, and these stones served as landmarks or boundaries. Their real date is ascertained to be about the year 1200 B. C. Around the upper surface of the cones are represented the symbols of the divinities whose names occur in the inscriptions, and together with these, various constellations that undoubtedly existed

---

[1] Vid. M. Obry, *Du Berceau*, pp. 19, 20.
[2] *L'Ancien Orient*, t. i. p. 341, text and note.
[3] Vid. Schlegel, *Uranographie Chinois*, p. 406.
[4] Vid. 1st Rawl. Pl. 70; 3d Rawl. Pl. 41-44.

upon the Babylonian sphere, some of them zodiacal, but the majority of them extra-zodiacal.[1] For myself, I do not hesitate to say that the upper surface of these cones was intended as an astronomical representation of the sacred mount of paradise, together with the primary divisions of the sphere having their generative point in the summit of this mountain. Thus, we have Aries and Scorpio, the pincers of the latter put for Libra, answering to the equinoctial signs, while the Eagle and Dog are certainly solstitial signs. Then we have Draco or the Serpent, the Tortoise, etc., which must appertain especially to the northern heavens. But the figures represented around the apex of the cones are to me the most decisive. There is the crescent and the eight-rayed star, varied to seven rays, inclosed in a circle; and what is especially remarkable, a four-rayed star, answering to the four cardinal regions, while four rivers are represented issuing from a common source in the centre, taking their courses toward the four intermediate points. It is impossible to take these intermediate radiations for anything but veritable rivers. The plate illustrating a similar document discovered by Mr. George Smith, accompanying his version of the same (Assyr. Dic. p. 236), where all these symbols are repeated in cruder form, shows these rivers with unmistakable plainness. Finally, we must consider that the conical stone itself was a temple, like the Betylus or Beth-el, was in fact a miniature Mt. Meru conceived as an immense conical hill.

Considering the facts above set forth, it is difficult to resist the conclusion that the system of land measuring, and of territorial divisions generally, both proceeding from the theory of the temple, was derived traditionally by the Babylonians from the sacred mount of the east, in which almost all their religious ideas seem primitively to have centred. The astronomical representation of the sacred mountain upon a boundary-stone can hardly be explained upon any other hypothesis. Another and very strong presumptive evidence of such traditional origin would be afforded if we could show that this system of measurements and divisions actually proceeded with the Babylonians from the theory of the temple, since the pyramidal temple was an artificial reproduction of the sacred mount. It will not be difficult, I think, to establish this point of

[1] Vid. 3d Rawl. Pl. 45. for cuts representing the upper surface of the cones. Cf. Rev. G. Rawlinson, *Five Monarchies*, ii. pp. 573, 574.

connection. In every one of the land grants, of which these conical stones were boundaries, there occurs, in the description of the territory conveyed, the Assyrian expression *Bit-as*, meaning "measure," thus: "fifty sekul in *measures* of great cubits." It is difficult to refer this Assyrian reading of *Bit-as* to any known Semitic root. The reason is, that it is in fact an Accadian term, so considered by M. Lenormant, to which he gives the reading *E-as*, these being the ordinary values of the two signs composing it. The etymology of the word is then apparent at a glance, namely, temple + measure, or to measure. This shows that the usage of the expression as denoting "measure" in land divisions was derived originally from the temple. Recall here, for a still further confirmation, the hieratic form of the character signifying "house, temple," etc., as heretofore explained, in which we find the arrangement of nine identical squares, so resembling a plot of ground cut up into sections. Finally, we know that the augurial temple of the Romans constituted the theoretical basis of their system of land measuring, and generally of all territorial divisions, while the origin of it has been already traced to the valley of the Euphrates, where the two orders of ideas must have been also associated. The sacred mountain of the east, then, was alike the birthplace of these doctrines of the temple and of territorial divisions. Indeed, the system of nine squares, corresponding to that of the "planetary seal" symbolizing the cosmos, was seen to have been definitely connected with the same traditional locality in a previous chapter.

SEC. 93. In the augurial temple of the Etrusco-Romans, according to the statement of Dr. William Smith previously cited, there was involved the notion of a terrestrial space, marked off and divided, and of a certain limited celestial space, corresponding to and placed in immediate connection with it; and all the divisions were located with special reference to the cardinal regions. Here was, then, literally a heaven + earth, an astronomical or celestial earth. The same doctrines and methods were applied in the partitions of the soil, the territorial divisions of the state, and finally in the laying-out of towns, the inauguration of cities, and probably in the foundation of the state itself. Everything was thus a temple, and at the same time a celestial earth. In each instance, the least unit of such a divided plot would be a geometrical figure, like that of the Spartan symbol of the Dioscuri, corresponding to the Chinese character *Tsing*, "a well," put also for the constellation of Gemini, or of the Dioscuri. It was shown in a previous chapter that this very symbol was associated also among the Chinese, with the divisions of the soil and formation of towns, thus: "Nine families constitute a *well*, four wells an *inclosure*, and four inclosures a *community*." Not only this, but a passage was cited to the effect that the constellation *Tsing* was taken for the symbol of territorial divisions, for the rules in laying out and founding a capital, and for the inauguration of the state. To these data should be added the fact heretofore insisted upon that the geometrical figure representing the symbol in question corresponds perfectly to the geographical scheme of five regions associated with *Akkad*, with the Biblical *Aram*, and various other countries previously named, and finally with the traditional mount of paradise itself. The Chinese character *Tsing* was also, as name of the constellation Gemini, connected with the same sacred locality, which constituted, in fact, the primitive celestial earth, the original cosmos, the first abode of humanity.

The statement has been made and repeated that the ancient civilizations and kingdoms had a genealogy which could be traced to the paradisiacal mountain ; and I believe the evidences now placed before the reader will be considered a sufficient verification of the fact. Savagism never gave birth, either immediately or remotely, to the religious and political institutions of antiquity. They were born of a *sacred science*, that had been transmitted from the earliest ages, to which the recollections of all the races reverted as the bright and golden era. We gather from the various peoples the most widely separated in antiquity, not merely general indications of such common inheritance, but the most circumstantial evidences tending to this conclusion. Thus, among the Etrusco-Romans, the temple, being literally an astronomical earth, and distinctly involving the notion of the cosmos, constituted obviously the germinal centre from which the entire state organization proceeded. The traditional inheritance of these notions, on the part of the Romans, from the Asiatic Olympus seems at first wholly incredible. Nevertheless, the proofs to this effect are quite conclusive. As stated by Dr. William Smith in a previous chapter, the gods were supposed to be seated in the *north*, which fact was thought to determine the direction of the main line in land surveying. This conception respecting the seat of the divinities finds its only explanation in the primitive tradi-

tions centring in Mt. Meru. Another evidence not less direct is
derived from some striking features of the Roman Pantheon. Herr
Nissen remarks relative to this renowned edifice : —

"The axis of the temple is only five degrees westward from the
pole. The seven gods of the Pantheon are the *septem triones*, to be
compared to the seven oxen (that is, the seven stars of the chariot)
which never disappear from the circumpolar region." "The position of the seven stars determined the location of the temple as
primal reason, while the reference to the dwelling-place of Jupiter
in the *eighth region* formed a second motive." [1]

These facts show with what remarkable fidelity the Etrusco-
Romans had preserved the ancient tradition, and clearly demonstrate
the ultimate origin of their sacred science ; though it appears to
have been received intermediately through the Babylonians. Like
evidences of a nature too circumstantial and direct to be explained
upon any principle of accidental causes or of normal development
have been drawn from the valley of the Euphrates, of the Nile,
from the Semitic populations generally of Western Asia, and from
the nationalities of the distant east, the Aryans of India and the
ancient empire of the Chinese. Dr. Schlegel notices a peculiar and
archaic form of the Chinese temple, the upper portion being in the
form of a dome to represent the heavens, the lower portion of a
square form to symbolize the earth ; conceptions whose exact analogy with the doctrines of the temple prevailing in other quarters of
the old world will be at once recognized. The connection of the
Chinese *Tsing*, also, with the fundamental ideas appertaining to
the state, compared with its obvious relation to the same order of
ideas among other peoples, is very remarkable, and the influences to
be drawn therefrom are quite obvious. Everywhere the temple is
conceived as a cosmos, a heaven and earth, an astronomical earth, in
fact ; and the expansion of these ideas forms the basis of the theory
of the state, which is itself a great temple, a terrestrial heaven, an
organized world redeemed from chaos, whose archetypal conception
has been inherited from primeval tradition, centring in that region
from whence the races first departed to people the earth.

SEC. 94. An effort to make a radical distinction between the
Hebrew conception of the tabernacle and temple, on one hand, and
that of the sacred edifices of peoples surrounding the Israelites, on

[1] *Das Templum*, p. 225.

the other, would ever be, in my view, unsuccessful. Originally and
fundamentally they were the same, and had been derived from a
common source. But in this conception the traditional element was
all important, a neglect of which would inevitably conduct to erroneous ideas. The aim was to represent the cosmos as the dwelling
of God. This, however, was not the cosmos in our modern sense,
except in a wholly subordinate degree. The reference was to the
particular heaven and earth known to the first men, and to which
all the traditions and all the cosmogonies especially related. An
expansion of this original idea to include the entire expanse of
heaven and earth would be natural and even legitimate, so long as
the primary reference to the sacred mount of paradise was held
most prominent and fundamental. To forget this primary reference was henceforth to convert the entire scheme into a philosophic
view of God and the universe, which could result only in pantheism. It is quite certain that, during the course of ages, the all-
important traditional element became itself, not merely subordinate,
but wholly mythical. The result was that the entire expanse of
heaven was taken as God's dwelling-place, while the entire earth
was assimilated to the earth goddess; and this is precisely the interpretation which modern investigators attach to-day to the original
doctrines of antiquity relative to these subjects. It appears to me
that the investigations contained in our sixth chapter, together with
various other considerations that might be urged, prove conclusively
that such interpretation is a misconception of the primitive idea.
The original reference was to a certain characteristic geographical
locality, and to a well-defined and limited celestial region placed in
immediate association with it. The traditional element referred to
these, and not to the entire heaven and earth collectively taken.

Professor Max Müller has shown that the Chinese word *Tien*,
and the Aryan *Dyu*, or *Dyaus*, denoted originally the sky, but that
they were employed likewise, by a natural association of ideas, as
titles of the supreme divinity; thus, the Greek *Zeus* and the Latin
Jupiter are names derived from *Dyu*, or *Dyaus*.[1] Nevertheless, the
divine being, thus conceived and invoked, was regarded as endowed
with strict personality. To these early titles of divinity is now to
be added that of the Accadian *An*, "elevated, heaven, god." This
term is otherwise employed for the personal pronoun, third person,

---

[1] *Lect. Sci. Language*, 2d series, pp. 456, 457, etc.

masculine and singular.   Hence, the conception was "heaven-he"
or "god-he," like the Aryan "Dyaus-he," a strictly personal deity.
But was the "heaven" which gave rise to the primitive conception
of a personal and divine being the entire expanse of the sky?
Professor Müller does not take note of this particular inquiry, but
it is very important for us in the present connection.   The simple
fact that, according to the earliest and most universal tradition, the
seat of the divine hierarchy was supposed to be the north, and the
extreme north, that is to say, the circumpolar region, the Su-Meru
of the Hindus, ought to be regarded as conclusive upon the point
before us.   There are any number of allusions in the sacred books
of antiquity, as well as traditionary notions, perfectly inconsistent
with the idea that the entire expanse of heaven was taken for the
Deity or the abode of the superior powers.   The Aryans of India
and Persia definitely located this abode in the particular celestial
region around the summit of Mt. Meru.   The same notions prevailed at Babylon, as evinced by the various expressions, "heaven
of Anu," "Bit-Kharris of the east," to which for proof of the ordinary Semitic conception may be added the prophet's allusion "the
stars of El," and the Haranite name Shemal, the Sumilu of the
cuneiform texts, but especially for the north and the great divinity
of the north.   Finally, I recall here the proofs previously introduced, that the Accadian An had particular and primary reference
to the same celestial region, and the passage cited from Herr Nissen
in the last section, showing that among the Romans the especial
seat of Jupiter was the "eighth region" in relation to the septem
triones, or seven stars of the great Dipper.   Thus, I submit it to
the judgment of the learned reader whether the data now before us
do not necessitate the conclusion that the primitive reference and
application of the notions and divine names referred to was not
limited, and that expressly, to the particular heaven and earth, constituting, as I have shown, the staple element of the traditions of
all Asia respecting the first abode of humanity; traditions which
had originally an actual historical and geographical basis.[1]

---

[1] Since the completion of the manuscript of the present work, I have more fully
realized the important bearing of the fact that primitive sacred tradition regarded
a particular heaven and earth, instead of the heaven and earth in general, upon
the labors of certain German authors in the field of comparative, especially Aryan mythology.   The same fact has a similar bearing upon Professor Müller's

SEC. 95. We return now to the Hebrew conception of the tabernacle and temple, as compared with that of surrounding nations, attached to structures of a like sacred character. I think it is true as regards the majority at least of these peoples that the original researches in this department. The German works to which special reference is had are those of Dr. Julius Grill (*Die Erzväter der Menschheit*, etc. Leipzig, 1875), and Dr. P. Asmus (*Die Indog. Religion*, etc. Halle, 1875). All proceed upon the supposition that the original heaven father and earth mother had reference to the entire heaven and the entire earth, or at least as taken in a general and indefinite sense ; upon the supposition also that the notions pertaining to both had a purely naturalistic development. I believe these are two fundamental errors, proved such by the present researches, and that they tend greatly to mislead the comparative mythologists of the present day. The notions of the heaven father and earth mother were rather traditional than naturalistic, and were inherited by widely different peoples. They were far more ancient than the Aryan development itself, and they obviously centred originally in the great Olympus of all Asia. The two authors, Drs. Grill and Asmus, represent the two opposite and extreme tendencies in Oriental research : 1st. That which would derive the religions of the Bible from the heathen systems. Dr. Grill attempts to trace the Mosaic history contained in the first chapters of Genesis to the Aryans of India and Persia. He forgets that the notions which he derives exclusively from these sources were far more ancient than either the Aryans or Semites, ethnologically speaking. They were common to all the cultured races of antiquity, and, as we have shown, derived originally from the primitive home of mankind. The Aryans on one hand, the Semites on the other, were equally indebted to ages long prior to them for those traditionary conceptions of which the author treats. Indeed, it seems to me that Dr. Grill's entire theory is shown to be erroneous by the fact developed in the present treatise. 2d. Dr. Asmus represents that tendency which virtually seeks to build up a contrast between the religions of the Bible as revealed, and the heathen systems as purely naturalistic. This standpoint, in my estimation, is just as erroneous as that of Dr. Grill. *The ground ideas of all the ancient religions were the same, and were a traditional inheritance from the original centre of humanity.* This proposition, so fully demonstrated in these pages, is sufficient to overthrow the two opposite theories of the authors named. The *naturalism* with which Dr. Asmus seeks to contrast the Biblical theism is in fact a decayed *symbolism*, in which the primeval doctrines were revealed. I do not propose here to return to the extremes in the use of the symbolic principle that characterized a former period in Oriental matters. On the other hand, I cannot adopt the other extreme, now so generally in vogue, which assumes practically that symbolic art never existed as an element of the ancient religions. The present volume contains ample evidence that symbolism was a vital principle in the primitive doctrines. Thus, generally, where Dr. Asmus sees nothing but pure naturalism, from the standpoint of the present researches I distinctly recognize the evidences of symbolic conceptions, whose original sense had been lost or perverted. Nevertheless, the author is to be excused, I think, for not being able to find the modern speculative doctrine of the

traditional element relating to the temple had come to be wholly
subordinate, and even mythical, and that the entire expanse of heaven and earth, as symbolized in the arrangement of such structures,
had been assumed as the exclusive standpoint. In such case a pantheistic view of God and the universe would be almost inevitably
the result. Stripped of its original historical and geographical
element, the conception became exclusively a philosophic dogma.
Heaven in general became assimilated to the male divinity, and the
earth in general to the female divinity. Everything tended towards
a pure *abstraction* speculatively speaking, while the original concrete personality became, in the popular conception and in the religious sense, broken up into numberless personified conceptions,
whose primitive identity was wholly forgotten.

The Hebrews preserved, or if not, Moses restored, the primeval
tradition, in its full symbolic import, respecting the artificial structure designed for the abode of divinity. The relation of the *Beth*-
absolute and unknowable in the ancient systems. For myself, I have not been
able to discover it in the religions of the Bible.

Yet more recent than the two treatises above referred to, and involving, although in a different field of research, the same fundamental error, according to
my judgment, is the work issued by Dr. Ignaz Goldziher, *Der Mythos bei den
Hebräern*, etc. Leipzig, 1876. The general tendency of the author is similar to
that of Dr. Grill's production, except that he does not attempt to derive the Hebrew religion from Aryan, but rather from Semitic mythology. Both Dr. Grill and
Dr. Goldziher deal largely in etymologies of personal names, etymologies in which
it is difficult usually to place much confidence. Certainly the two systems cannot
be correct, for substantially they disprove each other. As before intimated, both
proceeded upon the principle that the heaven father meant heaven in general,
instead of a particular and traditional heaven, etc. Another grave error in both
authors is the constant assumption that, if a personal name involves a mythological sense etymologically, then the personage so named must be regarded mythical,
and not historical. But this principle would consign nine tenths of the Assyrian
canon, as every cuneiform scholar knows, to the region of myth. The same of
the Chaldæan kings. Thus, the name *Lik-an-Bagas* means " light of the god
Bagas : " but was Lik-Bagas (or *Urukh*) therefore a myth ? Certainly not, for
we have many inscriptions bearing his name. Admit, then, with Dr. Goldziher,
that *Ab-ram*, " high father," really means " heaven father." This is no proof
that *Abram*, or *Abraham*, was a mythical personage. But when Dr. Goldziher
derives the name *Dan* from the Assyrian *Du-ni*, " to march, to go," we lose all
confidence in his etymologies even, and seek a firmer basis of conclusions (vid. p.
144. Cf. Norris, *Assyr. Dic.* i. 248). The syllable *Du* is Accadian, meaning " to
go," etc. A reading adopted into the Assyrian with the addition of *ni* as phonetic complement. It must be impossible, I think, to form out of such elements a
name *Dan*, " to go," etc.

Moad to the *Har-Moad* was fully recognized. The planetary regions and the zodiacal divisions were not excluded, but they were held subordinate. The planets and zodiacal constellations were supposed, in fact, to revolve around Mt. Meru as the common centre, just as we conceive the entire stellar world to turn on its axis, corresponding to the two poles. The Hebrew tabernacle, then, was the cosmos, the celestial earth, in its least form; and the twelve tribes of Israel, together with the territorial divisions allotted to them, being assimilated to the zodiacal constellations, completed the organization of the heavenly kingdom. In the end Mt. Sion simply replaced the traditional mount of paradise.

The blessings pronounced by Jacob at the time of his death upon his twelve sons (Gen. xlix.), together with the corresponding passages, have always appeared to me conclusive, to the effect that the intention was, in accordance with what appears to have been a prevalent custom, to assimilate each one to a particular zodiacal sign or constellation. Many of the older commentators, such as Dr. Clarke, who seems to have been guided very much by the views of Dr. Hale, did not hesitate to adopt this hypothesis; and it was certainly countenanced by some of the most respectable among the ancient authorities. But the construction put upon this theory by such writers as Dupuis of France, Nork of Germany, and others that might be named, have tended to make more recent critics a little cautious. M. Dupuis' statements of facts, so far as regarded the sources available at that time, appear to have been perfectly conscientious and usually correct, and he is being more and more cited by the best writers of the present day. But this, however, in no sense as adopting the author's constructions of the facts and the general hypothesis of his voluminous work. I think the author was justified in saying that the twelve tribes of Israel had for their ensigns the twelve constellations of the zodiac, but not in the conclusion that the origin of all religion was astrology. I wish to cite here the judicious remarks of Professor Bush upon Jacob's language to the effect: —

"That the peculiar phraseology in which the blessings are couched has, in most cases, a verbal allusion to the *nomes* bestowed upon the twelve phylarchs, or *princes of tribes*, at their birth — a circumstance not indeed obvious to the English reader, but palpable to one who consults the original" (Notes, Gen. xlix.).

It would be difficult to prove beyond doubt that the phylarchs, or phratries, princes of tribes, were associated in some way with the twelve zodiacal divisions, yet it seems to me quite probable. The Arabian tribes were thus associated, and it appears to have been an ancient Asiatic custom. The cuneiform *Dun-pa-uddu* is given as the name of Mercury, corresponding to the sign Aries, in a text already cited. In another text, the same name is put for the "prince of the men of Haran." an old city with which the Abrahamic race is especially connected in the account of Genesis. This is not conclusive as to the sons of Jacob, but it has a tendency in that direction. Really, the most direct evidence, and which seems to me almost positive, is the very language employed by Jacob, strikingly applicable in most instances to the ancient zodiacal constellations. Thus, Judah is "a lion's whelp"= Leo; "Simeon and Levi are brethren" = Gemini; "Dan shall be a serpent . . . that biteth the horse heels" = Scorpio, sometimes represented as a serpent biting the heels of Sagittarius. Such analogies carry with them the force of demonstration. The Hebrew camp, then, arranged in the form of a square, the sides facing the cardinal regions, and the tabernacle in the centre, was in every sense of the word an astronomical or a celestial earth ; and the temple on Mt. Sion, central point of the territory divided according to the number of the tribes, constituted a heavenly kingdom, whose primitive model was the traditional mount of paradise. Such was a perfected type of the politico-religious institutions of the old world.

SEC. 96. Professor Bush's allusion to the phylarchs calls to mind an investigation by Mr. Grote, quite critical and satisfactory, upon the primitive formation of Greek society, especially of Attica, from which the fraternal and religious character of these first organizations is plainly to be inferred. The phratries were sometimes identified with the tribes, at other times held as subordinate divisions. In practice the number varied, but theoretically appears to have been typical, twelve being the most frequent. In Attica we have four tribes, to each of which appertained three phratries, or phylarchies. Theoretically, all the members were regarded as descended from a common ancestral head, though in point of fact such was hardly ever the case. Each phratry had its "deme" or division of soil, like the tribes of Israel. The combination of the numbers four and twelve, like three gates to each side of a temple, was certainly not accidental, but had an obvious reference to some scheme founded on the zodiacal divisions. For the rest, I refer to Mr. Grote's own language as follows : —

" That every phratry contained an equal number of gentes, and every gens an equal number of families, is a supposition hardly admissible without better evidence than we possess. But apart from this questionable precision of numerical scale, the phratries and gentes themselves were real, ancient, and durable associations among the Athenian people, highly important to be understood. The basis of the whole was the house, hearth, or family, — a number of which, greater or less, composed the gens or genos. This gens was therefore a clan, sept, or enlarged and partly factitious brotherhood, bound together by common religious ceremonies," etc. " Each phratry was considered as belonging to one of the four tribes, and all the phratries of the same tribe enjoyed a certain periodical communion of sacred rites." "Such was the primitive religious and social union of the population of Attica in its gradually ascending scale, as distinguished from the political union, probably of later introduction." [1]

It is probable that the primitive social and religious organizations in the Nile valley, to which the Egyptian nomes appertained, were very similar to those of Attica, and it is safe to assume that all the ancient communities were at first of like character. In every case religious ideas seem to have exercised a controlling influence. But the two extracts, one from Mr. Grote, the other from Professor Bush, taken in connection, exhibit at once the vital relationship, the fundamental analogy, in fact, between the state as the greatest unit and the family as the least unit, in the ancient civilizations; between the temple as an image of the cosmos and the house as a human habitation; between the national divinity and altar and the divinity of the hearth. The reader will perceive at once, therefore, the connection of the present chapter with the second, which treats upon the divinity of the hearth. The ground thought, which unites the three notions of cosmos, temple, and house, is that of *dwelling*, to which appertains that of the paternal and filial relation, from whence the ideas of the "children of Israel," "children of the kingdom," the divine paternity, etc. We find here an explanation of the notion of cutting, of division, appertaining to the customs of the altar, of sacrifices, and the ratification of

---

[1] *History of Greece*, iii. pp. 54, 55.

covenants by sacrifices, as being fundamental in the doctrine of
the temple and of the cosmos, illustrated particularly in our fifth
chapter. The expression, "a new heaven and a new earth," is sometimes equivalent to "a new covenant" in scriptural usage. The
creation of one and the ratification of the other become strictly cognate conceptions; thus, the cosmos is a dwelling, the temple is a
house, and creation's altar is the hearth, where love first blossomed
and God first entered into covenant with humanity.

SEC. 97. The theory of the ancient civilizations was a divine
idea, and its origin and practical realization in the kingdoms of
antiquity, in the light in which we are now able to view the subject, may well excite our surprise and admiration. The two principal and component elements, as developed in our second chapter
and in the present one, are those actually constituting the primitive
stratum of conceptions upon which the religious, political, and
social institutions of the entire ancient world were founded, so far
at least as concerns all the cultured races. The Christian civilization, as embodied in the theory of its great Founder, is radically
different, and this difference proceeds mainly from the doctrine
that man is the real cosmos, the true temple. Religion, to be universal instead of national, must be released from the bonds of local
worship and the material temple. It cannot be confined to Mt.
Sion, nor to Mt. Gerizim. On the other hand, to lose sight of the
temple, and its prototype the cosmos, is to reject the traditional element and to convert religion into mere philosophy. The origin of
all religion, its historical phases and development through all ages,
must be preserved. These are the conditions of the problem of a
religion strictly universal, and of a civilization founded upon it.
The only solution of this problem under the circumstances is that
man, not in the individual sense exclusively, nor in the generic, but
in both senses, constitutes the cosmos, the temple, the abode of the
Divine Spirit. The church is then the celestial earth, the heavenly
kingdom, the temple, Christ being the corner-stone. This, however, was no after-thought, but had been anticipated from the beginning. As heretofore remarked, the Biblical psychology corresponds to the ancient cosmogony. Heaven answers to the *pneuma,*
or "spirit," the intermediate region or atmosphere to the *psyche,* or
"soul," and earth to the *soma,* or "organized body;" the *sarx,*
or "flesh," being nearly synonymous with "sin" in New Testa-

ment usage. The relation here between the macrocosm and the microcosm is exact, and the extreme antiquity and almost universal prevalence of the two co-related ideas are points impossible to be doubted. Nor was this doctrine at first merely speculative, which would be the same as to say pantheistic; it was strictly religious, for the reason that the fundamental conception was that of a dwelling, a habitation of divinity. As the cosmos is a dwelling of God, as the temple is such, so man as the real cosmos, the true temple, is a house of God, constructed of living stones, of which Christ is the chief corner-stone.

The world's history affords no example of a greater advance than that which presents itself here as between the ancient and the Christian civilizations, as the latter was conceived by its Founder. From the obstacles presented by the diversities and jealousies of different nationalities, a national and local religion could never become universal, unless the nationality itself became such, which was impossible. Besides this, the worship of the same Deity in different localities, and under various names, had been one of the principal causes of the growth of polytheism, from the inability of the human mind to retain the identity amid so many differences. It was thus necessary to destroy the ancient temples and to overthrow the civilizations centring in them. The advance, therefore, which here presents itself to view, was the complete redemption of religion from the bondage to material temples, to special localities, yet in a manner not to lose its essential characteristics derived from the temple. In the New Testament, everything that in the Old Testament centred in the temple is transferred to man, and this is the solution of the problem. Henceforth the language is, " Ye are the temple of God, and the Spirit of God dwelleth in you " (1 Cor. iii. 9). This is the new creation, the new heaven and earth, and thus, finally, do the two religions of the Bible flow from the first chapters of Genesis.

From whence came this sublime science, deeper and broader than all the philosophies? and by what extraordinary means had it been preserved in its integrity and purity, since those hoary ages to which we have traced its origin? In the midst of the dark periods of history, when old empires were being uprooted, and the light of former civilizations was going out, where did this ancient and sacred lore find an asylum? Who preserved the models that had been

brought from the primeval " mountain of the assembly " ?   Or who,
when they had been buried beneath the drifting sands of centuries,
recovered the lost treasures, restored them to new sanctuaries, the
centres of new civilizations?   I cannot answer these questions, nor
can they be answered, except it be permitted to reply: The priest-
kings of antiquity !

CHAPTER IX.

THE EARTH GODDESS.

SEC. 98. It was the opinion of many ancient mythologists, and it has been held by a large number of modern investigators, that all the gods and goddesses of antiquity ultimately resolve themselves into one original androgynous divinity; that is to say, a god and goddess essentially one, yet sufficiently distinct to be conjugally related. Another opinion extensively entertained both by ancient and modern authorities is that the male and female divinities thus conceived were assimilated to the two chief divisions of the cosmos, the male principle to heaven and the female to the earth. So far as my own investigations have tended to a definite conclusion on these two important points, the result has been decidedly in favor of the views just stated. The races whose pantheon exhibits the least appearance of having been derived from one androgynous personage are those whose antiquity is not nearly so great as that of others; while the most ancient nations who have preserved distinct traditions relative to their primitive worship furnish us with the best evidence tending to confirm the opinions to which we refer. As regards the ancient populations of Western Asia, the facts now known leave little room for doubt, and the same is to be said of the ancient Egyptians. Alluding to the Phœnician inscriptions, M. De Vogüe has the following observations: —

"It is the same with these texts as with the Egyptian inscriptions, which under the degenerated symbols of a gross polytheism have revealed the existence of veritable dogmas. The learned interpreters of these inscriptions have demonstrated, with the assistance of the formulas employed and the figured representations, that at the basis of the Egyptian religion, notwithstanding the contrary appearances, there existed the belief in one eternal God. Less personal than the divinity of the Bible, and above all less distinct from the created material, the Egyptian deity is nevertheless incorporeal,

invisible, without beginning or end; the innumerable divinities of
the Egyptian pantheon are only the personified attributes, the
deified potencies, of the incomprehensible and inaccessible being.
Cause and prototype of the visible world, he has a double essence;
he possesses and resumes the two principles of all terrestrial generations, the male and female principle; he is a duality in unity; a
conception which, in consequence of the duplication of the symbols,
has given birth to the series of female divinities. Such is the divinity revealed to us by Egyptologists. Less fortunate than M. de
Rougé and M. Mariette, we have at our disposition, in place of numberless pages covering the walls of the temples and the rolls of the
sacred rituals, only some rare and brief inscriptions; but these suffice to indicate the path to be followed, and to verify the numerous
and profound analogies that exist between Egypt and Phœnicia.
It has been already demonstrated that the worship of the Phœnician Baal implied the primitive belief in one God; the same also
as regards the worship of the Assyrian *Bel*, the Syrian *Hadad*,
the *Moloch* of the Ammonites, the *Marna* of the Philistines, etc.,
divinities whose very *names involve the notion* of unity and supre-
macy." [1]

Both M. Mariette and M. Maspero cite a passage from Iam-
blichus as expressing the actual truth, which reads thus:—

"The God of the Egyptians when he is considered as the hidden
force animating all things with light is called *Ammon;* when he is
the intelligent spirit who resumes all other intelligences he is
Emeth; when he is that which accomplishes all things with skill
and truth he is named *Phtah;* and finally, when he is the good
and beneficent being he is *Osiris.*" [2]

On the subject of the Babylonian pantheon Mr. George Smith
remarks:—

"At the head of the Babylonian mythology stands a deity who
was sometimes identified with the heavens, sometimes considered as
the ruler and god of heaven. This deity is named Anu, his sign is
the simple star (Accad. *An*), the symbol of divinity, and at other
times the Maltese cross (*Susru*). Anu represents abstract divinity,
and he appears as an original principle, perhaps the original principle of nature. He represents the universe as the upper and lower
regions, and when these were divided, the upper region or heaven
was called Anu, while the lower region or earth was called Anatu,
Anatu being the female principle or wife of Anu." [3]

---

[1] *Mélanges d'Archéologie*, pp. 50, 51.
[2] Mariette, *Musée à Boulaq*, p. 22; Maspero, *Hist. Anc.*, pp. 28, 29.
[3] *Chal. Acct. of Gen.*, p. 54.

Professor Max Müller has the following relative to the Chinese:

"In China, where there always has been a strong tendency towards order and regularity, some kind of system has been super-induced by the recognition of two powers, one active, the other passive, one male, the other female, which comprehend everything, and which, in the mind of the more enlightened, tower high above the great crowd of minor spirits. These two powers are within and beneath and behind everything that is double in nature, and they have frequently been identified with heaven and earth. We can clearly see, however, that the spirit of heaven occupied from the beginning a much higher position than the spirit of the earth. It is in the historical books only, in the Shu-king, that we are told that heaven and earth together are the father and mother of all things." [1]

SEC. 99. Professor Müller appears to doubt whether the two principles, male and female, assimilated to heaven and earth, were really primitive in the Chinese system. But this doctrine apper-tained to the cosmogony; and the recent investigations of Dr. Schlegel seem to me perfectly conclusive, that originally the Chi-nese system was strictly analogous to the doctrines held by other an-cient peoples. *Yang* and *Yin*, light and darkness, male and female, assimilated to heaven and earth, both proceeding from a primal an-drogynous principle, constituted the fundamental ideas which were truly archaic in China. I cite the following from among number-less testimonies, brought out by Dr. Schlegel's researches: —

"Before the two principles, or laws, were separated, their ether was a mixture like an egg, says the *Yi-king*. Thus, as they had assimilated *heaven* to the active and generative force of nature, or to *Yang*, and the *earth* to the passive force of nature which re-ceives and produces, or to *Yin*, they named these two forces *heaven* and *earth*, from whence was derived the supposition that at first heaven and earth were in a chaotic state." [2]

Human nature as the product of the interaction of both was sup-posed to involve completely the attributes of both, and was thus a microcosm. The original identity of this Chinese system with that of the Babylonians and Egyptians, and their derivation from some common source, are matters upon which, in the present state of knowledge, it is difficult to harbor serious doubts. Among the

[1] *Lect. on Sci. of Religion.* Vid. Littell's *Living Age*, Aug. 20, 1870, p. 486.
[2] *Uranographie Chinois*, p. 254.

more recent labors of M. Lenormant, devoted to the analysis of the
religious ideas prevalent in the valley of the Euphrates, is that relating to the primitive Accadian system, as *distinguished* from the
Chaldæo-Babylonian. This study, if my estimate is correct, is the
least satisfactory among the author's usually happy efforts and
admirable expositions of primitive doctrines, and needs a thorough
revision throughout. Among other points in relation to which the
author allows himself, as it appears to me, to be greatly misled, and
to which he attaches constantly an undue importance, is that indicated in the subjoined passage : —

" At the summit of the (divine) hierarchy, they (the Accadians)
admit, it is true, a certain number of divinities, as *An*, *Dingir*, or
Dimir. But their nature does not differ essentially from that of
the inferior spirits, the name given to these, *Zi* ('spirit'), being
the same as that applied to the gods. These (divinities) are beings
of the same essence, distinguished solely by a particular qualification ; their power is supposed to be greater than that of others." [1]

Thus, a fact which appears constantly in the books of the Old
and New Testaments, namely, the application of the same terms
(Heb. רוּחַ, Gr. πνεῦμα, "spirit") to the Deity, the Divine *Spirit*,
and to the created finite *spirit*, M. Lenormant interprets, in the
foregoing extract, as derogatory to the Accadian conception of the
Divine Being, as evidence of the want of systematic arrangement
and of a marked difference from the Chaldæo-Babylonian pantheon.
The grave and peculiar importance which he attaches to the simple
circumstance indicated is in my view wholly misleading in tendency. That, however, which more especially concerns our topic is
the following extract : —

" Two of the greatest of these gods, those who take rank before
all others, namely, *Anna* and *Hea*, have no higher titles than those
of 'spirit of heaven' (*Zi-anna*), and 'spirit of earth' (*Zi-kia*).
It is thus that they are addressed in the most solemn invocations,
and this characterizes purely their original and fundamental nature." [2]

These statements are eminently correct ; but the titles referred
to, which the author seems to interpret as of a low order, are actually those that tend to identify Anna and Hea, and the notions
thus attached to them, with the highest divinities and the most ele-

---

[1] *La Magie*, p. 139.　　　　　　[2] Ibid., pp. 139, 140.

vated order of conceptions among those nations the most ancient and renowned in antiquity. According to these titles, *Anna*, as "spirit of heaven," is one with the Babylonian *Anu*, the Chinese *Tien*, the Egyptian *Horus*, primitively put for heaven in opposition to *Set*, the earth. *Hea* also corresponds, as "spirit of earth," to his strictly Babylonian character as "lord of the earth" and of "mankind," and may be assimilated to the Egyptian *Tum*, who dwells alone on the "abyss of waters." In point of fact, Anna and Hea, both male divinities put in opposition, appertain to the second stage of development from the primal androgynous principle. The first stage is that in which the one proceeds to duality, male and female, heaven and earth, like Anu and Anatu. In the second stage the two principles proceed to duplicate themselves. Heaven as male takes a celestial goddess, and earth as female takes a male divinity, becoming the same as the Babylonian Hea and his wife *Nin-ki-gal*, who, in the second, replaces Anatu in the first stage. Anna and Hea, as male deities, assimilated to heaven and earth, appertain to the second stage. Their position is thus perfectly normal, and their qualifications very significant of the higher order of conceptions, common to the most ancient and elaborate systems.

SEC. 100. The subjoined extracts from M. De Vogüe exhibit in a very clear light the system, in its full development, whose fundamental conceptions were doubtless primitive and widely prevalent in antiquity : —

"In principle, two causes have presided at the formation of all things, the *father* and the *mother ;* the father is *light*, the mother is *darkness*. The subdivisions of light are *heat. dryness, rarity*, and *celerity ;* while the subdivisions of darkness are *cold, humidity, heaviness*, and *slowness*. To the first is assimilated the superior hemisphere, and to the second the inferior hemisphere. The four elements are divided according to the same order : fire and air appertain to the male principle, water and earth to the female ; but these four elements proceed in the formation of things according to the method of generation ; the two sexes reappear at each stage, from whence results this confusion (only apparent) that the air, male in relation to the two inferior elements, is female in relation to fire ; and that water, female in relation to the superior elements, is male in respect to the earth."

"In the mathematical order, the first principle is that of the monad and of odd or fortunate numbers ; while the second is that of the dyad and of even or unlucky numbers."

" In the moral order, life, justice, good, appertain to the first principle; to the second, death, injustice, and evil."

" In the theogony and the astronomical order, the sun appertains to the first principle, the moon to the second; the five planets range under the one or the other; the seven planets (including the sun and moon) contain the causes of all things, but these are subordinate to the influence of the superior world of fixed stars, or of the twelve signs of the zodiac. These twelve signs in their turn are apportioned between the two principles, accordingly as they are considered male or female; and the same is true of the thirty-six principal constellations, which preside, some over the celestial, and others over the subterranean world. This entire 'celestial army' is animated and active: each of those stars is a god or genius, and ranks in a divine hierarchy, at the summit of which appears the indeterminate notion of a supreme providence."

" The reciprocal action of all these things, their combinations and antagonisms, produce all the phenomena of the sensible world, since nature is composed of contraries, and 'harmony is born of the reaction of contraries.' We might almost add, from the *identity of contraries*, since this celebrated formula constitutes actually the basis of the entire system." [1]

The remarkable accuracy of the foregoing analysis of ancient doctrines, based upon a similar statement derived from a work of Origen, as the author remarks, " happily recovered to science by M. Miller" (not the author of the present treatise), is not only attested by the combined results of modern researches, but is admitted by European scholars generally who have devoted special attention to these subjects. In many of the details as here set forth, the system had submitted doubtless to additions and modifications; but the fundamental conceptions were truly archaic, and appertained to the primitive stratum of ideas constituting the oldest developed forms of religious belief of which we have any knowledge. That which here appertains to the superior and inferior hemispheres, or to the two chief divisions of the sphere, is otherwise predicated of the two principal divisions of the cosmos, heaven and earth, or the upper and lower regions, according to Mr. Smith, assimilated to Anu and the female principle Anatu. This equivalence of the two halves of the sphere to the two divisions of the cosmos is that upon which, in the previous chapters, we have so frequently insisted. In all the ancient systems, that which is astro-

---

[1] *Mélanges*, etc., pp. 57-59.

nomically the superior hemisphere is symbolically the heaven, and that which is astronomically the lower hemisphere is symbolically the earth. Consequently the goddess who presides over the inferior hemisphere is only another form of the earth goddess, and the same remark applies as well to the male divinity directly associated with this goddess, when, as in the case of Hea, such male personage is placed in opposition to another, as Anu, for instance, presiding over the superior hemisphere. Another principle whose importance is now well understood, and which is generally recognized among mythologists, may be thus formulated : the goddess who personifies the primal chaos in the cosmogony reappears in the theogony as great mother, and in the planetary system as Venus ; thus, Tiamat or Omarka is one with Bilat, and the latter is only another form of Ishtar, the Babylonian Venus. What we term the earth goddess, therefore, properly represents them all. M. De Vogüe observes in reference to the various forms under which this female divinity appears : —

" We are forced to recognize in these different personages the successive and local modifications of a single divinity, whose worship had spread through the entire eastern basin of the Mediterranean, and had ramifications even into the Indo-Persian world. This divinity is no other than the great goddess of nature, the great mother, designated by the vague title of the Oriental Venus, of whom Lucien affirms that she partakes alike of the character of Juno, of Minerva, of Venus, of the Moon, of Cybele, of Diana, of Nemesis, and of the Furies, thus adding testimony to the unity of the original conception." [1]

SEC. 101. The foregoing remarks will serve to convey a general idea as respects the character of the earth goddess, as well as in regard to the comparative position conceived to be occupied by her, in relation to the male divinities in the various systems of antiquity. We are prepared now to enter upon still further investigations, with a view to attain more precise and definite ideas respecting the various points already noticed. It is important to gain a better defined notion, if possible, respecting the relation of the goddess, so termed, to the male principle, or to the primal divinity. On this point the completely satisfactory results of M. De Vogüe's researches, as presented in the extracts cited in our second chapter,

[1] Ibid., p. 48.

should be here recalled. The Phœnician expressions, "face of
Baal" and "name of Baal," to which are compared the Biblical
phrases "face of Jehovah" and "name of Jehovah," since in the
original texts there is a direct analogy between the two series of
expressions, are regarded by M. De Vogüe as sufficiently determining the primitive conception which gave rise at a subsequent period
to the notion of a distinct goddess. But it is evident that originally the idea was held as in no sense inconsistent with a pure
theism, a proper monotheistic doctrine. There was not here an
absolute unity, nor were there two utterly distinct, separate person-
alities. It was rather the notion of an inherent, necessary relation
of two principles, like plus and minus, positive and negative. It
was the primal deity in his external manifestation; and as the process of creation according to all the cosmogonies, the Mosaic included, was regarded as in some sense a generation, the idea of a
conjugal relation was certainly very natural and almost inevitable.
That the sacred writers in their representation of the Divine Being
in his relation to the world, and especially to redeemed humanity,
ever intended wholly to exclude such a conception is to me more
than doubtful, and indeed I believe the contrary is susceptible of
the plainest demonstration. Both in the Old and New Testament
there are scores of passages distinctly involving this idea, and in
fact, one might almost say that they constitute the favorite mode
of setting forth the relation of divinity to humanity, especially to
the church. It would be, I think, a very singular principle of
exegesis that would construe this mass of scriptural phraseology as
merely a customary figure of speech. It was a form of ·speech, at
any rate, that gave color to a prevalent doctrine which had been
terribly perverted, and common prudence would dictate the employment of some other form of hyperbole, if this was all that was
intended. There was, however, a great difference between the Biblical standpoint and that of surrounding peoples, which will be
brought out in the sequel of these researches.

The goddess of the ancient religions, then, was not wholly an
invention, a simple product of the degeneracy of man. It was a
most lamentable misconception, a perversion of a sacred and elevated doctrine, which, owing to the gross passions of mankind, was
sure to be corrupted, at the same time that it was necessary to be
revealed. It is probable that the earth goddess was vaguely and

popularly assimilated to the earth in general, or to the entire earth. But there is much reason to believe that such a vague association was not primitive, nor wholly legitimate. The idea was rather, I think, that of a divided, a cultivated, an inhabited earth. The wild, uninhabited regions were more regarded as the abode of demoniacal powers, the imps of darkness, than as the domain of that goddess whose character better comported with the customs and conceptions of an industrial, civilized life, than with a state of savagism. In a political and religious sense, however, the earth in its uncultivated state, and peopled by uncivilized hordes, was precisely that primal chaos of boiling, surging waters, personified by the goddess Omarka, whom Bel, the organizer, divided with his sword when he set the world in order. This was the battle between Bel and the dragon Tiamat, who represented the chaotic abyss. But as M. Lenormant has correctly observed, that which in the cosmogony and in the process of creation appears as a battle reappears in the theogony under the form of the conjugal relation, and it is this stage to which the earth goddess pertains in the more complete development. She represents, then, a divided, a cultivated, an inhabited earth, the "celestial earth," in fact, in the sense which we have learned to attach to this phrase. The notion thus conceived and defined is precisely that involved by the Biblical writers in the phraseology just referred to, which, among the Phœnicians and other peoples, had given rise to the conception of a distinct goddess. The celestial earth, as we have seen, was a politico-religious organization among men, which really constituted the union of church and state, supposed to be under the divine sanction; it was, so to speak, a heavenly kingdom transferred to earth, conceived as the organization of divinity in humanity. But otherwise conceived this was the earth goddess, and the Accadian phrase *An-ki*, "a divine, a heavenly earth," in which sense alone it would be often necessary to interpret it, was otherwise put for the goddess earth, or the earth goddess. But we saw in the last chapter that a traditional and even geographical element was primitively inherent in the conception of the celestial earth, and that the subsequent divorce of the notion from this element resulted in a pantheistic view of God and the world. Exactly the same thing occurred in relation to the earth goddess, whose character in later periods was identified with this pantheistic philosophy. These are points to which our

attention is to be now directed. The aim will be to establish an
original traditional and geographical element in the notion of the
earth goddess.

SEC. 102. The entire evidence heretofore presented, tending to
show that the particular heaven and earth known to the first men
was really *the* heaven and earth to which all the ancient cosmogonies primarily related, has a similar bearing also upon the question now before us. It goes far to establish the conclusion that
the primitive goddess was none other than this primitive earth,
placed in immediate relation to the celestial space to which the original male divinity had been assimilated. The Vedic Ida, originally put for the "earth," was especially associated with the paradisiacal mount in Hindu tradition, from whence it might be inferred
that Ida as earth goddess was identified with Mt. Meru, to which
likewise the notion of "celestial earth" was attached. The fact that
the pyramidal structures of Babylon and Borsippa were regarded as
temples of *An-ki*, and were designed as imitations of the traditional
mount, affords another confirmation of our hypothesis, since the
phrase *An-ki* is equally understood as the earth goddess.

But M. Lenormant has developed a class of facts pertaining to
Is-tar, the Babylonian Venus, in his critical studies of the Himyaric
inscriptions, that have a direct bearing upon the particular topic in
hand, and which appear to be quite conclusive. The results arrived
at may be thus stated : —

1st. The god *Ath-tor*, so frequently called in the Himyaric inscriptions, is a masculine, or more correctly an androgynous form
of the Oriental Venus, like the Venus *barbata* of Cyprus. The
Canaanites gave a feminine termination to this divine name, from
whence the well-known *Ashtoreth*, Greek *Astarte*. But in the Assyrian *Ishtar*, which is the more ancient form of the name, the feminine ending does not appear. In the Himyaric *Ath-tar*, the *s* is
changed to *t*, to which corresponds the cuneiform *Atar-samain*,
"Atar of the heavens," proving the equivalence of *Ath-tar*, *As-tar*,
and *Atar*, to the Assyrian *Ish-tar*. The original sense of *As-tar* is
"the goddess," and it was only at a later period, when polytheism
arose, that the term came to denote one among many goddesses.

2d. *Ishtar* is proved by the cuneiform texts to be the same as *Assat*,
"the mistress," and this last is assimilated by M. Lenormant to the
Hebrew *Ishah* (אִשָּׁה), "woman," the feminine form of *Ish* (אִישׁ),

" man," the two being generic terms for " man " and " woman."
The form *Ishah* is the name first bestowed upon Eve by Adam
(Gen. ii. 23).[1]

The opinion was strenuously maintained by some of the older
mythologists, though not much importance has been attached to it
of late, that really the goddess primitive in the ancient religions
was no other than the Biblical Eve. In the sense that all the
mythologies were founded upon the Mosaic history, this opinion
was very erroneous. The great races diverging from the original
centre of human populations, each independently of the others,
naturally preserved some recollections of the primitive epoch.
Among these races, the Babylonians preserved the most distinct
recollections, perhaps, of any ; though there exists a remarkable
agreement between all the ancient traditions, even in those particulars and details where we should least expect to find it. But to
return to the name of *Ishtar*, as equivalent to the Hebrew *Ishah*, as
generic title of the " woman." Professor Bush has the following
remarks upon this name bestowed upon the Biblical Eve : " The
original word for ' woman ' is *Isha*, the feminine of *Ish*, ' man,' and
properly signifies, however uncouth the sound to our ears, *man-ness* "
(Notes, Gen. ii. 23). Compare with this Rev. Mr. Sayce's rendering of the cuneiform passage : " Venus is a *female* at sunset ;
Venus is a *male* at sunrise," to which the translator adds the note :
" The Assyrian word here is very remarkable, *Zi-ca-rat*, as if we
could coin a word like *male-ess*." [2] The identity in import between
the cuneiform and the Hebrew expressions, being generic titles of
womankind, their application to *Ishtar* as " the goddess" on one
hand, and to Eve as " the woman " on the other, must be regarded
as quite conclusive evidence that the really primitive goddess was
strictly a traditional character, and that this tradition was derived
from the mount of paradise, associated especially with the particular
earth known to the first men, and with the female personage styled
the " mother of all living."

SEC. 103. At the time of bestowing upon Eve the name of
Ishah, " woman," Adam addressed to her the following language :
" Therefore shall a man leave his father and his mother, and shall

---

[1] Vid. *Lettres Assyriologique*, t. ii. pp. 54–59, for M. Lenormant's investigations in detail on the points noticed in the text.

[2] *Jour. Bib. Arch. So.*, London, iii. pp. 196, 197, and note.

cleave unto his wife ; and they shall be one flesh." As already once
observed, this exact phraseology is interpreted of the church by the
apostle to the Gentiles, thus : " For this cause shall a man leave his
father and mother, and shall be joined unto his wife, and they two
shall be one flesh. This is a great mystery ; but I speak concerning Christ and the church " (Eph. v. 31, 32). The context further develops the same doctrine. We thus connect in the most
direct and reliable manner the celestial earth, the earth goddess,
and the church, tracing them all to the terrestrial paradise for their
origin. We assume, consequently, a traditional and geographical
element as actually the basis of the original conception of both the
celestial earth and the earth goddess. The subsequent divorce of
this element equally from both, as will be now apparent, gave rise
to the pantheistic view of God and nature with which both were
associated. The effect was to convert religious doctrines, sacred
traditions, into mere speculative principles, and to interpret these of
the entire heavens and earth, of the cosmos in the modern sense,
instead of its primitive and traditional sense.

We are prepared now to enter still more into the details of the
conception embodied in the celestial earth assimilated to the earth
goddess. The assimilation of the cosmos to the temple, the fundamental conception of both proceeding from that of *division*, was
sufficiently established in the fifth chapter of the present treatise.
Now the two typical instruments of division in antiquity were the
sword and the *plough*, to both of which, various symbolical ideas
were attached relating to " foundations," whether of the cosmos,
the temple, or the terrestrial kingdom, modeled after the order and
divisions of the sphere. Thus, we have seen in the " Fragments of
Berosus," on one hand, that Belus, when he commenced to set the
world in order, and as the very first act of creation, cut the woman
asunder who personified chaos with his " sword," forming the heavens and earth out of the two portions thus divided. Again, the sword
was the symbol of power, of that despotic force, in fact, by which
alone the rude and chaotic elements of human society were brought
into subjection, the principle of order introduced, and the foundations of the ancient kingdoms laid. On the other hand, we have
seen the " plough " introduced, to which a bull and heifer had been
yoked in the ceremonies of founding towns and cities, of dividing
off the plot to be occupied as the site ; in the partition of the soil

to be allotted to cultivators, also ; the divisions in all cases being located according to the method of forming the temple, and with especial reference to the cardinal regions, or the primary divisions of the cosmos. The plough was not less a symbol of a divided earth in general, a cultivated field, an inhabited district, and of a peaceful, industrial life. It was by virtue of these divisions, often marked by great highways leading off from the national temple to the extreme boundaries of the state, that the entire kingdom was conceived as a great temple ; and it was by virtue of their location in the direction of the cardinal regions that the state was deemed a cosmos, a celestial earth, a heavenly kingdom.

But the primitive and vital connection of the various conceptions here referred to will be better understood by a brief analysis of the sense attached to certain terms in common use among the ancients. The Hebrew *Bara* (ברא), denoting the first act of creation in the Mosaic account, employed also in speaking of the *creation* of a new heaven and earth, has the primary sense of "to cut, to carve, to form by cutting and carving." Dr. Gesenius observes substantially, in reference to this term, that "the notion of breaking, cutting, separating, is inherent in the radical syllable *par* (פר), as well as in the softened form *bar* (בר)." From these two forms proceed various notions and terms, among which I note the following : —

1st. The Hebrew verb *Pa-rar* (פרר), "to cut in, to plough;" in one of its modifications, "to split, divide, or plough the sea;" and the same verb affords the substantive form *Par* (פר), "a young bullock, a heifer," and from this last comes another substantive *Pa-rah* (פרה), "a young cow, a cow with calf, or a cow for the yoke." The Assyrian language has the form *Par* or *Par-ri*, "bullocks." We see here that the plough constitutes the special instrument for cutting, dividing, and it is worthy of note in this connection that the Hebrew *Kha-rash* (חרש), "to cut in, engrave, thence to form, make, create," etc., has likewise the sense of "to plough;" showing that the two notions of "fabricating, creating," and of "ploughing," proceeding from the one idea of cutting or of division, were habitually associated.

2d. Under the radical syllable *Bar* we may note the Accadian *Bar*, "a sword," otherwise put for "yoke, pair, double, half," etc. In the Hebrew we have not only *Bara*, and its cognate terms *Barah*, *Bereeth*, heretofore explained in these pages, but *Ba-raq*

(בָּרָק), "lightning, glittering sword." Then we have *Bar* (בַּר), "a son," from *Bara*, "to beget," and *Bar*, "a field, arable land," thence "corn, the fruit of the field." Here we have again the notions involved of plough, plough ox, or of taurus and the heifer, employed in agricultural labors.

SEC. 104. We come now to illustrate the existence of a similar class of ideas connected with the heavens. The sun was very frequently, as is well known, assimilated to the bull or taurus, and it is a remarkable fact that this celestial taurus was conceived to labor, to plough in the heavenly fields, the same as his earthly representative in terrestrial fields. The cuneiform texts afford direct proof of the prevalence of such notions. As shown by M. Lenormant, and as well known to Assyriologues, the generic name for the planets in the inscriptions is the Accadian *Lubat*, Assyrian *Bibbu*, both denoting primarily some animal whose species is not yet determined, though Dr. Delitzsch conjectures with considerable probability that the ram or he-goat is intended.[1] In any event, the application of these generic terms to designate the planets proves that at an extremely remote epoch they were likened to animals, wandering and grazing in the celestial pastures. But the Accadian name of the planet Jupiter is *Lubat-guttav*, where the element *guttav*, or *gut-tam*, needs explanation. This term, however, is well understood by its equation in the texts to the Assyrian phrase *pitnu sa same*, interpreted by M. Lenormant as "the furrow of heaven," that is to say, "the ecliptic."[2] Jupiter is thus the planet whose orbit agrees nearly with the sun's annual course, or the animal that follows in the celestial furrow. If, then, the royal planet follows this furrow, what animal ploughs it? The analogy of ideas, as well as the etymology of the Accadian term *Gut-tam*, will suggest the answer. The element *Gut* is the Accadian name of the bull or taurus (Rep. 246), while *Tam*, "day," is only another value of the sign *Ud*, "the sun" (Rep. 424). The phrase *Gut-tam*, or *Guttav*, then, literally means the solar bull, and the furrow followed by the planet *Lubat-guttav* is that ploughed by the celestial taurus. The existence of a profound symbolism in the astro-mythology of the ancients can be no longer doubted in view of these most singular conceptions.

---

[1] Lenormant, *Frag. de Bérose*, p. 371, note. Delitzsch, *Assyr. Studien*, pp. 47, 48.
[2] *Op. cit.* p. 373, note. Cf. 2d Rawl. Pl. 26, l. 26.

The sun as being the taurus that ploughs the celestial field was yoked to the moon as cow or heifer, and we know that the moon was usually represented by a cow or heifer in the symbolism of the ancient mythologies. From all antiquity, the sun and moon were taken as the chief dividers of time into seasons and months. The solar year and lunar year were equally ancient, perhaps, and both orbs were naturally taken as dividers of time. But astronomically, to divide the year into four seasons is to divide the celestial space into four principal regions, and a partition of the year into twelve months is to effect a partition of the celestial space into twelve regions, corresponding to the zodiacal constellations, or the divisions of the ecliptic. There was thus an inherent, necessary relation between the divisions of time and those of space. We have direct proof, derived from the cuneiform texts, that the principle of sex-ualism was applied even in reference to time and space, length and breadth. Thus, the Accadian *Us*, " phallus, male, length, time, a period of sixty years, or losse " (Rep. 161), constitutes a determi-native for length, while the Accadian *Luku*, " mother, breadth " (Rep. 148), constitutes a determinative of breadth, associated neces-sarily with the idea of space.[1] There was a perfect consistency in these notions with fundamental ideas. The heaven was assumed as male, the earth as female, the one as father, the other as mother. The divisions of time depended on the motions of the heavenly bodies, naturally conducting to the assimilation of time to the male principle, and by contrast the association of space with the female principle would be equally natural.

We have seen in the extracts cited from M. De Vogüe that the sun appertained to the principle of light, to the superior hemi-sphere, to heaven ; the moon, on the other hand, was connected with the principle of darkness, the inferior hemisphere, symbolically taken for the earth. The taurus, according to this system of ideas, was symbol of the heavens, especially of the sun, and the heifer or cow of the earth, particularly the moon as a celestial earth or ter-restrial heaven, offering thus an explanation of the facts to which Dr. Faber alludes, cited in a previous chapter, relating to the vision

---

[1] Vid. Norris, i. p. 74. Two sets of terms were employed denoting length and breadth, *buda* and *sakki* on one hand, and *us* and *luku*, "male" and "female," on the other. Usually the sexual idea was disguised by mixing the terms, as *us* and *sakki*, and *buda* and *luku*. But this was not always done.

of Timarchus. Primarily we have here the same circle of ideas symbolized by Taurus as involved in the conception of the god Anu among the Babylonians. At first the taurus represents the upper and lower regions, the same as Anu. Secondly, when these regions are divided, the taurus is put for the upper regions or heaven, the same as Anu, and the heifer or cow for the lower regions or earth, corresponding to Anatu, wife of Anu. M. Léon Carré has the following relative to the cosmical doctrines of the Persians: —

"The first of all the animals was the Taurus, which, after having existed for a long time alone, was killed by Ahriman (the principle of darkness and evil) ; but his soul became the principle of all animated nature, under the name of *Goschoroun ;* and from his purified seed were born two *Tauri,* one male and the other female, which produced the entire animal species. Finally, the first man was *Kaiomorts,* issuing from the sides of the Taurus under the form of a youth."[1]

The next stage of development would be the assumption of a celestial taurus and cow, the sun and moon, ploughing the celestial fields on one hand, these being represented by the terrestrial bull and cow, laboring in terrestrial fields on the other hand. The furrows mark the divisions of time and of the celestial space in one case, and those of terrestrial space, located with special reference to the former, in the other instance.

SEC. 105. It will not be difficult to apprehend the connection of these ideas with the general theory of the celestial earth, and with the character of the earth goddess. The two Accadian characters employed phonetically in writing the name *Is-tar,* Assyrian *Ishtar,* appear to have been selected as well in reference to their ideographic sense ; and as such, they involve the exact fundamental conception, according to my view, of the earth goddess, considered as "a divided, a cultivated earth." Thus, the Accadian *Is* means "earth" (Rep. 215) ; and *Tar* has among others the sense of "to cut, to separate, to divide" (Rep. 5), being similar to that involved in the Semitic radicals *Bar* and *Par* already considered. The literal Accadian meaning of *Is-tar,* the Babylonian title of Venus, would be, therefore, "a divided, a ploughed, a cultivated earth," and I believe that such was the absolutely primitive idea involved in the character of the earth goddess.

[1] *Ancien Orient,* etc., t. ii. p. 388.

It is necessary to recall here the facts verified in a previous chapter (Secs. 55, 56) relative to the augurial temple, the origin of the system of land measuring derived from it, as well as the singular custom of yoking a bull with a heifer, attached to a plough, in marking out the divisions of the soil and the sites of towns and cities, all directly associated with the ceremonies of founding cities and even states. The furrows thus ploughed in the earth were located with special reference to those ploughed by the celestial taurus and cow in the heavenly fields. The terrestrial city and kingdom were thus modeled after the celestial city and kingdom, peopled by the heavenly hosts, who were frequently compared to the inhabitants of the earthly kingdom. To complete the conception of a kingdom of heaven on earth, the king and queen were assimilated as human rulers to the sun and moon as celestial rulers; while the nobles and principal officers of state were associated with the principal constellations. In other cases, the zodiacal constellations, especially, were assimilated to the twelve tribes, to the twelve cities forming the state, or again to the twelve portions into which the public domain was partitioned. These notions were all fundamental in the theory of the ancient civilizations; and we have shown how widely prevalent they were in antiquity. Each kingdom, according to these conceptions, was a cosmos, a temple, a celestial earth, a goddess, and finally, in other terms, a church. The fundamental idea was that of division, the sword and the plough constituting the typical instruments employed. To the male personage, the organizer, the subduer of the primal chaos, appertained the sword. To the female personage, the goddess, appertained the plough, the harvest, the products of agricultural industry. Such especially was the character of Ceres. We see in the doctrines shown to have been fundamental in the general idea of a celestial earth, as well as in the character of the earth goddess, the ground of a multitude of comparisons employed by the sacred writers. The Saviour, even, frequently compares the kingdom of heaven to a cultivated field; and such comparisons were by no means accidental; they proceeded from fundamental notions, associated as well with the temple, the world or cosmos, as with a divided, cultivated earth. All such illustrations are the legitimate expansion of certain grand ideas whose origin in each instance we have traced to the traditional abode of primeval humanity. We see here the origin and import,

also, of that class of Biblical expressions in depicting the rise and
fall of empires, upon which, as previously cited, Bishop Lowth has
expressed some very correct views, yet without fully apprehending
the real significance of such language. The double conception involved in the Hebrew *Olam* and Greek *Aion*, finds a proper explanation likewise in the facts now before us. The sense of "world"
relating to the cosmos, on one hand, and that of "age, dispensation," relating to the renewal of things, to new creations, new heavens and earth, on the other hand, flow naturally from the principles
set forth in these pages relative to the cosmogony, the temple, the
celestial earth, the theory of the ancient civilizations. The foundation of the world and the beginning of time were the types of all
other foundations and cyclical periods, and it was according to the
strict analogy of ideas that the custom prevailed of identifying the
commencement of chronological eras with the founding of the state,
or of the temple conceived as the centre of the state.

SEC. 106. With the present chapter closes the first natural
and principal division of this entire treatise. Although the researches embodied in the succeeding chapters are strictly related to
those that have gone before, and are necessary to complete the view
of antiquity which it has been our aim to set forth, they appertain
to a class of ideas somewhat different, constituting, in fact, a separate division of our work. It seems proper, therefore, to introduce
here a brief review of our past labors, that we may better realize
and comprehend the standpoint to which these investigations have
conducted.

The founders of the Babylonian civilization, when they migrated
from the east to the plains of Shinar, were not the sole inheritors of
a sacred tradition and science which had been derived from that
primitive epoch and civilization of which the world to-day possesses but very little knowledge. They who had traveled eastward from the original and common centre of populations, settling
in the region now known as China, who had founded in fact the
celestial empire at a period whose chronology is almost unknown,
were not less the inheritors of a tradition and science which formed
the basis of primitive doctrines and of politico-religious institutions
in many respects strikingly similar to those that constituted the original stratum of the Hamite formations of Western Asia. The first
Hamite populations located in the valley of the Nile were by no

means ignorant of the same doctrines, the same theories, pertaining alike to the cosmos, the temple, the state, to the organization of terrestrial kingdoms after the model of the celestial. The Aryans of India and Persia, as well as those of Greece and Rome, who seem to have departed from the common home of all the races at a later epoch, who had each for themselves established independent types of civilization, although much indebted to those races who had preceded them in the path of human development and progress, seem to have preserved distinct recollections of the traditional abode from which all had departed, and of the ideas primitively associated with it. The same is to be said of the Semitic races. Thus, from the most widely separated nationalities of the old world we have gathered the proofs of the existence of primeval doctrines, theories of a cosmical, religious, political, and even social character, so similar in detail that the hypothesis of their common origin in some region that had been historically and geographically the centre of all these peoples seems to be completely established.

There was nothing crude, nothing that indicates a low order of development, in those ideas so full of symbolic import, according to which the primitive foundations were laid. Nothing more elevated and sublime, nothing more broad and comprehensive, has ever been put forth by any mind than those doctrines concerning the world assimilated to the temple, and the temple to man. The expansion of the idea of the family into that of the nation, of the hearth into that of the national altar, on one side; the construction of the ancient kingdoms after the model of the heavenly, as embodied in the notion of the celestial earth, on the other side, and the assimilation of all equally to the cosmos, to the cultivated field, and to the simple apartments of the private dwelling, — the one ground thought being everywhere that of *division*, — sufficiently evince the profound insight, the sublimity of genius, that presided over the beginnings of all human knowledge and development. It might well be said that all history, all historical progress, was born of the primitive cosmogony. It was from this as the original germ that everything in antiquity seemed to have proceeded. It was like the birth of the world from some mystic cave, and indeed it was such, — that cave in which the northern bear nestles her young by day, and leads them forth on the heights of the sacred mount by night.

But it is difficult to account for the origin of those primitive doc-

trines which all the **religions,** philosophies, and mysteries even had
taken for their base; to account also for their propagation in every
quarter of the ancient world, as well as their preservation in such
purity and detail, except on the hypothesis of our third chapter.
They who had constructed the first temples were the founders of
the ancient civilizations.   It was they who laid the corner-stones of
the pyramids of Borsippa and of Sakkara; who divided the sphere,
who invented writing, who wrote the sacred books, who taught
the ancient nations how to live and how to work.   It was these
mystic corporations of the early ages to which all antiquity attributed the first revelations of truth, the invention of the arts of civilized life, the foundation of beneficent institutions, the redemption
of humanity from savagism.   Everything tends to confirm this supposition, and I am constrained to believe it, even in spite of myself.

SEC. 107. In discussing the question of the actual geographical
locality of the *Gan-Eden* of Genesis, in our sixth chapter, much
stress was laid upon the necessity of adhering strictly to the essential conditions of the problem.   These conditions are substantially
as already defined : 1st. To discover a region on the earth's surface
whose peculiar geographical features correspond, or nearly so, to
those described in the Mosaic account of the Garden of Eden.   2d.
A region to which the traditions of various peoples point as having
been the first abode of humanity, and as that also from which their
ancestors had departed when they migrated to the countries occupied by these peoples in historical times. 3d. A region from which
it was physically and morally possible that the different races had
departed, according to all known facts concerning their origin, and
to which, by the aid of their traditions, sacred books, and the results of linguistic science, it was possible to retrace their steps along
the routes originally taken by them.   To these conditions I have
pointed out an additional one: 4th. The existence of a well-defined
stellar region, uniformly associated in tradition with the particular
terrestrial locality, assumed as the first abode of humanity.   Now
it may be properly and safely assumed that no two regions on the
earth's surface fulfilling all these requirements could by any reasonable possibility exist.   Consequently if one locality can be found
that does fulfill them, such must have been the primitive centre
from which the various races of men departed to people the earth.
As thus presented, the problem assumes a strictly scientific form,

and its solution, therefore, if properly worked out, must be regarded in the same light. The peculiar geographical features of *Gan-Eden* are themselves of a character quite extraordinary, and no other region on the globe known to geography in any way answers to them except the plateau of Pamir, which has been identified by scholars, as we have seen, with the Mosaic description of Eden. This region seems to form the great water-shed of all Asia, from which four large and navigable rivers take their rise, flowing from thence toward different quarters of the Asiatic continent. As for the other conditions named the reader has seen, from the facts detailed in the sixth chapter, a most remarkable concurrence of all the circumstances to satisfy them all.

We attain, then, a reliable standpoint historically, or at least traditionally and geographically, at the very dawn of human existence on earth so far as the earliest traditions known to antiquity afford any indications. If there actually was a period still earlier, no tradition relates to it. The evidences of its reality must be drawn from other sources. The question of the geological evidences of the antiquity of man, we are not as yet prepared to discuss, nor do we intend to do so, exhaustively at any rate, in the present treatise. We hope in future chapters to prepare the way for it, and to indicate generally our views. We assume, for the present, that *Gan-Eden* was the beginning of history; the proofs of this, drawn from ancient and universal tradition, as already before us, are entitled to great weight, but many additional facts are yet to be presented. But the locality in question was not merely the traditional and geographical starting-point of humanity. We have shown that it was the birthplace of the ancient religions, the sacred mysteries, and that the fundamental ideas forming the basis of the theory of ancient civilization had been inherited originally from the same locality. It was there on that mountain plain which overlooks all Asia that the physical, intellectual, and religious history of our race had its beginning. The first to rise above the dark waters of chaos and the deluge, it was there that the first civilization on the globe was planted. We know that its foundations were laid in wisdom, since it gave birth to those grand ideas whose influence extended throughout antiquity to the very latest epochs. How long it endured it is impossible to decide. It must have preceded the oldest civilizations known to us, and it is probable that it did

not endure much later than the period of their origin. It was at
all events the great foster-mother of all the brilliant epochs that
succeeded it. It was there that the first foundations were laid, the
first heaven and earth separated; and it was there that those models were perfected according to which the constructions of subsequent ages were scrupulously designed.

To be able to assign some definite and completely reliable chronology for this most mysterious epoch in human history would be a
result so exceedingly desirable and important as to justify even a
hazardous attempt to do so; and we confess in advance that such
will be one of the leading aims in the chapters to follow.

BOOK IV.

CHAPTER X.

ZODIACAL ARRANGEMENT OF THESE ASTERISMS.

SEC. 108. In the present chapter we enter upon a series of investigations of which a remarkable astronomical tablet constitutes the chief object of interest. Under the form in which it appears in the "British Museum Series," so called, this inscription has been for many years familiar to cuneiform scholars. I am not aware, however, that it has ever been regarded as of particular importance by Assyriologues, or that any one has devoted to it a special study. The greatest significance ever attached to it, so far as my information extends, is that by M. F. Lenormant, who cites it in illustration of the symbolical system of geography which prevailed in the times of Sargon the ancient, to which we have alluded in a previous chapter. In reference to this monarch, and as heretofore quoted, M. Lenormant observes : —

"He considers the country of *Akkad*, or Chaldæa, as situated in the centre of the universe, and as surrounded by four other countries, which correspond exactly to the four cardinal points : *Ilama* is east, *Martu* is west, *Gutium* is north, and *Subarti* is south. See particularly upon this system the fragment of a tablet, where twelve stars preside over the destinies of each of these countries, and where are described the influences exerted upon them during each month." [1]

The tablet here referred to is the one selected by us for special study. It is, as M. Lenormant intimates, quite fragmentary, cer-

[1] *Frag. de Bérose*, p. 321.

tain portions only being entire, particularly the one shown in our third plate, to which our chief interest attaches, and which we entitle "the twelve stars of Phœnicia," following the colophon at the foot of the list. For convenience I repeat the list below according to the Accadian values of the cuneiform characters interlined in the plate, adding some translations:—

1. The star *As-kar*.	7. The star *Su-qi*. The Rival.
2. The star *Sir*. The Serpent.	8. The star *Qaq-sidi*. Dog-star.
3. The star *Bar-tabba-galgal*.	9. The star *Bir*.
4. The star *Nin-Makh*. Great Mistress.	10. The star *Ungal*. The king. Jupiter.
5. The star *Nibe-anu*. Planet Mars.	11. The star *Al-lap*. The Taurus.
6. The star *Nam-makh*. Great Destiny.	12. The star *Lab-u*.
The twelve stars of	the country of Phœnicia.

The asterisms are numbered according to the order in which I suppose they were intended to be read, though Nos. 11 and 12 I conceive to have been purposely inverted, as this is not their proper zodiacal order. I had formerly regarded the true title of this list of asterisms to be "the twelve stars of Accad," being misled by Mr. Norris in his "Assyrian Dictionary." [1] He was doubtless misled by the colophon, which appears next above our list in the published text, and reads, *12 mul-mes mat Akkad-ki*, or "the twelve stars of Accad." He probably inferred likewise that the colophon, *12 mul-mes mat martu-ki*, or "the twelve stars of the west," of "Phœnicia," related to a series of asterisms named immediately below it in the text. This is an error, as is evident from Rev. A. H. Sayce's rendering of the list.[2] This author very properly applies to the series given in our plate the title, "the twelve stars of the west."[3] But the reference cannot be to the western quarter of the heavens, as he seems to suppose.[4] The phrase *mat martu-ki*, like *mat Akkad-ki*, can be interpreted only of a country. It is often put for the west country indefinitely; sometimes, as Dr. Schrader thinks, for Canaan.[5] Definitely, however, *mat martu-ki* is equated in the texts to *mat a-har-ri-i*, or "Phœnicia," as understood by Dr. Delitzsch and cuneiform scholars generally.[6] A very good reason for the supposition that in this case the phrase *mat martu-ki* does not

---

[1] *Assyr. Dic.*, i. p. 109. [2] Vid. *Jour. Bib. Arch. So.*. London, iii. p. 173.
[3] Ibid., p. 176. [4] Ibid., p. 167.
[5] *Keilinschrift. u. d. Alt. Test.*, pp. 14, 15.
[6] Vid. 2d Rawl. Pl. 50, Revs. l. 57. c. d. Cf. Delitzsch, *Assyr. Studien*, pp. 38, 139; Norris, *Assyr. Dic.*, i. 28.

PLATE III.

THE TWELVE STARS OF PHŒNICIA.

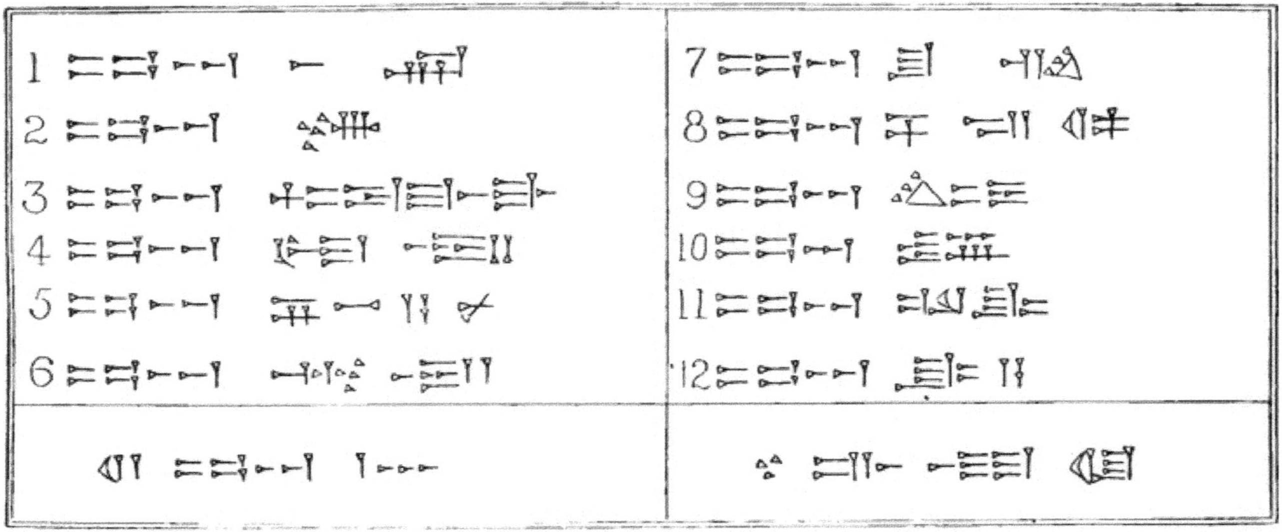

apply to the Canaanitish country generally is the fact that these
asterisms have in some sense an adjustment to the zodiac, leading to
the inference that a particular nationality, like the Phœnicians, had
appropriated them specially as such. The selection of twelve stars,
corresponding to the number of the zodiacal divisions, as presiding
over the destinies of the nation, seems to have been a common practice in those ancient times, and the various lists of asterisms originally contained in our tablet appear to have served a purpose of this
kind. As the population of Canaan was divided into many nationalities, some particular one in the present instance must be intended,
and thus *mat martu* should be understood here of Phœnicia. The
entire list of the twelve stars of Accad is wanting in this inscription, except the two names partially defaced in the first line of
the published text, one of which is obviously *Ni-bi-ru*, the seventh
name of Mercury in the series of twelve names applied to this
planet, answering to the twelve signs of the zodiac.

To facilitate the study of our tablet, I submit a comparative table
relating to the zodiac, including the list of twelve stars of Phœnicia,
showing their proposed adjustment to the zodiacal signs : —

MONOGRAMS.	ASSYR. NAMES.	NAMES OF MERCURY.	ZOD. SIGNS.	STARS OF PHŒNICIA.
1. *Bara.*	*Ni-sa-nu.*	*Dun-pa-uddu.*	*Aries.*	(11) *Lab-a.*
2. *Gut.*	*Ai-ru.*	*Ud-al-tar.*	*Taurus.*	(12) *Al-lap.*
3. *Uku.*	*Si-va-nu.*	*As-kar, Bab-ilini.*	*Gemini.*	(1) *As-kar.*
4. *Su.*	*Du-u-zu.*	*Da-pi-nu.*	*Cancer.*	(2) *Sir.*
5. *Ne.*	*A-bu.*	*Mak-ru-u.*	*Leo.*	(3) *Bar-tabba-galgal.*
6. *Kin.*	*U-lu-lu.*	*Sak-ve-su.*	*Virgo.*	(4) *Nin-makh.*
7. *Tul.*	*Tas-ri-tu.*	*Ni-bi-ru.*	*Libra.*	(5) *Ni-be-Anu.*
8. *Apin.*	*Arakh-Samna.*	*Rab-bu.*	*Scorpio.*	(6) *Nam-makh.*
9. *Gan.*	*Ki-si-li-vu.*	*Rabaz.*	*Sagittarius.*	(7) *Su-qi.*
10. *Ab.*	*Ta-bi-ta.*	*Sar.*	*Capricorn.*	(8) *Qaq-Sidi.*
11. *As.*	*Sa-ba-tu.*	*Gal.*	*Aquarius.*	(9) *Bir.*
12. *Se.*	*Ad-da-ru.*	*Kha Hea.*	*Pisces.*	(10) *Ungal.*

SEC. 109. The first column gives the names of the Accadian monograms for the months of the Babylonian year, the month *Bara*
being the first, corresponding to the zodiacal sign Aries. The
second column contains the Assyrian names of the months in their
regular order answering to the monograms and signs. The Hebrew
names were derived from these. Thus, the Assyrian *Ni-sa-nu* be-

comes the Hebrew *Nisan*, etc.[1] The third column gives the names
of Mercury corresponding to each month and sign. We have a
tablet showing the adjustment of these names to the zodiac, so that
no doubt exists in relation to this matter.[2] In the fourth column
the order of the zodiacal signs and their names are presented.
Finally, in the fifth column I have arranged the names of the stars
of Phœnicia according to my theory of their intended adjustment
to the zodiac. It is to be understood that there is no question here
as to the proper order of any list contained in the foregoing table
except in regard to the last. Our investigation has reference to this
alone, employing the other lists simply as aids in determining the
true arrangements of the Phœnician asterisms. We are prepared
now to enter directly upon this question.

1st. *The star As-kar.* The characters composing the name of this
asterism are read *Dil-gan* by Rev. A. H. Sayce, and they often take
these phonetic values. But the reading *As-kar* is equally legitimate, is that adopted by M. Lenormant, and, for reasons that will
appear in full in the next chapter, has been preferred here. It will
be seen that the name *As-kar* heads the list of Phœnician stars,
which circumstance, if any zodiacal arrangement is to be presumed
here, would lead naturally to the assimilation of *As-kar* to the
opening month of the Babylonian calendar, that is to say, to the
month *Bara*, Assyrian *Nisanu*, the sign Aries. Instead of this, as
shown in the table of proposed adjustments, I have assimilated
As-kar to the sign Gemini, Accadian month *Uku*, or "month of the
brick," the Assyrian *Sivanu*, being the third instead of the first
month of the Babylonian year. That which first determined me to
this arrangement was a comparison between the name *As-kar* and
that of Mercury as it attends the sun in the sign Gemini, or during
the third month. The title of Mercury in this zodiacal division is
the Accadian *An as-kar kâ an-mes*, or the Assyrian *il as-kar bab-
ilini*, the sense being the same in both cases, namely, "the god *As-
kar*, gate of the gods." Thus, the name *As-kar* is simply the title
of Mercury in the sign Gemini, or during the third month of the
Babylonian calendar, from which it follows that the zodiacal arrangement of the twelve stars of Phœnicia does not correspond to

---

[1] Vid. Norris, *Assyr. Dic.*, i. p. 50, for a table of monograms, names of the
months, etc.
[2] Vid. 3d Rawl. Pl. 53, No. 2, Obs.

the Babylonian year opening with the sign Aries, but that the whole series is to be moved forward on the zodiac two entire signs, so that the star *As-kar* shall fall in the sign Gemini. Accordingly, it will be seen from the table of adjustments given in the last section that I have arranged the whole series of Phœnician names with reference to this one fact.

The phrase *bab-ilini*, "gate of the gods," is evidently a mere qualification of *As-kar ;* and his being considered a god in one case, and a star or constellation in the other, is in no sense material, as all the stars were considered divinities. The evidence then appears to be conclusive to the effect that the twelve stars of Phœnicia present a zodiacal scheme, which opens with the sign Gemini, and concludes with that of Taurus. Nevertheless, there is an objection to this hypothesis which it is necessary to set aside before we can safely rest in the conclusion seemingly so obvious at first view; and it arises from the statement of M. Lenormant, in which he is supported by Rev. A. H. Sayce, that the title *As-kar* is a generic one, applied equally in the texts to the planets Mars, Jupiter, and Mercury.[1] If such is really the fact, we cannot assume here that the name in question is necessarily to be assimilated to the title of Mercury in the sign Gemini. The two texts cited by M. Lenormant in proof of his statement are the same as those relied upon by Rev. Mr. Sayce. The application of the name *As-kar* to the planet Mars is predicated upon a bilingual passage, which reads as follows: . . . *As-kar = il Ni-be-anu ;* " . . . *As-kar* is the god *Nibeanu*, or Mars." [2] The character which precedes *As-kar* in this phrase is almost wholly defaced; but enough remains to show that it terminated with *two parallel* lines, in which case it was neither *Mul*, " star," nor *An*, "god." It is evident, then, in this instance, that *As-kar* is not considered a star nor a divinity; and the text affords no proof of M. Lenormant's statement.

The text supposed to demonstrate the assimilation of *As-kar* to Jupiter is a list of twelve asterisms, or divinities, six of them equated in some sense to Jupiter, and six to Saturn, and may be presented in the following form so far as relates to the point before us : [3] —

[1] Vid. Lenormant, *Frag. de Bérose*, p. 375, note ; Sayce, *Trans. Bib. Arch. So.*, iii. pp. 167, 176.

[2] 2d Rawl. Pl. 39, No. 5, l. 64.

[3] 2d Rawl. Pl. 57, Revs. Col. 1, ll. 44–47.

The star *Gan-gusur* = the star *Lubat-guttav* = Jupiter.
The star *Mar-duk* = the star *Lubat-guttav* = Jupiter.
The star *As-kar* = the star *Lubat-guttav* = Jupiter.
The star *Qaq-sidi* = the star *Lubat-guttav* = Jupiter.

M. Lenormant has seen here an ordinary bilingual equation of certain names to others; in this instance, of *Gan-gusur*, etc., to *Lubat-guttav*, or the planet Jupiter. Such a view was, to say the least, very natural; but it has seemed to me erroneous, and I find that Rev. Mr. Sayce perceives here a serious difficulty, and thus he observes: —

" The scribe, therefore, who wrote the passage in question must have misunderstood his copy, and have identified with Jupiter a group of stars which were coupled with it in consequence of their proximity to the ecliptic." [1]

The author shows that this list contains the names of asterisms that cannot by any possibility be identified with either Jupiter or Saturn. The same may be said with reference to *Qaq-sidi*, which occupies the fourth position in this series. The text, therefore, affords no evidence of the assimilation of *As-kar* to Jupiter; and we return to our original supposition that the name is exclusively a title of Mercury; of Mercury, however, only as it attends the sun in the sign Gemini, or during the third month of the Babylonian year. The twelve stars of Phœnicia, then, or of the west, present a zodiacal scheme that opens with this sign.

SEC. 110. But we cannot agree with Rev. Mr. Sayce respecting the text last cited, that the scribe has " misunderstood his copy." Is it not more probable that our author has misinterpreted the scribe? The twelve names of Mercury, as we have seen, afford a regular zodiacal scheme. In a certain sense, as shown by the text, every one of these names is equated to the "god Marduk." Yet these are not properly titles of Marduk, as Rev. Mr. Sayce appears at times to suppose. *Dun-pa-uddu, Da-pi-nu, Sak-ve-sa*, etc., are titles of Mercury beyond question, and never of Marduk. What, then, is the meaning of Marduk's association with all these names? It is simply, as Assyriologues have often suggested, that " Marduk rules the entire year" according to the parlance of astrology. Now in the series of twelve asterisms before referred to, in which our English Assyriologue believes the scribe to have made a mistake, is

[1] *Trans. Bib. Arch. So.*, iii. pp. 170, 171.

it not probable that we have another zodiacal arrangement, not in ordinary practical use, perhaps, but one in which Jupiter is supposed to rule the first half of the year, while Saturn rules the last half? Such, at any rate, is my view of the import of this text; and it offers a satisfactory explanation of the apparently anomalous facts which had given rise in Rev. Mr. Sayce's mind to the suggestion made by him. But there seems to be some positive evidence that the text in question was intended for a zodiacal arrangement.

The star *Qaq-sidi*, as will be seen in the extract given, occupies here the fourth position, which zodiacally answers to the sign Cancer, where the sun attains its highest exaltation at the summer solstice, a period universally connected in antiquity with the heliacal rising of the dog-star. Now Mr. Norris identifies *Qaq-sidi* with the dog-star, and cites in proof a cuneiform passage which is thus rendered by Rev. Mr. Sayce: "In the days of variable storms (and) heat, in the days of the rising of *Qaq-sidi*, which (is) like bronze," etc.[1] This language can have reference only, it would seem, to the burning season of the summer solstice, when the sky is like bronze; thence also to the heliacal rising of Sirius which was supposed to mark that period. The fact, then, that this star assumes the fourth position in the list referred to, being thus perfectly normal in any zodiacal scheme, constructed with reference to its heliacal rising, is another very significant circumstance tending to the conclusion that we actually have here a zodiacal arrangement. In addition to this it is to be noticed that the star *As-kar* holds the third position, answering to the sign Gemini, thus corresponding to the order of the twelve names of Mercury where *As-kar bab-ilini* is put for the third sign. On the other hand, we have here a confirmation of our theory: 1st. That the "star *As-kar*" and the "god *As-kar*, gate of the gods," refer to one and the same asterism. 2d. That the asterism thus denoted is to be assimilated zodiacally to Gemini.

We find ourselves under the necessity of removing another doubt which arises from the following statement by Rev. Mr. Sayce: "*Dil-gan* (*As-kar*) was the patron star of Babylon . . . just as *Mar-buda* was of Nipur, and this fixes its identity with the star of Merodach."[2] In the two texts cited by the author, I fail to see any evidence supporting the opinions here expressed. The third

[1] Norris, *Assyr. Dic.*, i. p. 109; Sayce, *Trans.*, etc., iii. p. 174.
[2] *Trans.*, etc., iii. p. 171.

name of Mercury, or *As-kar bab-ilini*, is taken as one of his proofs;
but he falls into a double error with regard to it, unless I am myself
very much deceived. In the first place, he interprets this entire
series of names, usually regarded as titles of Mercury, and by himself so understood in other places, as being really names of *Marduk*.
or Merodach, the patron divinity of Babylon. But the fact is that
these are not titles of *Marduk*, who merely rules the year in this
case, as already explained in this chapter. Hence, *As-kar* or *Dil-
gan* is not here identified with *Marduk*. It would be impossible to
show, from any other text in existence, that *Dun-pa-uddu, Da-pi-nu*,
to which we may add *As-kar*, were ever applied to *Marduk*. They
are the most ordinary epithets of Mercury. Again, the author interprets the phrase *bab-ilini*, "gate of the gods," as being identical
with *bab-ilu*, "gate of Ilu," name of Babylon. It is perfectly correct that Babylon was styled the "gate of Ilu, or El," but this is
quite different from the "gate of the gods" of the entire pantheon,
obviously relating to the traditionary mount of the east, conceived
as the seat of the great divinities, and regarded in all Asia, as M.
Renan has stated, as "the gate of the universe." The other text
cited is equally defective, in my view, as affording any evidence
that *As-kar* or *Dil-gan* was the patron divinity of Babylon, and thus
one with the god *Marduk*. The text is much broken up, but the
following exhibits its main features:[1] —

1. The star *Al-lap* = *Sip-par-ki* = city of Sippara.
2. The star *Mar-buda* = *Bil-kit-ki* = Calneh, or Nipur.
3. . . . . *kar* = *Bab-ilini* = "gate of the gods."
4. . . . . = *As-sar-ki* = Asshur.
5. . . . . = *Su-si-ki* = (?).

Doubtless, the middle or third line of this text had the name *As-
kar, bab-ilini*, in its perfect state. But there is no proof, so far as I
am able to perceive, that it was assimilated to *Marduk*, nor that it
represented the city of Babylon. The point which this writer
assumes, as before intimated, is the equivalence of the phrase *Bab-
ilini* to *Bab-il*, name of Babylon; but there is no proof of it in the
two texts cited, nor am I aware of any others tending to support
this view. Thus, we return again to our first position, namely, that
the star *As-kar* is Mercury in the sign Gemini, which accordingly
opens this old Phœnician calendar, if indeed this arrangement of
asterisms was ever employed for such a purpose.

[1] Vid. 2d Rawl. Pl. 48, Col. 1, ll. 55-59.

SEC. 111. 2d. *The Star Qaq-sidi.* It has been shown that this name denotes the dog-star, whose heliacal rising with the sun in Cancer marked the period of the summer solstice. In any zodiacal scheme constructed with reference to its heliacal rising, this star would hold the fourth position, agreeing with that of Cancer, and as seen in the text already noticed (Sec. 109). In the series of twelve stars of the west, or of Phœnicia, *Qaq-sidi* occurs in the *eighth* position. Now the eighth zodiacal sign is Scorpio. It would be difficult to assign any valid reason, either astronomical or astro-mythological, for assimilating the dog-star to the sign Scorpio. The dog-star had always in antiquity a solstitial character, connected thus with either the sign Cancer or Capricorn. A word of explanation here respecting the different risings of the stars. A star is said to effect its *cosmical* rising when it comes to the eastern horizon exactly at sunrise; but owing to the superior light of this luminary, it is not visible to the naked eye. A star is said to rise *heliacally* with the sun when it mounts the eastern horizon a little time before the sun, or so as to render it visible to the unassisted eye. The sun gains two hours during each month on the risings of the fixed stars, hence at the end of six months it has gained twelve hours. If, then, the dog-star rises heliacally with the sun in Cancer at the summer solstice, six months after with the sun in Capricorn, it will have gained twelve hours, and will set in the west about the time the dog-star rises in the east. The *acronycal* rising of a star is that when it comes to the eastern horizon a few moments after sunset, or a sufficient time after to be visible to the eye without the aid of instruments. The sun attains the winter solstice in Capricorn six months after the summer solstice, which occurs with the sun in Cancer. The dog-star, therefore, by its heliacal rising marked the period of the summer solstice, and by its acronycal rising that of the winter solstice. Although the two risings might not correspond exactly to the two solstices, they appear to have been so considered for a long period in antiquity.

We return now to the consideration of *Qaq-sidi*, holding the eighth position in our list of asterisms. With *Qaq-sidi*, or the dog-star, in the eighth sign, or Scorpio, we are at once in the midst of difficulties. It must fall in Cancer, the fourth sign, answering to its heliacal rising, or in Capricorn, the tenth sign, agreeably to its acronycal rising, or it is impossible to attribute any really zodiacal character to its position. According to our theory of adjustment of these twelve stars to the zodiac, it is necessary to move them all forward two entire signs as compared with the Babylonian system; in other words, we must move *Qaq-sidi* from Scorpio, the eighth sign, into Capricorn, or the tenth sign. By reference to our table of adjustments (Sec. 108), it will be seen that *Qaq-sidi* falls in the sign Capricorn, as if to mark, by its acronycal rising, the period of the winter solstice. The question arises, then, whether the ancients ever associated a dog-star with Capricorn and with the solstice of winter. Professor A. Romieu, in his critical treatise on the "Egyptian Decans," already referred to in these pages, cites a remarkable statement from Firmicus: "On the right of Sagittarius rises the ship Argo, and on the left the dog." "We recognize in divers authors," adds the professor, "the notion of two dogs, guardians of the limits of the sun's course; and in this case, the southern dog would be found in the region of Capricorn precisely to the left of Sagittarius."[1] Now, whether these two dogs were really separate asterisms, or merely the two phases of the heliacal and acronycal risings of one and the same star, it is certain that the Hamites of Egypt primitively located a dog-star exactly in the position in which, according to our method of adjustment, the table of the twelve stars of Phœnicia places *Qaq-sidi*, that is to say, in the sign Capricorn. A coincidence so remarkable as that here exhibited is exceedingly rare in antiquarian researches; and it is of a character to add much strength to our hypothesis relative to these Phœnician asterisms.

SEC. 112. 3d. *The star Nin-makh.* The meaning of this name is "great mistress" or "divine lady;" and the reference is undoubtedly to the mother goddess, probably to *Ishtar*, the Babylonian Venus. This star is the fourth in the series of Phœnician asterisms; but it is plain that it has no connection with the fourth zodiacal sign, or Cancer. According to our method of adjustment, it should fall in the sixth sign, and this is Virgo, or the Virgin. The Accadian name of the corresponding month is interpreted by M. Lenormant as "month of the message of Ishtar."[2] In the absence of all other proofs tending to the support of our hypothesis, the evidence afforded by the facts here presented would be regarded

---

[1] *Sur un Décan,* etc., p. 35.
[2] *Premières Civilisations,* t. ii. p. 71.

justly as very direct, although not absolutely conclusive. It would still be a question whether the proof was not merely contingent and accidental. But taken in connection with the data previously introduced, it becomes difficult to resist the conviction of the existence of a veritable law in the method adopted for the adjustment of these stars. In each case, we have only to move the Phœnician stars forward two signs on the zodiac, as compared with the Babylonian system, in order to perceive at once that the asterism has found its normal position. Thus, *Nin-makh*, the fourth star, is seen at a glance to appertain to the sixth sign, or to Virgo.

4th. *The star Nam-makh.* The Accadian *Nam* has the sense of "destiny" (Rep. 82); and *Makh* signifies "great, very great" (Rep. 68). The Star *Nam-makh*, therefore, was supposed to exercise a preëminent influence over destinies, according to the notions of the ancient astrologists. It was the opinion almost universally held that the sign Scorpio was supreme in this respect. Mr. Burritt remarks, "Scorpio was considered by the ancient astrologers as a sign accursed. The Egyptians fixed the entrance of the sun into Scorpio as the commencement of the reign of Typhon." [1] It was held also to exercise a great influence in fixing the destinies at birth. The correspondence, then, between the meaning of *Nam-makh* and the traditional character of Scorpio is very exact: and the fact that, according to our theory of adjustment, the star *Nam-makh* appertains to this sign constitutes another reason for the supposition that this theory rests upon a firm basis.

5th. *The star Ungal, or the King.* We shall find in reference to this asterism apparently a very marked discrepancy as regards our hypothesis. An inspection of the table of adjustments, as given in the 108th section, will show that *Ungal*, the Assyrian *Sar*, "king," is the *tenth* Phœnician asterism, and the *tenth* name of Mercury. This appears to be at first a striking coincidence between both name and number. But while the tenth name of Mercury agreeably to the Babylonian system falls in the tenth zodiacal sign, or Capricorn, the tenth star of Phœnicia having the same name, according to our method of adjustment, falls in the *twelfth* zodiacal division, or in Pisces. Apparently, therefore, it has been carried forward from its normal position to the extent of two signs into an abnormal one. If, in moving the Phœnician asterisms forward two

[1] *Geog. Heavens*, p. 102.

signs in relation to the Babylonian system, it brought the Phœnician star of the King into Capricorn, where we have the same
name as a title of Mercury, it would be an instance like that of
Nim-makh; the asterism would be brought *into* position by the
operation.   But instead of this, it seems to take the Phœnician
Ungal directly *out* of position, carrying it forward into the twelfth
sign.   This appears to be a serious objection to our theory of adjustments, although it admits of an explanation in perfect accord
with it.

It is obvious that the King, holding always the tenth position,
must be assimilated to the tenth antediluvian king, the *Sisithros* or
Xisuthrus, of Berosus, the Babylonian Noah.   These ten kings of
the antediluvian period, answering to the ten patriarchs of the
Mosaical record including Noah as the last, had been assimilated
by the Babylonians to the ten zodiacal signs, beginning with Aries
and ending with Capricorn.   The fact of such assimilation was
long since established by Dr. Movers, more recently confirmed by
M. Lenormant, and of this the principal proofs will be presented
in the chapter immediately following.   The tenth king of Berosus,
being *Sisithros,* was associated accordingly with the tenth zodiacal
division, with Capricorn, to which the Assyrian title of Mercury,
Sar, "the king." is seen to correspond.   This tenth personage of
the antediluvian genealogy appears to have been regarded traditionally by the Babylonians as *the·king·par excellence.*   He was
Sydik, "the just," like Noah, who is styled in Genesis "a just
man," etc.; and the same title *Sydik* was very extensively known
as the name of the planet Jupiter, to whom this tenth king appears
to have been assimilated.   Thus much for the Babylonian system.

The Phœnician system appears likewise to have been connected
in some way with the antediluvian dynasties, and indeed our next
chapter will contain ample proofs of it.   The position of the star
Ungal, therefore, as assimilated to the tenth king, corresponding to
Sisithros and Noah, has to be preserved intact.   As this scheme
commences with Gemini, the star *Ungal,* or "the king," was necessarily transferred to the sign Pisces, to which corresponds the singular title of Mercury that reads, *Kha an e-a,* "Fish of the god
Hea."   Until quite recently M. Lenormant has not hesitated to
identify the name *Hea* with the Biblical *Noah,* a fact which, if real,
would be demonstrative of the accuracy of these explanations rela-

tive to the star *Ungal*. It would show definitely a design, based upon traditional ideas, of connecting the star of the King with the sign Pisces, as assimilated to the Biblical Noah. This whole matter will receive full attention in the next succeeding chapter, there being no opportunity to discuss it properly in the present chapter. We assume, then, provisionally, partly on the basis of M. Lenormant's previous views, that *Hea* is to be considered the same as the Biblical *Noah*. Hence, while Sisithros, as the tenth king, should be associated with Capricorn in the Babylonian system, the star *Ungal*, in the Phœnician system, was purposely assigned to Pisces, in order to connect it with the god *Hea* identified with *Noah* of Genesis.

SEC. 113. If there has seemed to be, in view of the evidences thus far presented, still some shades of obscurity and doubt remaining as to the actual scientific character of the method of adjustment proposed for these asterisms, I believe the data to be now introduced will tend to remove such doubt, and settle the question of the zodiacal arrangement of the "12 stars of the west," an expression that in the present instance must relate definitely to Phœnicia. It will be necessary to place before the reader's eye an accurate representation, so far as possible, of the published text of the inscription being studied, at least so far as relates to our twelve stars. The following table gives the Accadian values of the characters composing the text, the dotted lines showing those portions that are wanting: —

1st Col. Months.	2d Col. Months.	1st Col. Stars.	2d Col Stars.
1.	.... kû.	Mul. .... Se.	Mal Ni-bi. ...ru.
2.	.... ar-a.	12 Mul-mes.	Mat Akkad-ki.
3.	Ab. G m-gan-au.	(1) Mul As-kar.	(7) Mul Su-gi.
4.	Ab. Abba-uddu.	(2) Mul Sir.	(8) Mul Qag-sidi.
5. Ab. Si....an-bit-ti.	Ab. As-a-an.	(3) Mul Bar-tabba-gulgal.	(9) Mul Bir.
6. Ab. Kisal-vri-ba-a.	Ab. Se-kin-tar.	(4) Mul Nin-makh.	(10) Mul Ungal.
		(5) Mul Ni-be-anu.	(11) Mul Al-lap.
7. Ab. Bara. Mat.,	Ab. Gut. Mat Rama-ki.	(6) Mul Num-makh.	(12) Mul Lab-a.
8. Ab.	Ab. Mar-ta-ki.		
9.	.... Gu-ti-i.	12 Mul-mes.	Mat Martu-ki.

The Accadian *Ab* signifies "month," and *Mul* "star." The inscription as represented shows two columns of months and two of names of stars; these last constituting the special object of our study. The first column of months, very defective, seems intended to set forth the astrological influences exerted by each, the names attached not being regular denominations of the months. But the second column of months, down to the sixth line of the text, contained the regular Accadian names in full; four of which remain entire, the two upper ones being defective. Above all, the tablet contained much more, but it is now lost.

It must be apparent, I think, that the two columns of months involved in some sense a regular adjustment to the two columns of stars. But of the first column of months we can decide little more than this: it contained names purely mystical, and not in ordinary use in the calendar. Such is not the case as regards the second column; it shows even now several names well known as belonging to the Accadian calendar. The four names, the months, *Gan*, or *Gan-gan-na*, *Ab*, or *Abba-udda*, and so the month *As*, and *Se*, are seen in full, and it is to be especially noted that they close the Babylonian year. The proposition, then, which I have to submit is as follows: *The second column of months was intended, as showing the proper adjustment of the second column of asterisms to the Babylonian calendar and the zodiacal signs.* A completely satisfactory proof of this is seen in the very first line of the text, although so defective. The syllable *kû* in the second column is a part of the Accadian name of the seventh month, *Tul-kû.* The corresponding asterism in the second column of stars is *Ni-bi-ru*, and this is the seventh name of Mercury, which answers to the month *Tul-kû* in the ordinary lists. In our text, *Ni-bi-ru* is given as one of the stars of Accad, the only one in fact whose name remains in a tolerable state of preservation. It has this great importance to us: it proves beyond question that the second column of months shows the zodiacal arrangement of the second column of asterisms. Compare, then, this table with our table of adjustments (Sec. 108), and it will be seen that I have equated the month *Gan* to the star *Sugi*, the month *Ab* to the star *Qaq-sidi*, the month *As* to the star *Bir*, and the month *Se* to the star *Ungal*. This closes the Babylonian calendar, as indicated by the heavy line beneath, yet there are two months remaining to be adjusted. and two asterisms in the second column that remain unadjusted. We have first the "month Gut," which means the Taurus; and to correspond is the star *Al-lab*, or *Al-lap*, evidently the same as *Alap*, ordinary Assyrian name for the Taurus. Then there is the *month Martu-ki*, "month of the west," of Phœnicia, but without stating what particular month is to be understood. Probably we should assume the month *Bara*, the sign Aries, to which must be adjusted the star *Lab-a*, since all the other months and signs have been properly accounted for.

SEC. 114. Nothing more is required, according to my judgment, to establish the correctness of the theory which has been put forth relative to the true zodiacal arrangement of these asterisms. My aim has been to place a matter of such importance, considering the obscurity still existing relative to Babylonian uranography, upon a strictly scientific basis, and so far as concerns these asterisms I believe the effort has been quite successful. Before resting fully in this conclusion, however, it is necessary to notice briefly an astronomical tablet recently discovered and partly translated by Mr. George Smith, as seen in his "Assyrian Discoveries." According to this tablet, the star *Ni-be-anu*, or *Nibat-anu*, is assimilated to the sign Sagittarius, corresponding to the month *Kisi-livu*, or *Kisler*, instead of to Libra, as in the scheme here proposed.[1] But as *Nibe-anu* is regarded as a planet, and not as a fixed star, nothing results from a variation of this kind. Yet, owing to a certain erroneous statement by the author, the circumstance here noted might be interpreted as a serious objection to my hypothesis. Mr. Smith remarks: " The star Nibat-anu has hitherto been erroneously supposed to be a planet." [2] Naturally, we look for some proof of a declaration so important, and one that is opposed to the universally received opinions of Assyriologues for many years past ; but the author affords us none, except that which seems to have deceived him, namely, the fact that Nibat-anu is assimilated to a particular month and sign in the tablet translated by him, the month Kisler and sign Sagittarius. He has inferred from this, apparently, that Nibat-anu was the name of a constellation or fixed star, and not of a planet. But this term denotes a planet beyond question, and in all probability Mars. Two complete lists of the seven planets are given in the published texts, and in both Nibat-anu is included among them.[3] These lists as cited below leave no doubt upon the matter. Hence, we reaffirm our belief that the zodiacal arrangement of the twelve stars to which we have devoted our attention, and as determined by us, is scientifically correct.

---

[1] *Assyr. Dis.*, p. 408.   [2] Ibid., p. 407.
[3] Vid. 2d Rawl. Pl. 48, Obs. Col. i. ll. 48–54.   Cf. 3d Rawl. Pl. 57, No. 6. ll. 65–67.

It is necessary to offer briefly some explanation here relative to
the month *Gut*, equated to the country *Mat ilama-ki* and to the
star *Mul al-lap*, also relative to the month *Martu-ki*, equated to *Mul
lab-a*, the star Lab-a.   The month *Se-kin-tar* that immediately precedes these closes the Babylonian year, as is well known, and the
heavy line beneath is designed to indicate this fact.   Yet two asterisms remained to be adjusted to their respective zodiacal positions,
and in the two months here alluded to the scribe has undertaken to
do so, but he has introduced a mystical element that needs some
elucidation.   I shall endeavor to place this matter in a clear light
in another chapter devoted to the subject of the *Cherubim*.   For
the present it will suffice to say that the asterism *Gut*, being the
constellation Taurus, is here put for the country of *Ilama*, the Biblical Elam, modern Susiana.   In the symbolical system of geography,
of which our tablet offers a practical illustration, and in which the
country of Accad holds the central position, *Ilama* represents the
east, and *Martu*, or Phœnicia, the west.   The peculiar arrangement of the two months in question is due to these considerations.
The month *Gut*, set to Ilama, represents the east.   In other terms,
the celestial Taurus is transferred to the east quarter of the heavens, while Aries retains its normal position in the west quarter,
being put for the west country, that is to say, for *Martu*, or Phœnicia.   The month Aries, or *Bara*, opens the Babylonian year.
This accounts for the fact that the natural order of the two stars
Al-lap and *Lab-a* is inverted in the text, but I have restored them
to their normal position in the table of adjustments.   As the Phœnician stars represent a calendar that opens two signs after that of
Babylon, that is, in Gemini, it results that Aries and Taurus, commencing the Babylonian year, are made to close the Phœnician
year.   These remarks, I think, are entirely sufficient in elucidation,
and as an analysis of the text forming the subject of the present
study.

Sec. 115.   The question naturally arises here whether this arrangement of the "twelve stars of the west" was ever employed
by the Phœnicians, or by any other people, for the practical purposes of a calendar?   If we speak of historical times I do not think
any satisfactory evidence could be adduced, showing that such was
the case.   In Western Asia, and since a very early epoch, the Babylonian calendar seems to have been very generally adopted.   In

some instances two systems were employed, a civil and a sacred or religious year, which differed from each other as to the period of commencement. We know, too, that the order of the zodiacal signs, beginning with Aries and terminating with Pisces had no controlling influence, outside of the valley of the Euphrates, in determining the commencement of the calendars. They frequently varied in respect to the particular sign with which they opened. Still, I am not aware of the existence of any calendar in practical use since the opening of the historical period that commenced with the sign Gemini. If this arrangement of twelve stars which we have studied was ever in use for such purpose, it must have been in an epoch extremely remote, and it had afterwards been superseded for practical use by others, being itself exclusively retained subsequently merely as traditional, and as belonging to the sacred science inherited from a former age. Nevertheless, it is evident that a much greater importance was attached, in very early times, to these twelve stars of Phœnicia than the remarks just submitted would lead one to suppose. Ample evidence that such was actually the case will be produced in the chapters immediately following.

Our tablet establishes one important fact, to which allusion was made in a previous study, and which our present researches have tended to impress still more strongly upon the mind. It is that the habit of associating certain asterisms with such and such peoples or countries was quite prevalent in antiquity, and this proves that, in such case, they were regarded as celestial earths or terrestrial heavens, conformably to the traditional notions attached to the primitive abode of mankind. The frequent custom of dividing the population into twelve tribes, and the national domain into twelve districts, or cities, evidently arose from the same order of conceptions. In every instance the notion was a heaven + earth; the notion on a large scale corresponded to the two divisions of the temple itself, one representing heaven, the other the earth, and the original idea had been derived from the particular heaven and earth constituting the celestial and terrestrial paradise.

In the symbolical geography of Sargon the ancient, there can be no doubt that the country of Accad replaced the sacred mount of the East, the *Kharsak-kurra*, or "mountain of the world," as Rev. Mr. Sayce rather loosely interprets this phrase. The element *Khar* signifies "circle, bracelet" (Rep. 440), and *Sak* has the sense of

"head, front, summit" (Rep. 136). The syllable *Kur* is only another value of the Accadian *Mat*, "country," and has the meaning
of "to elevate, mountain, east" (Rep. 419). Finally, *Ra* signifies,
among other things, "inundation, deluge" (Rep. 303). "The
bracelet or circle of the summit of the diluvian mountain" may
be taken for the traditional conception, which has been crystallized
in this cuneiform expression. Mt. Meru, to which the phrase obviously relates, was regarded as the primitive Ararat, and its summit was sometimes known as the "circle or bracelet of Ida," the
mountain goddess, of whom some notices have been heretofore introduced. We say that Accad, or *Akkad*, a name signifying
"mountain or highland," according to M. Lenormant, replaced
the *Kharsak-kurra* in the symbolical system of the ancient Sargon.
Aram signifies "highland," also, and we have seen that among
the Semites a system precisely like that of which Accad formed
the centre was primitively connected with this country. The Biblical *Elam*, likewise, the *Ilama* of our tablet, has been interpreted
as meaning the same thing, that is, "highland." It is quite probable that originally certain asterisms were associated with each
one of these "highlands," always in imitation of that divine summit around which rolled the flaming chariot of the seven stars. It
is impossible not to see in these traditional ideas the origin of the
"high places" of worship to which the Hebrew Scriptures make
so many allusions. Indeed, the sacred summits of all antiquity
were but projecting spurs, so to speak, from the mount of paradise.
The sentiment that embodied itself in these sacred elevations was
perfectly human and natural, for as the immortal Goethe says: —

"On every height there lies repose."

CHAPTER XI.

ZODIACAL ARRANGEMENT OF THE ANTEDILUVIAN GENEALOGIES.

SEC. 116. In reporting the Chaldæo-Babylonian traditions pertaining to the first ages of the world, Berosus gives a list of ten kings supposed to have reigned during the antediluvian period. It was in the days of the last of these kings, named *Xisuthrus*, Greek *Sisithros*, that occurred the deluge. Biblical scholars have very naturally seen in this Babylonian genealogy a confirmation of that given by Moses, extending from Adam to Noah. In some sense it is proper to regard it as such. I think it could be shown that at Babylon this succession of primitive rulers was considered as real, and hence as historical. But certain facts have been developed by modern criticism tending to throw doubt upon the actual historical reality, at least of a portion of the list handed down by Berosus. The etymology that has been discovered of some of these names renders it almost certain that they were simple designations of zodiacal signs, or constellations, to which unquestionably they had been assimilated. It would be impossible to prove, at the present day, that all of them were of this nature; but those which undoubtedly were, forcibly give rise to suspicions as respects the historical character of the others. Hence, what was formerly supposed, and with much reason, to confirm the actual chronological succession of the Mosaical line of patriarchs, extending from Adam to Noah, appears now in a measure as an obstacle to considering it in this light; for one is strongly inclined to infer, if the Babylonian list was purely zodiacal, that the Mosaic genealogy was of like character, especially as recent developments in cuneiform science tend more and more to establish the fact that the Hebrew and Babylonian traditions relating to primeval times had a common origin. Undoubtedly, as we shall be able to show, both these genealogies were in some sense zodiacal. But the question arises, Were they

merely zodiacal? It does not appear to me possible, in the present
state of our knowledge, to offer a completely satisfactory reply to
this inquiry. That the two genealogies had been, at an extremely
remote epoch, definitely connected with the zodiac will be readily
conceded, I think, in view of the evidences that are to be presented
in this part of our studies. That they had in addition to this a
genuine historical character, I am not able to prove or disprove,
though I am inclined, from certain considerations that will gradually develop themselves as we proceed, to adopt the traditionary
belief respecting this matter. We give, in the first place, our consideration to the Babylonian genealogy, which is reported by Berosus as follows : (1) Alorus ; (2) Alaparus ; (3) Almelon ; (4)
Ammenon ; (5) Amegalarus ; (6) Davonus ; (7) Edoranchus ; (8)
Amemphsinus ; (9) Otiartes ; (10) Xisuthrus.

It was stated in a previous chapter that the twelve zodiacal constellations had been employed at Babylon for the purpose of marking the divisions of three distinct periods of time. 1st. They
marked the twelve hours of the day, these being double hours and
termed *Kas-bu* in the inscriptions. 2d. They corresponded to the
twelve months of the year, a fact familiar to every one. 3d. They
were associated to the twelve divisions of the great cosmical year,
which was supposed to have commenced at the dawn of creation.[1]
In each case the starting-point appears to have been Aries, or the
Ram. There can be no doubt of this, except as regards the hours,
and since the custom was so widely prevalent of beginning the day
at evening, there is every probability that the hours likewise were
reckoned from Aries. It was with these great cosmical periods that
the reigns of the ten antediluvian kings were habitually identified.
It was in this general sense without doubt, and, so far as I am aware,
in no other, that the ten kings had been associated with the zodiacal signs, beginning with Aries and terminating with Capricorn.
Alorus, the first ruler, was connected with Aries, Alaparus with
Taurus, and thus on to Capricorn, to which Xisuthrus appertained.
There are indications in this arrangement that these ten kings were
at least believed to be historical, though the basis of such belief, as
we shall see, was not wholly sound. The twelve hours were certainly time periods, and so were the twelve months of the year.

[1] Vid. Lenormant, *Frag. de Bérose*, pp. 184–240. Cf. Movers, *Die Phœnizier*.
i. pp. 164–166.

There is no reason to doubt that the cosmical periods were equally regarded as such. But while the first two appertained to the calendar simply, the last was properly speaking chronological. It is evident that these chronological periods were not correctly historical, yet they appear to have been assumed as such, and this shows that the ten kings were taken in some sense as historical. Thus, if we accept the Babylonian belief as a proof, the genealogy transmitted by Berosus must be regarded in this light. But the facts to be now presented tend wholly, or nearly so, in the opposite direction.

SEC. 117. The following extract contains Dr. Movers' proposed etymology of the two names, Alorus and Alaparus, standing at the head of the Babylonian genealogy : —

"It is without any hesitation, since the year commenced with Aries and Taurus, that we interpret the name of the king *Al-or-us* as *aries lucis* (אֵיל אוּר), and *Alap-ar-us* as *taurus ignis* (אֶלֶף אוּר)." [1]

The term *Alap* is an ordinary Semitic expression for the Taurus, and *Ail*, to which Dr. Movers assimilates *Al* in the name *Al-or-us*, is equally common as a designation for *Aries*, or the *Ram*. We should hesitate in a matter of such great importance to rest wholly upon the authority of so learned a critic even as the one here cited. But M. F. Lenormant, whose critical ability is not less marked and extensively acknowledged, has recently given the same matter a careful consideration, and thus expresses his views relative to the names of these kings : —

"The first two are certain, and their forms have submitted to but little alteration, for we find the same words which in the Assyrian language denote the first two zodiacal signs, *ail*, 'the Ram,' *alap*, 'the Taurus.' As to the final elements, *orus* and *arus*, it seems that we have a right to recognize them in the word *Ur*, 'light.' We have, then, *Ail-ur*, 'Aries of light,' and *Alap-ur*, 'Taurus of light,' and these two names are, alone considered, decisive for the character of the list." "Thus, the ten antediluvian kings are personifications of the solar houses (מַזָּלוֹת), which the idolatrous Hebrews, during the period of Assyrian influence, adored with the sun, the moon, and all the celestial hosts." [2]

The author interprets the third name with almost equal confidence as denoting " the sons," that is, " the twins," answering to

---

[1] *Die Phœn.*, i. p. 165.
[2] *Frag. de Bérose*, pp. 235, 236, 238.

Gemini; but here there is, to say the least, an element of uncertainty. So far as concerns the first two names it must be admitted, if I am able to judge, that they are simple designations of the two signs which correspond to the commencement of the Babylonian calendar. I am not prepared to adopt, in its entire consequences, the inferences drawn by the author that these two names "are decisive for the character of the whole series." That all the kings had been assimilated to the zodiacal constellations appears to admit of no question. But I think that, beginning with Gemini, there are many indications of a proper historical element as constituting the basis, and these will appear as we proceed.

We turn our attention now to the Mosaic genealogy of the patriarchs, beginning with Adam and terminating with Noah. It will not be difficult to show here at least an indirect relation to the zodiacal signs. Some of the names very clearly indicate to what particular signs they are to be assimilated. But as for identity of names, such as appears in those of Alorus and Alaparus, it will be difficult to prove it, except in one or two instances at most. The proofs of an association with the zodiac consist, for the most part, in the direct analogies of ideas existing, as denoted by the Hebrew names, and as compared with other titles whose reference to the zodiacal divisions is not a matter of doubt. But the Mosaic list presents, in some particulars, a system quite distinct from that of the Babylonians; and any attempt to connect the Hebrew names directly with the Babylonian would necessarily prove, as always heretofore, a complete failure. The real secret of the adjustment of the Mosaic genealogy to the zodiac is precisely that which has been demonstrated with respect to the "twelve stars of the west," or of Phœnicia: the whole list must be moved forward two entire signs in relation to the Babylonian system. Adam, the same as the star *As-kar*, corresponds to Gemini; and Noah, the tenth patriarch, answers to the sign Pisces, the star of the *King*. The moment we assume this standpoint of investigation, the analogies between the ideas involved in the Hebrew names and various others whose zodiacal character is fixed become very striking and numerous. The names and positions of the asterisms which occupied our attention in the last chapter will be found to contribute largely, though not exclusively, to the elucidation of the zodiacal character of the Mosaic genealogy. These asterisms will serve frequently as interpreters between the known and unknown. With these preliminary remarks, we enter directly upon the investigations before us.

SEC. 118. 1st. *As-kar and Adam.* The Accadian *As* has the sense of "happy, propitious" (Rep. 1). The element *Kar*, as previously explained, means "summit;" but this is only another value of the sign *Gan*, "inclosure, garden," identical with the Hebrew *Gan* in the phrase *Gan-Eden*, or "Garden of Eden." Whether we select the interpretation, then, of "propitious summit," of "happy garden," or "garden of delights," the reference of the term *As-kar* to the mount of paradise, the traditional abode of primeval humanity, is a point upon which not much doubt can be entertained. But it is apparently not less certain that *As-kar* was a name of the first man, the same as Adam is supposed to have been; for Professor Max Müller, in some remarks upon the "Northern Mythology" appertaining to the ancient populations of the north of Europe, states the following, which evidently justifies our opinion : —

"The second son of Mannus, *Isco*, has been identified by Grimm with *Askr*, another name of the first-born man. *Askr* means likewise ash-tree, and it has been supposed that the name *ash* thus given to the first man came from the same conception which led the Greeks to imagine that one of the races of man sprang from ash-trees (ἐκμελιᾶν). Alcuin still uses the expression, son of the ash-tree, as synonymous with man." [1]

As given by Professor Müller, this name of the first-born man is without a vowel in the last half of the word. It is sometimes written *Askur*, of which *Askar*, phonetically identical with the cuneiform *As-kar*, would be but a simple variant. The two names may be taken, therefore, as originally the same. The legends respecting *Askur*, as contained in the "Northern Mythology," are thus variously related by different writers: —

"As Bör's sons were once walking on the seashore, they found two blocks (or trees), of which they created a man called *Askur* (*ash*), and a woman *Embla* (*alder*)." "Three gods issuing from Asgarthr found two trees, the *ash* and *alder*, upon the borders of the sea, which were without force and posterity. The gods had compassion upon them, and formed of one a man and of the other a woman." [2]

---

[1] *Lect. Sci. Language,* 2d series, p. 476.

[2] *Encyc Americana,* vol. ix., art. "Northern Mythology." Cf. Le Bas, *L'Univers: Suède et Norwège,* p. 9.

It is probable that the name *Askar* gave rise to the term *Asgard* or *Asgarthr*, applied to the residence of the Scandinavian divinities, or those of the northern mythology, whose chief seems to have been the god *Odin*. The writers just cited have respectively the following notices of this locality : —

"The residence of the gods is *Asgard*, a fortress of heaven, whence the bridge *Bifrost* leads to the earth. *Asgard* contained the palaces of the gods. . . . In the centre of *Asgard*, in the valley of *Ida*, was the place of meeting, where the gods administered justice." " Odin, Nile, and Ve constructed upon earth an immense fortress in defense against the giants, and called it *Midgard* (the world). At the centre of this fortress is found *Asgarthr*, the residence of the gods. . . . It is there that the twelve great gods are assembled." [1]

The followers of *Odin*, according to all accounts, came from Asia from beyond the Caspian Sea; and everything tends to identify the ancient *Asgard* with *Meru* of the Hindus, *Albordj* of the Persians, the *Eden* of Genesis. Compare the "valley of Ida" with the goddess Ida, associated with Mt. Meru. Compare, also, the place of the assembly of the gods in *Asgard* with the mount of assembly, to which Isaiah alludes. But recent writers do not hesitate to identify these two localities; and this point may be assumed without special labors to establish it. That to which we wish to call attention is the relation between *Askur*, *Asgard*, and the cuneiform *As-kar*. The permutation of the vowels *u* and *a*, and the consonants *k* and *g*, is so frequent in nearly all languages that it needs no explanation in the present case. The Accadian *g* is often changed to *k*, and *vice versa*. We hold, then that the name *Asgard*, as locality, is simply *As-gar-d*, or the name As-kar taking a locative termination, like *Akka-d* for *Akkad*, according to M. Lenormant's etymology. Thus, *Askur*, name of the first man, gives rise to *Asgard*, applied to the first abode of man, and both are to be assimilated to *As-kar*, relating to the same abode, being employed also as a name of Mercury. The classic authors identified Odin with Mercury, and it is remarkable that, like Mercury, Odin had twelve chief names. It is difficult to resist the conclusion in view of these facts: 1st. That the Accadian *As-kar* was the name of the first human abode derived from the title of the first man. 2d. That the Scandinavian *Asgard* was the name of the same locality, derived from *Askur*, also a title

[1] *Enc. Amer.*, ix. p. 319 ; Le Bas, *L'Univers: Suède et Norwège*, p. 9.

of the first man. 3d. That the two sets of terminologies were derived from the same source. 4th. That the notions involved in these terms offer an exact analogy with those connected with the name *Adam*, traditionally associated with the term *Adama*, signifying "earth," here the particular earth constituting man's primitive home.

SEC. 119. In the last chapter the star *As-kar* was identified with the "god *As-kar*, gate of the gods," as name of Mercury, both being assimilated to the zodiacal sign Gemini. It has been stated heretofore, and it is a fact well known, that the Oriental zodiacs, especially the Hindu, represent a man and woman in the sign Gemini, instead of two male figures as in the Greek system. The man and woman thus represented appear to have been identified with the Hindu *Yama* and *Yami*, reputed first human pair. Professor Whitney, a portion of whose language has been previously cited, has the following upon Yama: —

" His name does not come, according to the usual interpretation, from the root *yam*, 'subdue, repress;' it is radically akin to the Latin *gem-ini*, etc., and means 'twin.' In him and in his sister Yami are conceived the first human pair, parents of the whole following race; he is therefore, as expressly stated in the hymns, the first who made his way to the skies, pointing out the road thither to all succeeding generations, and preparing a place for their reception ; by the most natural transition, then, he becomes their king. It is in entire consistency with this that, in Persian story, where he appears as Yima (later Jem-shid), he is made ruler of the golden age and founder of the Paradise." [1]

In his interpretation of the character of Yama and Yami, Professor Whitney adopts the views of Professor Roth. Professor Müller rejects the interpretation entirely, regarding Yama wholly in the light of a solar deity. After citing various texts from the hymns, he observes : —

" These indications, though fragmentary, are sufficient to show that the character of Yama, such as we find it in the last book of the Rig-Veda, might well have been suggested by the setting sun, personified as the leader of the human race, as himself a mortal, yet as a king, as the ruler of the departed, as worshiped with the fathers, as the first witness of an immortality to be enjoyed by the fathers," etc.[2]

---

[1] *Orient. and Ling. Studies*, p. 45.
[2] *Lect. Sci. Language*, 2d series, p. 535.

The same writer has the following remarks, quite important to our purpose : —

" As the east was to the early thinkers the source of life, the west was to them *Nirriti*, the *exodus*, the land of death. The sun, conceived as setting or dying every day, was the first who had trodden the path of life from east to west, — the first mortal, — the first to show us the way when our course is run, and our sun sets in the far west. Thither the fathers followed Yama; there they sit with him rejoicing, and thither we too shall go when his messengers (day and night) have found us out." [1]

Finally, the author concludes : —

" Professor Roth has pointed out some more minute coincidences in the story of Jem-shid, but his attempt at changing Yama and Yima into an Indian and Persian *Adam* was, I believe, a mistake. Professor Kuhn was right, therefore, in rejecting this portion of Professor Roth's analysis." [2]

I freely acknowledge myself greatly indebted to Professor Müller in his researches upon the subject of comparative mythology, in respect to which he is in some sense the founder of a school. But it has always seemed to me that two fundamental errors may be pointed out in his system, tending to vitiate almost all his interpretations of the ancient legends. But these are not so much errors as mistaken points of view. In the first place, he occupies too exclusively the standpoint of the *Rig-Veda.* Secondly, he confines himself too exclusively to the *diurnal aspects* of nature. In relation to the first point indicated, it is to be said that the legends connected with Osiris by the Egyptians are strictly analogous, in almost all the features pertaining to his character, to those relating to the Hindu Yama. Osiris is the sun, yet he formerly reigned on the earth. He is god of the dead, like Yama, and his descent into the dark regions of the west has given rise to the same conceptions that among the Hindus were associated with Yama. The fundamental idea in both characters has been correctly pointed out by M. Mariette, as cited in a previous chapter (Sec. 85), although exclusively interpreted by him of Osiris. *The sun's course had been taken as a symbol of human existence*, a fact that appears in nearly all the ancient religions. It will be seen from this ground principle that the two characters of Yama as sun-god and as the first man, representative of the whole human race, are perfectly con-

[1] *Lect. Sci. Language,* 2d series, p. 534.   [2] Ibid., p. 542.

sistent with each other, and doubtless did appertain to this Vedic personage. I think, therefore, that Professor Roth's interpretation is perfectly correct, and that Professor Whitney was fully justified in adopting it. At the same time Professor Müller is fully supported in assimilating Yama to the sun's course.

With regard to the second point alluded to, it may be observed that in general the same opposition which exists in the diurnal aspects of nature, as between day and night, morning and evening, repeats itself as between summer and winter, spring and autumn, and finally as between the primordial night and the morning of creation. These three distinct phases apply to the sun, and are distinctly to be traced in the character of the Egyptian Osiris. He is the nocturnal sun, the winter's sun, and the sun of the original night of chaos. His birth into the light under the form of *Ra* is the birth of humanity represented in the Adam of Genesis. That Yama's character included these three phases admits of little doubt. His mother is *Saranyû*, the "dark storm cloud which in the beginning of all things soared in space."[1] This is Professor Roth's interpretation of the myth of *Saranyû*, which appears to me perfectly correct, assimilating this "cloud," as he does, to the primeval chaos. Professor Müller rejects this view also, but I think without sufficient reason. The relation of *Saranyû* as mother of Yama is precisely like that of Osiris to Ra, considered in their cosmical aspects.

SEC. 120. The three phases of character appertaining to the sun-god, as symbolizing the life of humanity, correspond to the three periods of time whose divisions were marked by the twelve signs of the zodiac, namely, the day, the year, and the great cosmical year, which opened at the dawn of creation. Aries, the Ram, was the "lamb slain from the foundation of the world." Taurus, who appertained always to the watery element, personifies the watery chaos, the Taurus, in fact, from whose purified seed springs a male and female taurus, as in Persian legend, or is killed by the lion when the sun reaches its highest exaltation at the summer solstice. In all cases the life of humanity, history properly speaking, begins with Gemini, so that Aries and Taurus, in the Babylonian scheme, appertained strictly to the creation week of Genesis. Thus, Gemini as zodiacal constellation, and *As-kar* as leading star of

[1] Vid. *Lect. Sci. Language.* 2d series, p. 503.

Phœnicia, whose assimilation to each other was shown in the last chapter, really represent the paradisiacal man, the progenitor of the human race.

2d. *Unyal and Noah.* Having treated upon the first personage in the antediluvian genealogy, we proceed to consider the last, the hero of the deluge, the Biblical Noah. It is not intended here to discuss the subject of the deluge, except in so far as necessary to establish the zodiacal relation of the hero of it to the sign Pisces. The Hindu tradition of the deluge, as contained in the "Satapatha Brahmana," will be of much service to us, and not having the English version by Professor Müller, I present a rendering from the French of M. Lenormant : —

" One morning they carried some water to Manu with which to bathe himself, and while in the act of doing so he discovered a fish in his hands. It addressed him in these words: ' Protect me, and I will save you.' ' From what will you save me?' ' A deluge is about to destroy all creatures; it is from this that I will save you.' ' How shall I protect you?' The fish replied, ' While we are small we are in great peril, because the fishes devour each other. Guard me at first in a vase. When I shall become too large for it, construct for me a basin. When I am become still larger, carry me to the ocean. Then I will preserve you from destruction.' Soon he became a great fish, and he said to Manu, ' In the year in which I shall attain my full growth the deluge will take place. Construct then a vessel, and adore me. When the waters begin to overflow, enter into the vessel, and I will save you.'

" After having thus guarded it, Manu carried the fish to the ocean. In the year indicated he constructed a vessel, adored the fish, and when the deluge began, he entered into the vessel. As he navigated the ship, the fish came to him, and Manu attached the cable of the vessel to the horn of the fish, and by this means it was made to pass to the mountain of the north. The fish then said: ' I have saved you ; attach the vessel to a tree, so that the waters may not carry it away whilst you remain upon the mountain : and as fast as the waters retire, descend.' Manu descended with the waters, whence it is called ' *the descent of Manu* ' of the mountain of the north. The deluge had destroyed all creatures, Manu alone being saved." [1]

M. Obry states on the authority of M. Neve that the seven *Devas*, that is to say, the seven stars of the great Dipper, accompanied Manu in the vessel : and this would constitute the typical number

[1] *Premières Civilisations*, ii. pp. 124, 125.

of eight persons saved in the ark.[1] M. Lenormant sees in these
traditions the origin of the notion of *Saviour;* that is, he who pre-
serves from the destruction of the deluge. It is remarkable that
the god Hea, who appears as god of the deluge in the " Chaldæan
Account" rendered by Mr. Smith, takes very often in the texts the
title *Salmanu,* "Saviour," which recalls the expression of the fish
to Manu: " I have *saved* you." One of the most notable epithets
of Hea in the inscriptions is "sentient, or intelligent fish." But
that to be particularly considered in the present connection is the
title of Mercury in the sign Pisces: *ul kha an E-a,* "star of the
fish of the god Hea." It is to this sign that the star *Ungal,* Assyrian *Sar,* "king," appertains, as shown in the last chapter. If we
compare these facts with the Hindu legend, it will be plain at once
that both the Hindu and Chaldæan traditions of the deluge, associated with Manu and Hea, appertain zodiacally to the sign Pisces.
The matter is so apparent, in fact, as to exclude all doubt.

SEC. 121. Upon the subject of the identification of the Biblical
name *Noah* with that of the god *Hea,* M. Lenormant as late as
1874 published his views as follows: —

" The habitual Assyrian translation of the Accadian *É-a* in the
bilingual documents is *bit,* ' house; ' but it is certainly not thus that
we should read the name of this divinity. I believe it to be necessary to select for this reading a derivative of the root *Navah,* ' to
reside, to abide; ' that is to say, the word *Nuah, Nua,* ' abode, residence,' which in some Assyrian translations corresponds to the Accadian *É-a.* To the same root appertains likewise the name *Ninua*
(Nineveh), signifying equally ' abode,' and which is not only the
appellation of the city of Nineveh, but that of a goddess, daughter
of the divinity of which we seek to determine the name. Thus, I
believe that it is necessary to adhere to the reading *Nuah,* already
proposed by Dr. Hincks, but without being able at the time to render sufficient proof for it. That which fully determines me in
favor of this reading *Nuah,* which is the exact equivalence in sense
of the Accadian *É-a,* is the part played by the deity in question in
the (Chaldæan) account of the deluge, and the analogy of *Nuah*
with the Biblical *Noah.*"[2]

As I have already intimated, the author has corrected himself to
some extent in a more recent work. Not having it in possession at
the present writing, I am not able to state precisely the nature and

---

[1] *Du Berceau,* etc., p. 6.
[2] *Prem. Civilisations,* ii. pp. 130, 131.

effect of his correction, but hope to present the matter hereafter
in the form of a note.   It is to this extent, at any rate, that the
Accadian reading *Hea has been adopted into the Assyrian.*[1]   Mr.
Smith has proceeded on this principle in his rendering of the " Deluge Tablets," and, as it now appears, was wholly justified in doing
so.   But is the reading *Nuah* to be abandoned as wholly incorrect?
Dr. Hincks read *Nuah*, Sir Henry Rawlinson *Nuha*, adopted by Mr.
Norris, long before the discovery of the " Deluge Tablets."   It is
hardly possible that they had not some good reason for such readings,
both being similar to the Biblical *Noah*.   The literal sense of *Éa*
is house + water, "abode of the waters," so striking a reference to
the *ark* that it is difficult not to admit here an express intention.

[1] My recollection as expressed in the text was correct.   M. Lenormant doubts
the reading *Nuah* as the Assyrian for *Hea*, and thinks that Dr. Schrader and the
English Assyriologues have had reason for the supposition that the Assyrian retains the Accadian form of the name (*La Divination*, Paris, 1875, p. 89, note 3).
But the identity of the two personages, *Hea* and *Noah*, is placed beyond doubt by
two important facts to which the author alludes.

1st. The god *Hea* often takes the name of *Nisroch* in the texts, upon which the
author observes : " This enables us to comprehend how the Jewish tradition has
always associated the name of Noah with that of the Chaldæo-Assyrian deity
Nisroch — one of the appellations of Nuah (*Hea*) — in a manner which was till
now inexplicable. ' Nisroch,' said the celebrated Rabbin Raschi (Is. xxxvii. 38),
' is a plank from the ark of Noah.' "   (*La Magie*, p. 119.)

2d. In reference to the ship of Hea, the author again remarks : " One of the
hymns of the collection upon magic (vid. 4th Rawl. Pl. 25), extremely difficult
to comprehend since we have only the Accadian text without an Assyrian version, and since it is full of technical terms unexplained, turns entirely upon this
ship of Hea, ornamented with ' seven times seven lions of the desert,' and in
which Hea navigates : ' Hea, the master of destinies, with his wife Davkina,
whose vivifying word, Silik-Mulu-khi, prophesies the favorable renown of Mun-
abge, he who conducts the lord of the earth, and *Nin-gar*, the great pilot of heaven.' "   (Ibid., pp. 149. 150.)

These extracts prove not only that Hea had an ark or vessel in which he navigated the waters, but that, under the name of Nisroch, he was identified with
Noah in Jewish tradition.   The assimilation of Noah, therefore, to the sign Pisces,
to which appertains the name, "fish of the god Hea," must be considered as a
settled point.   The facts relating to Adam and Gemini on one hand, to Noah
and Pisces on the other, tend to support each other.   If Adam be assigned to
Gemini, Noah, as the tenth from Adam, must be associated with Pisces.   Or, if we
first determine Noah's position in Pisces, this fixes Adam's in Gemini.   The two
series of facts reduplicate their force in each other.   I see that Hea takes especially the title of *Nun*, "prince," *Ungal*, "king," in the texts; and this confirms
our remarks respecting the star *Ungal*, and its connection with Pisces.

Taken as the name of a personage, it would mean "he who abides upon the waters," an allusion to *Noah* so direct as almost to force us to assume it as such.

For the present, then, I am not able to identify etymologically, either the name *As-kar* with *Adam*, or that of *Hea* with *Noah*. But I strongly suspect such a connection between *As* and the Hebrew *Esh* or *Ash* (אֵשׁ), the old form of *Ish* (אִישׁ), "man" definitely applied to *Adam;* also between *Hea*, under the form of *Nuah*, and the Biblical *Noah*. If *As* meant "man," then *As-kar* denoted the "man of Eden," or garden of Eden, for the Accadian *Kar*, "summit," with the other value *Gan*, "inclosure, garden," evidently contains a crystallized tradition of paradise. But aside from these considerations, it seems to me the demonstration is complete that zodiacally the antediluvian genealogy of Moses commenced with the sign or constellation Gemini, and terminated with Pisces; and it follows that the scheme of Genesis was different in this respect from that of Berosus. In the hieratic form of the monogram for Nineveh, having the value *Ninua*, we have the image of a fish inclosed in a basin (Rep. 191). Compare with this the fish in the Hindu legend of the deluge, and the fish of Hea, the god of the deluge, appertaining to the sign Pisces. Compare, again, the account of Jonah's mission to Nineveh, his being three days in the belly of the fish, and finally the Saviour's allusion to this last circumstance, in view of his descent into Hades, over which, according to the Babylonian mythology, the god Hea was ruler. Noah was a type of the Saviour, the ark of the church, and the terrestrial paradise had been transferred to Hades, as previously shown.

In relation to the star *Ungal*, or of the king, it is unnecessary to add much to what was said in the last chapter. I think that the hero of the deluge was considered the king *par excellence*. According to the Babylonian system, this name is assimilated to Capricorn; but upon the hypothesis assumed by us, it should be connected with Pisces. But having shown that, zodiacally speaking, the antediluvian history began with Gemini, and terminated with Pisces, we proceed to the examination in brief of some of the intermediate names and periods.

Sec. 122. 3d. *Sir and Seth.* The Accadian *Sir* signifies "serpent;" as a star of Phœnicia, and as shown in the last chapter, it falls in the sign Cancer, which is the next after Gemini. It is per-

feetly obvious, according to my view, that we have here an express
allusion to the serpent of Eden.   If Gemini and *As-kar* represent
the paradisiacal man, then Cancer and *Sir* denote man after having
fallen under the destructive wiles of the serpent.   The Hebrew
term denoting the serpent of Eden is *Na-khash* (נחש), and it is
well understood by Hebraists that astronomically it relates to the
constellation Draco, or the Dragon, coiled around the north celestial
pole, the very region, as we have seen, associated in tradition with
the mount of paradise.   But this is, so to speak, the cosmical serpent, appertaining to the sun's course, as symbol of the life of humanity during the great cosmical year, already explained.   The
serpent that appertains to the sun's annual course, equally a symbol
of human life, is the constellation Hydra, and this corresponds well
to the sign Cancer.   Finally, the sun in its daily course encounters
a like enemy, the constellation Serpens, the Ophiuchus, often substituted for Scorpio, to which Jacob compared Dan : "Dan shall be
a serpent (Heb. *Na-khash*) by the way" (Gen. xlix. 17).   In each
of these three phases of the sun's course, of which, as before observed, the twelve signs mark the twelve divisions, this luminary
encounters an enemy in its path, almost at the opening of its career.
Thrice repeated, therefore, is the starry record of those primitive
traditions that relate to the temptation and fall.   It is hardly necessary to insist here, in view of the facts previously placed before
the reader, that these traditions are not purely astronomical.   The
sun's course was taken as a type of human existence ; the fact is
evinced by numberless expressions that might be gleaned from the
sacred writings of antiquity; and it is this very fact, all important
to be considered, which affords the true standpoint for the interpretation of the ancient legends relating to the death and resurrection of the sun-god.

The Assyrian name of the month answering to Cancer is *Du-u-zu*,
for which the Hebrew nomenclature substitutes *Tammuz*, name of
the Syrian Adonis, the sun-god who suffers a violent and untimely
death.   All these names refer to one and the same personage, to
which similar ideas were attached in each case.   The cuneiform
Tur-zi, "son of life," and apparently *Tam-zi*, "sun of life," are to
be added to the list of expressions involving the same conceptions.
In *Tur-zi*, "son of life," it is possible to see an allusion to *Abel*,
son of *Eve*, "mother of all living," since the name *Eve* is derived

from a word signifying " life," like the Accadian *Zi*. It is not only possible, but quite probable, I think, that such a traditional reference was intended. The death of Abel seems to have given to his name, in all antiquity, the significance of the beloved son, who meets with a violent and untimely death. It is not without especial meaning, as it appears to me, that the name of Mercury corresponding to the sign Cancer and thus to the star *Sir*, or of the Serpent, is *Da-pi-nu*, a title interpreted by M. Lenormant as signifying " protector, covering," similar thus to the notion of a protection, covering, a hiding-place from sin, the consequences of the fall.[1] This title is frequently applied to other divinities as protectors, but it is the especial title of Mercury as he attends the sun in the sign Cancer, corresponding to the month in which the sun-god dies by violence.

In consequence of the substitution of Seth for Cain, in the line of Adam's posterity, he occupies the second place in the antediluvian genealogy, and, according to our theory of adjustments, must be assimilated to the sign Cancer. Considering all the circumstances, the idea of a *substitution* here is very remarkable. The serpent, the temptation and fall, the promised seed that should bruise the serpent's head, the death of Abel as type of the great sacrifice, embodied in the legends of the dying sun-god, the conception of a covering, a substitution for sin, — all these ideas seem to centre in this one zodiacal division, receiving double significance from the coincidences that have developed themselves so unexpectedly under our hands. It is difficult to admit all these implications, yet the simple facts appear forcibly to suggest them, and we shall find hereafter nearly the same circle of ideas grouped around the sign Scorpio, as well as the extra-zodiacal constellation Draco. In such case a conscious design in each instance cannot well be denied. If, then, the star *As-kar* and the sign Gemini marked the period of the paradisiacal man, the star *Sir* and the sign Cancer embodied all those traditions relating to the fall, the promise, and the events immediately succeeding the expulsion from Eden.

SEC. 123. 4th. *Bar-tabba-galgal and Enos*. The expression *galgal* is a simple reduplication of the Accadian *Gal*, Assyrian *Rab*, "great," its repetition having the sense of "very great," or the superlative "greatest." Respecting the other elements in the name

---

[1] Vid. *Frag. de Bérose*, p. 251, for M. Lenormant's definition of *du-pi-nu*.

of this asterism, we have a bilingual phrase in which the god *Bar-tabba* is explained by the Assyrian *Ilu Kilalin*, the last term being defined "wholly," or "entire," by Mr. Norris.[1] The god *Bartabba* is thus the god "wholly divine," and the very great *Bartabba* is the "divinity supreme, highly exalted, all-powerful." All this applies with great force to the sun-god in his supposed extreme exaltation in the sign Leo, when the power of his rays is greatest. It calls to mind the expression "Lion of the tribe of Juda," applied by the Revelator (Rev. v. 5) to the Hebrew *Yahveh*, or Jehovah, in his manifestation in the Saviour. As appears from a remark by Rev. A. H. Sayce, Dr. Oppert considers the stars *Bar-tabba-galgal* and *Bar-tabba-dūdū* as epithets of the sun, in the two senses of "doubly great" and "doubly little," thus referring doubtless to its greatest power at mid-summer and its least power in mid-winter.[2] The sign of the Lion, then, symbolizes the sun in its supreme energy and force. If Gemini be put for the solar twins, Cancer represents the death of the twin sun by fratricidal hands. The Lion, then, typifies his resurrection and triumph over his last enemy, and the name of the fifth Arabian Lunar Mansion designates the power by which he was raised. This "Arabic title is *Dhira*, 'the paw,' that is, of the Lion, which the Arab astronomers stretch out over a much larger region of the sky than he occupies with us."[3]

The name Enos (Heb. אֱנוֹשׁ), applied to the son of Seth, is only another form of *Ish*, signifying "man," from which is *Ishah*, "woman." According to the theory of some exegetes, Eve supposed that her first-born, Cain, was the promised one, or he who should remedy the sad consequences of the fall, bruising the serpent's head. Hence, on the birth of Cain she exclaimed, "I have gotten a man, even *Yahveh*," this being considered the literal sense of the phrase, "I have gotten a man from the *Lord*." *Yahveh* was supposed to be the promised one, he who should come, the deliverer. The second occurrence of the divine name *Yahveh* in the Scriptures is in the singular expression connected with the name of Enos: "Then began men to call upon the name of the *Lord*" (Gen. iv. 26). The original has *Sam-Yahveh*. Much uncertainty exists as to the literal sense to be attached to the language quoted. Under any

---

[1] 3d Rawl. Pl. 68, No. 2, 1. 68. Cf. Norris, ii. p. 558.

[2] *Trans. Bib. Arch. So. Lond.*, iii. p. 167.

[3] Professor Whitney, *Orient. and Ling. Studies*, 2d series, p. 352.

circumstances we show a direct association of ideas as connected with the sign Leo, the star *Bar-tabba-galgal*, the names Enos and *Yahveh*, and the " Lion of the tribe of Juda."

5th. *Ninmakh and Cainan.* The star *Ninmakh* is Venus, and corresponds to the sign Virgo, or the Virgin. The name Cainan (Heb. קֵינָן) signifies " a smith," and is equivalent to *Cain* (Heb. קַיִן), according to Dr. Fürst, hence denotes " a lance, spear, originally that which is pointed." It is the name of a male personage, while *Ninmakh* denotes a female character, the goddess Venus. At first view there appears here an insuperable objection to our general theory respecting a zodiacal arrangement of the antediluvian genealogy. But underlying it there will be found an actual confirmation. Cainan, as just observed, is equivalent to Cain, signifying on one hand a lance, spear, arrow, or anything pointed, as a weapon of war ; on the other hand denoting a possession, he who was gotten, referring to the expression of Eve, " I have gotten a man," etc., on the birth of Cain. There is a double reference involved, therefore, to the great mother and to war, the weapons of war. Now Ishtar assimilated to Virgo was both the great mother and the war goddess, represented as an archeress. M. Lenormant observes : —

" The sixth month is called ' the month of the message of Ishtar,' and a passage of the prism of Assurbanipal formally attests that the archeress became for us the virgin, who corresponds to this month in the zodiac, and is no other than Ishtar." [1]

The connection of the bow and arrow with the very name Cainan is very direct. In addition to this we have seen in a previous study that, under the title of *Athtar*, Venus was worshiped in Arabia as a male personage, or at least as an androgynous divinity ; and furthermore, as war goddess, Ishtar certainly assumes a male character. Thus, the analogy of ideas is very striking in this instance, while the difference in sex offers no serious objection when all the facts are considered. It amounts, in fact, to no difference.

SEC. 124. 6th. *Nibe-anu and Mahalaleel.* Libra is here the corresponding zodiacal division. The star *Nibe-anu* is supposed to be the planet Mars, who had his domicile in Aries and Scorpio. But primitively the pincers or claws of Scorpio covered the entire zodiacal space now occupied by Libra. It was shown in a former study that the Egyptians figured a solar mountain in Libra, associated

[1] *Prem. Civilisations*, ii. p. 73.

with the mother goddess and her child on one hand, and on the
other with the god Tum. The personage last indicated, although
conceived as male by the Egyptians, appertained to the lower hemi-
isphere, and it is only natural, therefore, that the same personage
becomes female in the cuneiform texts. We have previously cited
the phrase in which the sublime mountain of Tum is said to be
Ishtar, calling to mind the Vedic Ida definitely associated with Mt.
Meru. The Accadian name of the month answering to Libra is
Tul-kû, which I interpret "sublime mountain." Thus, everything
indicates a specific design on the part of both the Egyptians and
Chaldæans to locate zodiacally the traditional mount of paradise in
Libra, considered accordingly as in the east. Of this sacred mountain, as we have shown, the temples were conceived as imitations or
artificial reproductions. These remarks now will help to reveal a
definite relation of *Mahalaleel* to the zodiacal Libra. The meaning
of this patriarchal name is "praise of God." The Assyrian name
of the month corresponding to Libra is *Tas-ri-tu*, which Rev. A. H.
Sayce explains thus: " *Tasritu* or Tisri is a tiphel form of *Esritu*, a
'sanctuary.'" [1] The connection, then, between *Mahalaleel* and this
zodiacal sign is that between "praise of God" and "sanctuary,"
which is, of course, sufficiently direct and striking.

7th. *Nammakh and Jared.* We have seen that according to
uniform tradition the diluvian mountain was the same as that of
paradise. It has been seen also that this sacred mount was located
zodiacally in the sign Libra. In Hindu tradition, as shown from
the legend of the deluge already quoted in this chapter, the mount
of the deluge was called the "descent of Manu," and it seems that a
precisely similar notion had been localized in the vicinity of Ararat
in Armenia. The diluvian mount of the east was thus the "mount
of the descent." Zodiacally speaking this idea had been associated
with the sun, which makes its *descent* from the solar mount located
in Libra as it passes from the superior into the lower hemisphere.
This descent takes place definitely when the sun passes from Libra
into Scorpio. These remarks will suffice to illustrate the connection between Scorpio and the patriarchal name *Jared*. — it signifies
literally "the descent," or "he who descends." Two facts appear
to conflict with the explanation here offered. First, Jared is in no
way associated traditionally with the deluge. Secondly, we have

seen that really the hero of the deluge was definitely connected
with the sign Pisces. These circumstances admit of an explanation
in harmony with our theory of adjustments, but certain very important principles are involved in the matter which cannot well be set
forth at present, and we must await further developments in these
researches. It will suffice to say now that everything in this case
is subordinate to the sun's annual course as representative of human life. Annually the sun makes its descent in passing from
Libra into Scorpio, but cosmically it would be different, — a matter
that we are not yet in condition to explain. We shall return to
it, however, in a future chapter. The connection of the star *Nam-
makh*, or " great destiny," with Scorpio has been already sufficiently
set forth.

8th. *Sugi and Enoch.* We labor under the same disability here
as in relation to Jared. Certain principles are yet to be established
before the significance of Enoch's assimilation to the sign Sagittarius can be realized. Enoch, as previously shown, is a name that
signifies " the initiated." The name *Sugi* means " a rival," according to Dr. Delitzsch.[1] The rivalry here is between Gemini and
Sagittarius, placed in opposition on the sphere. It is otherwise a
rivalry between the twins, like that of Jacob and Esau, Cain and
Abel, etc. This was one of the subjects pertaining to the ancient
mysteries, and hence the connection of Enoch, " the initiated," with
this zodiacal sign. The sun in the upper hemisphere, particularly
in Gemini, was one of the solar twins, and in the lower hemisphere,
in Sagittarius, the other solar twin. They were placed in opposition, in rivalry. Confirmatory facts and additional explanations
relative to these points will be hereafter submitted.

Sec. 125. 9th. *Qaqsidi and Methuselah.* The proposed assimilation of these names is to the sign Capricorn. The star *Qaqsidi*, as
before shown, is to be identified with the Southern Dog, which
anciently marked the extreme limits of the sun's course in the
lower hemisphere at the period of the winter solstice, when the
solar power becomes almost completely exhausted. This is, in fact,
the old sun conceived by the ancients as a man bowed down with
extreme age, who goes to renew his life in the dark waters of death.[2]
At other times the sun at this period was likened to a man afflicted

---

[1] Vid. *Assyr. Studien*, II. i., pp. 120, 121.
[2] Vid. Brugsch, *Matériaux*, etc., pp. 44, 45.

with leprosy, who goes to bathe in these waters to renew his health.
The name *Methuselah* signifies "man of the dart, or sword, a warrior." This, according to Hebrew lexicography. Professor Bush
remarks on the authority of the marginal reading: "The import of
this name in the original is ' He dieth, and the sending forth,' as if
it were an intimation of the sending forth of the waters of the deluge" (Notes, Gen. v. 21). According to the well considered opinions of M. Lenormant, Rev. A. H. Sayce, and Sir Henry Rawlinson,
the "Izdhubar or Deluge Tablets" discovered and translated by
Mr. George Smith, consisting originally of the number twelve, are
definitely to be connected with the twelve signs of the zodiac. Mr.
Smith himself rejects this view, but it is probable that very few
Assyriologues could be found to agree with him. In the system of
the Izdhubar series, the hero of the deluge appertains to the sign
Aquarius, and his father *Ubara-tutu* to Capricorn, to which also we
have assigned Methuselah. In M. Lenormant's view, *Ubara-tutu*
represents the old sun, and I think that Methuselah as the oldest
man is to be regarded in the same light. For the rest, it will be
convenient to introduce M. Lenormant's own language : —

"The moment when the sun commences to renew its force, and
enters upon its ascending path, the moment when it is cured of its
annual malady, and the fear of death, is precisely the eleventh
month (Aquarius) of the Chaldæo-Assyrian year, the month following that of the solstice of winter (Capricorn). Such being the case,
the culminating epoch of decadence, and of the leprosy of the igne-
ous and solar deity, ought to be in the preceding month, or at the
solstice. It thus appears in the poem. The symbolic malady of Izdhubar, who undertakes a voyage in search of Sisithros, appertains
to the tenth tablet (and sign). At the same time, in the Accadian
names of the months the tenth is called ' the month of the spot
(leprosy) of the declining sun.' "[1]

Rev. A. H. Sayce remarks: "The tenth month is termed *abba
uddi*, or *abba uddu*, the meaning of which is difficult to determine.
Abba signifies ' father,' also ' old ' and ' hollow,' and in the latter
sense is joined with *a*, ' water,' to denote the sea."[2] In relation to
Ubara-tutu, M. Lenormant again remarks : —

"That the name of the father of Sisithros, in the cuneiform documents, *Ubara-tutu*, a name taken from the Accadian tongue, signi-

---

[1] *Prem. Civil.*, ii. pp. 77, 78.
[2] *Trans. Bib. Arch. So.*, iii. p. 164.

fies 'the golden splendor of the setting sun,' and that the name of the father of Enoch, in the Bible, Yirad (*Jared* in the Vulgate), signifies 'descent, setting.' The Babylonian tradition unites with the personage Sisithros the facts which the Bible distributes between Enoch and Noah, and the name of the father of Enoch corresponds in Hebrew, in its signification (of 'descent'), to that of the father of Sisithros in Accadian." [1]

According to the Babylonian tradition, Sisithros (Noah) is translated. Hence, in accord with the same scheme the father of Sisithros has a name signifying "descent," the same as that of Jared, father of Enoch. There is confusion in the Babylonian scheme; the Mosaic is consistent throughout, and our theory of adjustments will help to bring order out of chaos.

(*a*) Izdhubar, whose name signifies "mass of fire," is the sungod. His resemblance to Nimrod arises from the fact that the latter was taken for the sun-god, also, both being war-gods likewise, as were nearly all the solar deities. In the character of warrior both corresponded exactly to Methuselah, "man of the sword, warrior."

(*b*) In the tenth sign, or Capricorn, the sun either contracts a fatal malady, like that of Izdhubar, and as denoted by the Accadian name of the corresponding month, or attains an extreme old age, like Methuselah, and finally dies at the winter solstice, to be renewed under the form of a little child, as held by the Egyptians.[2] It is amid the waters of the deluge, answering to Aquarius, that the sun renews his health, or attains a renewed life. Methuselah dies at the going forth of the waters.

(*c*) The exact analogy exhibited here between the ideas connected with Methuselah on one side, and the legendary conceptions centring in the sign Capricorn on the other, seems to me fully conclusive as to the soundness of our theory of adjustments for the Mosaic genealogy.

SEC. 126. 10th. *Bir and Lamech.* The anonymous author of "Palmoni," on the authority of Dr. Ewald's researches, has the following: "Lamech means 'the destroyer.' The connection of this name with the introduction of wickedness, and the superinduction of the condemnation of the world, has been shown" (p. 73). The Accadian *Bir* has the sense of "to crumble, to fall in ruins, ruin"

---

[1] Op. cit., pp. 59, 60.
[2] Vid. Brugsch, *Matériaux*, etc., p. 9.

(Rep. 441).   Both the star *Bir* and patriarch Lamech appertain to the sign Aquarius.   This zodiacal division properly represents the destructive period of the deluge, when the old world fell to ruins. As Noah was saved from this catastrophe, being preserved to renew the period of human existence, he is naturally assimilated to the sign following, or to Pisces.   The correspondence of ideas associated with the names *Lamech* and *Bir* and with the sign Aquarius could hardly be more direct and perfect, and it adds another powerful support to the general hypothesis with which we have been occupied in the present chapter.

The evidences tending to connect the beginning and the end of the Mosaic genealogy of the antediluvian period with the signs Gemini and Pisces respectively were seen to be of a nature very difficult to set aside, and almost to force conviction.[1]   But the question would naturally arise, notwithstanding, whether an attempt to associate the other names of the list, according to the same scheme, with their respective zodiacal divisions, would not be attended with so many obstacles and develop so many objections as actually to counterbalance the force of facts previously presented.   It seemed proper, therefore, that we should make such an attempt, and the reader is now aware of the result.   We have encountered no serious obstacle, and on the other side have discovered much additional proof.   The evidences have not only been cumulative in their nature, but they have almost constantly multiplied upon our hands. According to all the principles of induction, it seems to me the hypothesis assumed ought to be considered as established.

As before observed, it is perfectly plain that the Babylonian adjustment of the ten antediluvian kings to the signs of the zodiac was radically different from that of the Mosaic genealogy.   The inference is thus quite natural, that the two schemes appertained to different, or at least independent, traditions respecting the period before the deluge.   Under any circumstances the fundamental difference here shown to exist is of great importance to the Biblical critic.   It places such critic on an independent footing so far as concerns the Babylonian traditions, an object very desirable to be

---

[1] In point of fact, although the identity of *names* still remains in some doubt, the absolute identity of *personages*, that is, of *As-kar* with *Adam*, and of *Hea* with *Noah*, and their assimilation to the zodiacal Gemini and Pisces, is a matter placed beyond all doubt by the data submitted.

attained, I think, in view of the tendency of recent discoveries to make Babylon the exclusive source of the Mosaic history pertaining to the primitive ages. These discoveries, properly interpreted, afford important confirmations of the Hebrew narrative; but it is well to be assured of the fact that the Hebrews were not necessarily indebted to the Babylonians for the Book of Genesis.

The question of the chronological succession of these genealogies, in view of the facts now before us, still forces itself upon our consideration. Were they not purely zodiacal, having no historical basis? That they were supposed to be chronological, that the cosmical year was regarded as a genuine time period, and thus in some sense historical, does not, in my mind, admit of serious doubts. It is not necessary that the cosmical year be scientifically accurate. It was supposed to be and was practically thus assumed. Did this cosmical year correspond to the precession, or retrograde movement, of the equinoctial points on the ecliptic, comprehending about twenty-five thousand years in each revolution? Such should have been the nature of it in order to give it a scientific character, and possibly it was so designed. But the reigns of the kings before the deluge, according to Babylonian estimates, had no correspondence with the period of precession appropriate to each sign. This would be about two thousand years to each of the ten kings, or some twenty thousand in the aggregate. On the one hand, the reigns of these kings were supposed to be far greater, and on the other it will not be difficult for us to show that actually the astronomical period to which the earliest traditions pertained was much less than twenty thousand years, even reckoning from the present time. There was, then, so far as we know, nothing of scientific chronology in the assumed cosmical year, nor in the assimilation of these kings to the zodiacal signs. But, as before remarked, everything seems to have been assumed as correct. For the rest, it will be necessary for us to attain some fixed standard date in high antiquity before we can treat this matter intelligently, and this is the ultimate result to which these investigations are directed.

CHAPTER XII.

THE CHERUBIM.

SEC. 127. The reader is requested now to turn to the transcription of the tablet upon which we have been so long engaged, as given in the 113th section, and to recall the explanations there submitted. In the second column of months, the seventh line, we have the Accadian phrase, *Ab Gut Mat Ilama-ki*, or "the month of the Taurus, the country of Ilama," the Biblical Elam, modern Susiana. Following this equation of the month of the Taurus to Ilama, in the eighth line, is the phrase *Ab Martu-ki*, "the month of the west," or Phœnicia; but what *particular* month is intended is not here indicated. Still below these, in the ninth line, we have the name of the country *Gu-ti-i*, or *Gutium*, to which obviously some month had been assigned by the scribe, but the name of it is entirely defaced. The reader sees in the extract from M. Lenormant (Sec. 108), that in the symbolical system of geography of which Akkad formed the centre, and of which this tablet constituted an exposition, the country of *Ilama* was put for the east, *Martu* for the west, *Gutium* for the north, and *Subarti* the south. It is obvious that to each of these four countries a special month was assigned, or rather a particular zodiacal constellation. To *Ilama*, put for the east, we know from the phrase just cited that the Taurus was assimilated. Beyond this, we are left almost wholly to conjecture, owing to the mutilated condition of the published text. It is true that Rev. A. H. Sayce has undertaken to complete the scheme, assigning Aries to Akkad, Taurus to Elam, Cancer to Martu, the west, or Phœnicia, Virgo to Gutium, without suggesting any equation for Subarti, representing the south.[1] So far as concerns the published text I am unable to find any sufficient basis for this scheme, and am persuaded that it is erroneous

[1] *Trans. Bib. Arch. Society*, London, vol. iii. p. 172, and note 2.

in respect to some of its proposed assimilations. The principle established in the 113th section, namely, that in the published text the second column of months offers an adjustment of the second column of stars to the ordinary Babylonian calendar, this principle, I say, enables us to assume with safety that *Mat Ilama-ki* has for its asterism the "star *Al-lab*," which is thus shown to be one with *Alap*, or the Taurus; and again that *Mar-tu-ki* is presided over by the "star *Lul-a*," or *Lab-a*, as I prefer to read the name. This must be Aries, the Ram: 1st. Since some *animal* is denoted by this name, as shown by Rev. A. H. Sayce in the place just cited; 2d. For the reason that all the other stars have been already adjusted to the zodiac, leaving only the *star Lab-a* for the *sign Aries*. In such case Aries is put for *Mar-tu-ki*, the *west*, just as Taurus is put for *Ilama-ki* and the *east*. The two zodiacal divisions are assigned to exactly *opposite* points of the compass, east and west, although normally they follow each other in the zodiac. This one fact proves a mystical intent on the part of the scribe, which we have now to study.

The fact has been heretofore sufficiently verified that the symbolical geography of Sargon the ancient, of which scheme the country of Akkad formed the centre, had been a traditional inheritance from the sacred mount of the northeast, the Gan-Eden of Genesis. Such being the case, Akkad replaces in this system the sacred mount itself, with which a like geographical scheme was associated. Now to the east of Akkad, as especially representing this cardinal region, was placed the country of Ilama, and to this region is assigned also the star *Al-lab*, evidently the same as the Assyrian *Alap*, applied to the Taurus, and a frequent designation of the man-headed bulls, whose connection with the Biblical Cherubim has been often suggested by scholars, and which will receive ample proof in the sequel of the present chapter. Bearing the two facts in mind, then, 1st, that Akkad here replaces the Gan-Eden of Genesis; 2d, that the star *Al-lab* or *Alap* is definitely put for the country of the east, compare the following statement of the Mosaic text; "So he drove out the man: and he placed at the east of the garden of Eden cherubim, and a flaming sword which turned every way, to keep the way of the tree of life" (Gen. iii. 24). In view of these considerations it will be difficult to resist the conclusion that the mystical intent of the scribe was, in this geographical symbolism tradition-

ally inherited from Eden, and in definitely assimilating the Taurus
to the country of the east, to incorporate in this arrangement the
tradition of the cherubim placed to the east of the Garden of Eden.
To this end he has put the two zodiacal constellations, Aries and
Taurus, for the two opposite points of the compass, — Aries for the
west, Taurus for the east, although they directly succeed each
other in the ordinary zodiacal order. Thus, as Aries answers to
the west, we must conceive Taurus in direct connection with the
opposite sign of the zodiac, which is Libra. But of this hereafter.

SEC. 128. We must study now the actual form of the cherubim.
That the *body* of these symbolic figures, to which four faces were
attached, according to the Scriptural accounts, was that of an *ox* or
bull admits of the clearest demonstration. One proof is derived
from the double description of these symbolical animals found in
the prophecy of Ezekiel. This writer proceeds thus : " And every
one had four faces, and every one had four wings. And their feet
were straight feet; and the sole of their feet was like the sole of a
calf's foot." " As for the likeness of their faces, they four had the
face of a man, and the face of a lion, on the right side: and they
four had the *face of an ox* on the left side ; they four also had the
face of an eagle " (ch. i. 6, 7, 10). The " face of an ox " and
the " feet of a calf " indicate quite clearly that the body of the
cherubim was modeled after the form of the taurus. This supposition is fully established by the description of the same figures in
the tenth chapter, where instead of the expression, " face of an ox,"
we have substituted " the face of a *cherub*," this term being the
singular of which the word *cherubim* is the plural. " And every
one had four faces: the first face was the face of a *cherub*, and the
second face was the face of a man, and the third the face of a
lion, and the fourth the face of an eagle " (x. 14). The four
faces of our description are evidently identical with those of the
other ; but the term *cherub* (כרוב) takes in the tenth chapter the
place of the *ox* (שור) in the first description. This proves that the
two terms were considered by the prophet as designations of one
and the same animal, — in other words, that the ox or taurus was regarded characteristically as the *cherub*, a term from which the plural
cherubim is formed. Dr. Faber, in the great work already cited in
these pages, regards the data furnished by these two descriptions of

the symbolic animals as wholly conclusive upon the point before us, and M. Lenormant adopts the same view.[1]

Another evidence not less conclusive that the body of the cherubim was that of the taurus is derived from the fact, now definitely settled, of the direct connection of the Scriptural cherubim with the man-bulls of Assyria. Exegetes had long since conjectured that there was such a connection, and now M. Lenormant produces the satisfactory proof of it in the following extract: —

"M. de Saulcy, with the ingenious and fine archæological tact that distinguishes him, has labored to establish the identity of the cherubim of the ark of the covenant, in the Holy of Holies of Solomon's temple, with the winged bulls placed at the gates of the Assyrian palaces. . . . A comparison of verse 14, 10th chapter of Ezekiel, with the 10th verse of the 1st chapter, proves that the prophet employs the word *Kerub* (*cherub*) in the sense of 'taurus.' The question is to-day decided, conformably to the opinion of the learned Academician, by the testimony of the Assyrian inscriptions, where we find the winged bulls with a human face designated alternately by the words *alapi* (bulls) and *kirubi* (ditto), and where the expression *kirub* is extended afterwards by catachresis to the gate itself flanked by these symbolic animals. Thus we see that we have not to search, as many scholars have done, such as Tuch, Gesenius, and Renan, for an Aryan origin of the Biblical name *cherubim*, comparing it to the Griffons (Γρύφες), located in India by the Greek legends, because *kirub* is a word perfectly Semitic, and is often applied to the ox as an animal employed in agriculture. The poetic imagination of the Hebrews represents the cherubim as guarding

[1] Faber, *Pagan Idolatry,* i. p. 421. Lenormant, *Frag. de Bérose,* pp. 137, 138. Exegetes have often maintained that the substitution here of *Kerub* for *Shar* is not conclusive as to the form of these figures. Admit that the demonstration is not mathematical; yet it is a moral certainty, considering the facts: 1st. That *Kerub* is known to have designated an ox, independently of Ezekiel's language. 2d. That the four faces of one description are the same as those of the other. 3d. That instead of the "face of an ox" in one instance, we have the substitution of the "face of a cherub" in the other. This proves that Ezekiel was aware of the application of the very term "cherub" to designate the "ox," and thus that the substitution of one for the other was perfectly legitimate. Now, if the singular "cherub" means an ox, certainly the fundamental idea, and thence the ground form, of the "cherubim" must have reference to this animal. However, other evidences not less conclusive follow in the text. The fact that the cherubim of Eden are placed in direct relation to the "tree of *life*," and that the taurus was universally the symbol of nature's generative forces, ought to suggest the leading idea of these symbolic animals.

the gate of the terrestrial paradise, just as their analogues did the Assyrian palace." [1]

The term *Alap*, or *Alapi* for the plural, preceded by the determinative of divinity, is quite frequent in the cuneiform texts as designation of the man-headed bulls. I have never noticed an instance where the word *Kirub* was applied to the same symbolical figures, and was inclined to doubt such usage for this term; but M. Lenormant's language, although he cites no inscription, shows that he is certain of the existence of texts of this character. Thus, the two species of evidence now before us leave no room for doubt that the taurus was characteristically the cherubic animal. It is this animal that the arrangement of the tablet of twelve stars forces us to locate in the zodiacal sign Libra, as intimated already, in which sign the Egyptians placed their solar mountain, evidently to be identified with the mountain to which Ishtar of Babylon, and Ida in India, had been associated, that is to say, the mount of paradise.

SEC. 129. It is quite evident that there existed on the part of the scribe, in the peculiar arrangement of our tablet, another design accessory to the one already pointed out. It is important to note the fact, in the first place, that the sign of the Balance, or Libra, was unknown to the Babylonian sphere. It is correctly stated by M. Lenormant, on the evidences furnished by the inscriptions, that the Pincers of the Scorpion occupied the zodiacal space, where we find the Balance in the Greek as well as in our modern zodiacs.[2] Another fact equally significant in its bearings, as we shall soon discover, relates to the symbolical representation of the sun in its annual course under the form of the taurus. A proof of this was presented in a previous chapter, as derived from the Accadian term *Gut-tam*, or *Gut-tav*, explained by the Assyrian *pitnu sa same*, "the furrow of heaven," that is, the ecliptic. The element *Gut* denotes the "taurus," and *Tam* is one of the values of the Accadian character designating the "sun." Thus, the sun in his zodiacal journey was deemed the solar bull, which yoked with the heifer, as personifying the moon, ploughed the celestial furrow. But the taurus in this case was likewise a representative of the generative force of nature, a force that was considered as appertaining to the sun, especially in connection with the moon. Bearing these facts in mind,

---

[1] *Frag. de Bérose*, pp. 80, 81.
[2] Vid. *Prem. Civilisations*, ii. pp. 68, 69.

we proceed to still another. Upon the ancient art monuments, designed to represent the annual sacrifice of a taurus, we see frequently figured a serpent and scorpion attacking the prostrate victim; and uniformly the scorpion is seen in the act of "seizing, holding, griping" with destructive force the generative organs of the victim.[1] There can be no mistake as to the design of these artistic representations. The taurus symbolizes the productive power of the sun in its annual course. The sacrifice of the taurus is then a type of the decay and gradual destruction of that force, as the sun passes from the upper into the lower hemisphere; and it is precisely in the signs Libra and Scorpio, both spaces formerly occupied by Scorpio alone, that this "descent" of the sun into the lower regions takes place. Naturally, owing to the circumstances here stated, the Taurus had been associated in conception with the sign of the Pincers, or Libra, denoting that period when the sun's productive power began to decay, and when this animal symbol of nature's generative force fell a victim to the power of darkness. M. Dupuis makes the following correct statement relative to the sign Scorpio: "The Egyptians placed here the seat of the evil principle Typhon, who kills Osiris, or the god with the head of an ox, as Plutarch observes, during the month when the sun passes through this sign."[2] The god under the human form, represented with the head of an ox, is simply another mode of expressing the notion involved in the man-headed bull, the symbolical animal of the Assyrians, corresponding in name and character to the Hebrew cherub or cherubim. The symbolical design of the scribe, then, in the peculiar arrangement of our tablet of twelve stars is manifestly that which we have supposed. There is a reference to the sun's descent into the lower hemisphere, and to the sacrifice of the taurus, at that period when this luminary passes from the sign of the Pincers into Scorpio itself. But there is obviously a primary reference to the tradition of the cherubim of Eden; since the solar mountain of the Egyptians, and the mountain to which Ishtar was assimilated, relating doubtless to the mount of paradise, were alike located zodiacally in Libra.

SEC. 130. We are prepared now to entertain the question upon

---

[1] Vid. Lajard, *Culte de Venus*, Pl. XVI. Cf. De Hammer, *Culte de Mithra*, Pl. XXIII., etc.

[2] *Origine de tous les Cultes*, iii. p. 239.

which critics have held so many different and contradictory opinions, — the question which regards the etymology of the word cherubim.　Dr. Kurtz, in the article on the "Cherubim" in Herzog's
"Real-Encyclopädie," as found in the condensed American translation of this work, opposes the opinion of Dr. Delitzsch, as seen in
the subjoined extract : —

"The explanation of Delitzsch is equally unsatisfactory, in which
the word *Karab* (כרב), according to the rule of radically related
verbs, *Qa-rab*, *Tra-rab* (קרב, צרב), and according to the analogy of
the Sanskrit *Gribh*, the Persian *Giriften*, and the Gothic *Gripan*
(Gr. Γρίφες), has as its ground the sense of *seizing, taking, holding*,
and the cherubim are supposed to be those who hold and sustain
the throne of God." [1]

Dr. Kurtz very pertinently objects, I think, that the Biblical
cherubim are never represented as seizing, holding, or sustaining
the throne of God. Nevertheless, the etymology of the Hebrew
Kerub, from which the plural *cherubim* is formed, as given by Dr.
Delitzsch, is supported by Dr. Fürst, and is undoubtedly correct.
The writer last named (vid. כרב) gives the following derivation :
"*Karab* (כרב), not used, equivalent to *Qarab* (קרב), in *Akrab*
(עקרב, 'scorpion'), 'to seize, to lay hold of, to grip,' the same as
Garuph (גרף), 'to hold fast.' From which *Kerub* (כרוב)." Dr.
Fürst (art. כרב) agrees with Dr. Delitzsch in construing the fundamental notion of seizing and holding, with reference to sustaining
or upholding the chariot or throne of Jehovah. But it appears to
me that Dr. Kurtz's objection to this view, as stated above, holds
perfectly good. One still more serious, I think, is derived from
the ground idea itself, as involved in *Karab*, equal to *Qarab*, in
Akrab, "the scorpion." The sense is not "to sustain, to uphold,"
as the throne of God, but "to seize, grip, hold," as the scorpion
with its pincers or claws, or as the bird of prey — the eagle for
instance — with its talons. That the conception last indicated is
really that from which the notion of the *cherub* has proceeded is
proved by two important facts : —

1st. Precisely the same sense of "to seize, to grip, to hold,"
involved in the Hebrew *Qarab* and *Akrab*, attaches to the Aryan
Merops, according to Professor Curtius (Grundz., p. 456), and
Merops was identified with the eagle by the Greeks. The eagle is

[1] *Prot. Theolog. Encyc.* i. ; art. "Cherubim."

sometimes substituted for the serpent in the Biblical descriptions of the four faces of the cherubim. The Garuda, or eagle, in the Hindu legends was directly associated with the sacred tree as guardian genius, and was otherwise conceived as one of the griffons, supposed to guard hidden treasures, just as the winged bulls protected the treasures of the Assyrian palaces. The etymology of the term *Kerub* proposed by Tuch, Gesenius, Renan, and others, from the Sanskrit *Gribh*, Greek *Gruphes*, was thus so far correct as pertains to the ground notion involved.

2d. The identical term *Ki-ru-bu*, assumed by Dr. Fried. Delitzsch as one with the Hebrew *Kerub*, thus confirming the views of M. Lenormant already cited upon this point, is definitely applied to designate a bird of prey, thought to be the vulture or hawk by the German Assyriologue just named.[1] Thus, although the Assyrian *Kirub* was otherwise employed with reference to the man-bulls, the sense involved was not that of "sustaining, upholding," but that of "to seize, to grip," like the scorpion or a bird of prey, and the evidence to this effect is the fact that the term was sometimes employed to designate the vulture, hawk, or a bird of like ferocious habits.

The data thus presented must be considered as perfectly conclusive, in my estimation, that the ground thought involved in the Hebrew term *cherub*, from which *cherubim* is formed, was that of "to seize, grip, hold," as the scorpion with its pincers or the eagle with its talons. But in fact, as already established, the cherubic animal is the *taurus*. By what possible association of ideas, then, could the notion of "seizing, griping, holding," be applied to the taurus? The peculiar and anomalous arrangement of our tablet, which has been described, taken in connection with the representations upon the art monuments, already alluded to, pertaining to the annual sacrifice of a taurus, affords us a most complete explanation as regards the rare combination of ideas such as that involved in these symbolic animals. The taurus to which more immediate reference is had is the solar bull, which personifies in his sacrifice the orb of day as it makes its annual "descent" from the sign of the Pincers, or Libra, into the lower hemisphere, significant type of

---

[1] *Assyr. Studien.* II. i. pp. 107, 108. Cf. 2d Rawl. Pl. 37, No. 1, Revs. l. 17. *Ku-ru-bu* is here taken by the author as radically one with *Ki-ru-bu*. Dr. Schrader is cited for the assimilation of the term to the Hebew *Kerub*, and the etymology of Dr. Franz Delitzsch already shown is adopted.

the destruction of nature's generative force. But we have seen, on one hand, that the sacred mount of the east was located zodiacally in the sign of the Pincers, where the Egyptians were accustomed to represent the great mother, who corresponds so exactly with the "mother of all living," and, on the other hand, it has been shown that the sun's course was regarded in antiquity as *a symbol of the life of humanity* (Sec. 85). This "descent," therefore, to which the name *Jared* referred, of which the sun's passage into the lower regions was a symbol, was the *fall of man!* The terrible loss attending that sad event, hereditarily transmitted to Adam's posterity, was like that of the sun as its celestial fires are quenched in the dark waters into which it descends at midwinter, or like that of the celestial taurus, attacked by the serpent and scorpion, or, finally, like that of the ox which has been houghed! The inheritance of an incapacity by the sons of Adam, such as is here so plainly indicated, is simply frightful to contemplate, but it accords perfectly with the dogmas of the old theology, and gives an intense significance to many a passage of the sacred text.

Sec. 131. The views set forth in the last section respecting the symbolical ideas fundamental in the conception of the cherubim find obviously a direct confirmation in the facts established in the second chapter (Sec. 21), with reference to the "tree of life," as well as in the various notions traditionally inherited by different Asiatic peoples pertaining to the sacred tree. The investigations referred to sufficed to show: 1st. That the *palm* had been taken for the tree of life, or at least as a type of it, even by the inspired writers; 2d. That the palm, and thus the tree of life itself, was directly associated with the calendar, being taken, in fact, as a symbol of the year, derived from its twelve fruit-harvests. As stated by the author of Genesis, the cherubim were placed to the east of the Garden of Eden, "to keep the way of the tree of life." Thus, if this tree was itself connected with the calendar, with the sun's annual course, it is necessary to assume that the cherubim were likewise thus associated. Such an inference, of course, would tend to confirm the opinion long since put forth by M. Dupuis that the cherubic figures, with their four faces, were closely related in origin to the four zodiacal constellations, or animals having similar faces, namely, the ox or taurus, the lion, the serpent or eagle, substituted for the scorpion, and finally the water-bearer, having the face of a man. In view of all the facts, I feel compelled to adopt substantially M. Dupuis' suggestion. But instead of seeing here any proof of the astrological origin of all religions, as this author does, I recognize in these symbolic figures, in connection with the data already introduced, an astronomical record, so to speak, confirming the Mosaic account of the fall of man. For the rest, the cuneiform texts afford the evidence that two distinct classes of *Alapi*, or *Kirubi*, were well known to the Chaldæo-Assyrians; that is to say, the celestial and the terrestrial. Thus we have the following phrases : [1] —

Il Alap an-ta, "the god Kirub of heaven."

Il Alap ki-ta, "the god Kirub of earth."

In this case, it is quite certain that the celestial *Alapi* were the prototypes of the terrestrial; and as the Biblical cherubim have been shown to have had a direct connection with one of these classes, it is necessary to infer a primary reference also to the other. According to the mystical arrangement of the scribe in our tablet, we must conceive the celestial Kirubi as located zodiacally in Libra, or the sign of the Pincers of the Scorpion. In this sign, as we have seen, the Egyptians placed the solar mountain, together with the great mother and her child, an obvious reference to the seed of the woman which should bruise the serpent's head (Gen. iii. 15). Now Joseph, symbolized by the ox, was evidently a type of Christ, the veritable seed of the woman. Joseph's descent into Egypt was like that of Christ into Hades, a region definitely assimilated to the lower hemisphere, as shown in our seventh chapter. Not only this, but the terrestrial paradise had been transferred to the same region ; and to this the Saviour alludes in the language : "This day shalt thou be with me in paradise." All this was represented in the annual course of the sun, which, as a symbol of the life of man, and especially of the God-man, goes to illustrate the significance of the notion fundamental in the conception of the cherubim, as already explained in the previous section.

It is necessary to take into consideration here another order of facts, pertaining to the tree of life as typified by the palm, and as embodied in nearly all the Asiatic traditions respecting the sacred tree. There was the male-palm and the female-palm, or date-tree. It was the practice in Babylonia, as stated by Herodotus, to tie the branches of the two species together, for the purpose of the produc-

[1] 3d Rawl. Pl. 66, Obs. Col. 3, ll. 26, 27.

tion of fruit. Rev. Mr. Rawlinson adds to this statement the following in a note: " All that is required for fructification, they tell us, is that the pollen from the blossoms of the male-palm should come into contact with the fruit of the female-palm or date-tree." [1] The name of the sacred tree answering to the tree of life, among the Aryans of India, is explained by M. Lenormant as the " tree of desires or of periods." [2] According to the same author, the representations of this tree upon the art monuments from the Euphrates valley sufficiently indicate the traditionary idea of its sexual character. If we connect these various notions inherited from tradition by nations widely separated with the data previously presented, especially in the twenty-first section, it will be difficult to avoid the conclusion that the tree of life had reference, not to any abstract principle, of which the first men had little conception, but to life in the concrete, and under its ordinary manifestation as a reproductive force. Such being the case, the symbolism which we have found fundamental in the conception of the cherubim, and even in the origin of the term, can hardly fail to be regarded as embodying a correct interpretation of the entire subject. The progenitors of mankind had sustained an irreparable loss, which had been entailed upon all succeeding generations by the natural laws of hereditary transmission. From that hour, the human race had ceased to be able to reproduce the paradisiacal man. The record of this terrible calamity had been symbolically represented at an unknown period in the two divisions of the zodiac formerly occupied by the Scorpion and its Pincers. For myself, I can make nothing else out of the data which have been submitted to the reader. Two other records, having the same import, are yet to be traced in the heavens.

SEC. 132. We have not yet exhausted the symbolism inherent in the cherubim. They have often been interpreted, in fact, as types of superhuman strength, of divine power; and their direct connection with the god Hercules in the Asiatic systems can be readily shown. The Assyrian Hercules takes usually in the cuneiform texts the name which is read Nin, Nin-ib or Nin-ip by the Messrs. Rawlinson, Adar by M. Lenormant. The simple *phonetic* values of the characters employed in writing the name of this di-

[1] Rawl. *Herod.*, i. p. 259, and note 7. On the sexual character of the palm, consult also Robinson's *Calmet*, art. " Palm-tree."
[2] *Frag. de Bérose*, p. 325.

vinity would be *Nin-ib*, with the permutation of the last element to *ip*, if one chooses. The ideographic value of these two signs would be *lord + generation*, or "lord of generation." It is well known that this personage was identified with the planet Saturn at an epoch quite early, although primitively, as M. Lenormant shows, he was the sun of the lower hemisphere. It is a fact well established that *Nin*, as Hercules-Saturn, had for his symbol the man-bull, already identified with the Biblical cherubim. One proof of this is the very name of Saturn, *Lubat-sak-us*; thus, *lubat*, "animal;" *sak*, "head;" *us*, "male, man;" hence, "the animal with the head of a man." *Nin* was likewise assimilated to the fish-god, to the god Hea, in fact, as we shall show. The extracts which we proceed to introduce will sufficiently verify the different statements just made. Rev. Mr. Rawlinson observes: —

"Many classical traditions, we must remember, identified Hercules with Saturn; and it seems certain that in the east at any rate this identification was common. Nin, in the inscriptions, is the god of strength and courage," etc. "In many respects he bears a close resemblance to Nergal or Mars. Like him, he is a god of battle and of the chase," etc. "He is the true 'fish-god' of Berosus, and is figured as such in the sculptures. In this point of view he is called 'the god of the sea,' 'he who dwells in the deep,' etc. Nin's emblem in Assyria is the man-bull, the impersonation of strength and power." [1]

That *Nin* or *Adar* was the true fish-god, identified with Hea and associated with the man-bulls, will appear for the rest from the following: 1st. The name of Mercury answering to the zodiacal sign Pisces is *Ka an e-a*, "fish of the god Hea." 2d. The Assyrian name of the month corresponding to Pisces is *A-da-ru*, or *Adar*. 3d. But the god Hea, in the distribution of the divinities according to the zodiacal signs, is otherwise assimilated to that of Taurus.[2] 4th. In his relation to this sign, Hea takes the title of *Alap Shamas*, "Taurus of the Sun." [3] But I desire to add here the important testimony of Sir H. Rawlinson, who has the following with reference to *Nin* or *Adar:* —

"In the stellar tablets it is clearly established that the god in question must represent the constellation Taurus, in virtue proba-

[1] *Five Monarchies*, etc., i. pp. 132, 133.
[2] Vid. Sayce, *Trans. Bib. Arch. So.*, iii. p. 148, note.
[3] Vid. Lenormant, *Frag. de Bérose*, p. 114.

bly of his connection with the man-bull, which, as the impersonation of strength and power, was dedicated to him." "M. Raoul
Rochette, in his elaborate memoir on the Assyrian Hercules, . . .
viewing the subject from a classical rather than an oriental point
of view, has accumulated abundant evidence to show that Hercules
was commonly confounded in the east with Saturn.   Damascius
thus quotes a tradition on the authority of Hellanicus and Hierony-
mus, the Peripatetic: 'That from the two primitive elements, water
and earth, was born a dragon, who, besides his *serpent's* head, had
two other heads, those of a *lion* and a *bull*, between which was
placed the *visage of god*' (Θεοῦ πρόσωπον, Ὠνομᾶσθαι δὲ Χρόνον ἀγήρατον
καὶ Ἡρακλῆα τὸν αὐτόν)." [1]

The Greek term *prosōpon* (πρόσωπον), employed above in reference to the "visage" of god, is the ordinary expression denoting
the "human countenance" as distinguished from those of animals.
It was termed in this case "the visage of god" instead of a man,
since the "dragon" alluded to was taken for a divinity.   But nothing is plainer than the fact that this "dragon" is to be identified
with the Biblical cherubim.   The proofs are: 1st.   That the cherubim of Scripture are to be identified with the man-bulls of Babylon
and Nineveh, as already fully shown.   2d.   That *Nin* or *Adar*, as
Hercules-Saturn, had for his emblem the man-bull, a fact also
abundantly established.   3d.   That this "dragon," having the four
faces of a *man*, a *lion*, a *bull*, and a *serpent*, substituted often for
the *eagle*, a description identifying this composite animal with the
Biblical cherubim in a manner not to be mistaken, was taken for
the god Hercules, at the same time for *Chronos* or *Saturn*, being in
this sense a time god, and thus answering exactly in character to
the Assyrian Hercules-Saturn.

SEC. 133. The data presented in the last section must be regarded as conclusive, I think, to the effect that the symbolic animals, termed *Kirubi* in the cuneiform texts, *Cherubim*, or *Kerubim*,
by the Biblical writers, were definitely associated with the Hercules,
or Hercules-Saturn of the Asiatic religions.   But it remains to show
that this personage was primitively the sun.   Alluding to *Nin-ip*,
whom he calls *Adar*, M. Lenormant has the following : —

"This entire side of the character of *Adar* is a remnant of his
ancient nature as solar deity, and it shows that when he possessed
this character he was the personification of the sun of darkness and

---

[1] Vid. Rawl., *Herod.*, i. pp. 504, 505, and note 9.

night, of the sun in the inferior hemisphere. This is indicated also by his essentially funeral nature, evinced in the ceremonies in honor of the Assyrian Hercules, most thoroughly studied by M. Rochette, — ceremonies where he figures as a god who dies, to be afterwards raised again, who is burned upon the funeral pile, and whose tomb is also shown."[1]

Thus, the primitive character of Hercules was that of the sun of the lower hemisphere, being therefore identical with Tammuz-Adonis, Osiris of Egypt, etc., that is to say, with the sun-god who suffers a violent death, but is afterwards raised to life again. All this goes to show that the Asiatic Hercules really answers in character to the "seed of the woman which should bruise the serpent's head," and his definite connection with the symbolic figures, termed the *Kirubi* or *Cherubim*, forces us likewise to identify him with the Jehovah of the Old Testament as manifested in the Christ of the New, who also suffers a violent death, descends into Hades, or the lower hemisphere, but is raised again on the third day. That which confirms this assimilation to *Yahveh*, or Jehovah, of the Old Testament is the fact, on one hand, that *Hea* is to be identified with Hercules, as already shown, while, on the other hand, he takes the name of *Auv Kinuv* (הוא כינא), the "Existent Being," a title whose first element is identical with the Semitic term from which the name *Yahveh* is formed.[2]

It is necessary now to connect all these facts with the zodiacal signs Libra and Scorpio, together with the ideas previously found centring there. Having done this, it will be impossible not to recognize there at once a combination of conceptions as remarkable in themselves as they are important in their bearings. If we are not essentially misled respecting the import of the data that have been presented, we must admit the existence here of a series of symbolical representations, embodying not only the main facts pertaining

---

[1] *Frag. de Bérose*, p. 113.

[2] Vid. Lenormant, *Frag. de Bérose*, p. 65, and *Lettres Assyriologiques*, t. ii. p. 194. M. Lenormant, in the place last cited, referring to הוא כינא as a title of *Hea*, observes, "This name signifies the 'Existent Being,' or the 'Absolute Being;' the sense is precisely the same as *Scayambhu* of the Hindus. One remarks, likewise, its direct relationship to יהוה, since it has almost exactly the same meaning and is derived from the same root. It is the 'Aω, which at Cyprus was one of the principal titles of Adonis." This Adonis was a character of Hercules, especially as the dying sun-god, who is raised to a new life.

to the fall of man, but those also relating to the restoration of humanity to the paradisiacal state. The only correct standpoint from which to interpret these extraordinary and symbolic combinations is the fact that the sun's course was held to be a type of human existence, a point which will be sufficiently verified in the course of the present treatise.[1] But the sun was considered uniformly also as a divinity ; a divinity, however, who shares the lot of man, and in whom the divine and human characters perfectly interpenetrate. Such was Osiris, unquestionably the sun of the lower hemisphere, judge of the dead, yet being supposed to have ruled on earth as a civilizer of men. Such was Yama of the Hindus, Yima of the Persians, undoubtedly the sun, yet reputed first man and founder of paradise. Such was the Syrian Adonis, whose solar character is well known, although in his human character he meets with an untimely death. Such was Nin or Adar, the Assyrian Hercules, primitively a solar deity, yet as a man supposed to have suffered death, and afterwards raised to life. This violent death and subsequent resurrection is almost always attributed to the sun-gods, particularly those appertaining to the inferior hemisphere. In most instances, also, they have their sepulchres, or tombs, regarded with especial sanctity. The ancient pyramidal temple of Babylon was thought to be the tomb of Bel, or Belus. It was one of the primitive structures of the country. There were two principal classes of pyramidal temples in the valley of the Euphrates, those whose *angles* faced the cardinal regions, and those whose *sides* faced these cardinal points, and, as M. Lenormant states, the latter were regarded as divine tombs,[2] that is, sepulchres of the dying sun-god, who thus himself shared the lot of humanity. The extreme antiquity of these conceptions applied to the sun, extending even into the night of ages, is a matter susceptible of abundant proof drawn from modern researches, and this is now generally admitted by Orientalists. But it is perfectly obvious that the notion of the death and resurrection of the sun-god was only a symbol of that

---

[1] In addition to the proofs submitted in the 85th section, the reader is requested to consult those introduced in Section 143. If the sacred books of antiquity were collated for the purpose, a mass of expressions could be found involving the notion in question ; but an attempt to exhaust this subject would lead us far away from our present purpose.

[2] *Frag. de Bérose*, pp. 357, 365.

of man, and was founded upon the notion yet more fundamental which regarded the sun's course as a type of human existence.

SEC. 134. The theory advocated by the school of M. Dupuis cannot be admitted here as a basis of interpretation of the facts which have been presented in the present chapter. In substance this theory assumes that all the ancient religions, those of the Bible among the rest, had their origin in astrology, or in astro-mythology, being based at first upon purely physical and material ideas. Thus, in accordance with these views, the account of the death and resurrection of Christ had no historical basis, but was derived from the ancient legends relating to the dying sun-god. The long priority of these legends, if such it is proper to regard them exclusively, to the opening of the Christian era admits now of no doubt. Hence, the question here forced upon the consideration of the Christian apologist is a very serious and important one, and it is one moreover that cannot well be set aside, if we have regard to the facts now being brought to light by modern research. It was many years since that Sir J. G. Wilkinson felt compelled to make the following remarks : —

" The sufferings and death of Osiris were the great mystery of the Egyptian religion, and some traces of it are perceptible among other people of antiquity. His being the divine goodness, and the abstract idea of ' good,' his manifestation upon earth (like an Indian god), his death and resurrection, and his office as judge of the dead in a future state, look like the early revelation of a future manifestation of the deity converted into a mythological fable." [1]

I believe Sir J. G. Wilkinson has correctly seen here that the question lies between the theory of M. Dupuis on one hand and a *primitive revelation* on the other. For myself, I adopt the hypothesis of a *primitive revelation* and one *written in the heavens!* Indeed, I submit the investigations of the present chapter as affording very striking and convincing indications of it. The principle of the gradual and normal development of religious ideas, to which the comparative mythologists of our day attach so much importance, does not account for the facts here presented to view. We have here conscious design and symbolism, and these of a most remarkable character. The notions which we have found centring in the zodiacal signs Libra and Scorpio had an existence, a conscious

[1] Vid. Rawl. *Herod.*, vol. ii. p. 219, note 3.

existence, before they were symbolically represented there. The
simple natural phenomena could never suggest these ideas, even
after the zodiac had been formed, for we need the written revelation now to enable us to comprehend the symbolism here expressed.
It was only the initiated who could understand its import without
such aid. With respect to the sun-god, especially, it would be
impossible to conceive the sun as dying and as being raised again,
until the sun's course had been taken for a symbol of the life of
humanity. But once conceived as such, it could then, with the
consummate ingenuity and wisdom which we see displayed here, be
made the vehicle of past history, and even of prophecy. I attribute the origin of this symbolism to those primitive "corporations," to that ancient order of priest-kings, who formed the subject
of our third chapter, and I believe that this view is yet to be vindi-
cated. At the least, a very high order of intelligence has presided
over those rare combinations of ideas, which we have seen centring
in the two zodiacal divisions named. The simple union of the
four principal constellations of the zodiac, forming the composite
figure of the cherubim with four faces, for whatever purpose we
assume it to have been done, implies a state of culture of no inconsiderable advancement. But when we realize the very great importance and extreme veneration attached to them by peoples widely
separated from each other, it is safe to infer that from the first they
embodied conceptions the most elevated in their character. Considering all the facts now before us, I feel justified in the conclusion
that both history and prophecy entered originally into the conception of these symbolic animals, — history of the fall and expulsion
from Eden, prophecy of the final redemption of man and restoration
to his first estate. In this view I have not hesitated to interpret the
Hercules of the Asiatic systems, of the promised seed of the woman.
These opinions will appear, doubtless, somewhat extravagant to
many of my readers, and devoid of sufficient support. But the
chapters which are to follow will present much additional proof, so
that it will be difficult to resist the combined force of the whole.

SEC. 135. In the tradition reported by Damascius respecting the
"dragon" with four faces, these serving to identify the figure as a
whole with the Biblical cherubim, the god thus symbolized is identified not only with Hercules, but also with the time-god or Chronos.
It is a matter almost of necessity to infer from this fact that the
cherubim were in some way connected with chronology, evidently

in this case a zodiacal chronology. This suggestion receives some confirmation from the fact that the god *Hea* must be identified with Chronos as well as Hercules. Thus, as Mr. George Smith has well observed, it was Chronos who gave warning of the deluge, according to the tradition preserved by Berosus, while in the "Deluge Tablets" it is *Hea* who foretells this catastrophe, proving that the two personages must be assimilated to each other. We are not prepared at present to explain in what sense, if any, the subject of chronology is connected with the cherubim, but we hope to throw some light upon this question in the three chapters following. It seems obvious, however, if Hercules and Chronos are one, if both are associated with the cherubim, and finally if these symbolic figures are to be located zodiacally in Libra, where we find also the sacred mount of paradise, that it is in this sign we should expect to find localized some traditionary idea affording us a key to the subject to which we refer. Nevertheless, it is a matter as yet involved in obscurity, and we pass it by for the time being.

The present section closes our studies upon the tablet of the "twelve stars of Phœnicia." We have shown that they had a definite zodiacal arrangement, although different from the Babylonian calendar. We have shown, also, that they exhibit the order in which the Mosaic antediluvian genealogy was adjusted to the zodiac. This, too, was different from the method adopted in the case of the antediluvian kings of Berosus. Finally, we have discovered in this tablet a manifest reference to primitive traditions centring in the *Gan-Eden* of Genesis, and especially a mystical allusion to the cherubim of Eden, that has led the way to the fundamental notion involved in these symbolic animals. But we have not discovered any positive proof that this arrangement of twelve stars had ever been employed for the practical purposes of a calendar. Everything indicates, however, that such had been the case at a very early epoch, and that it was the calendar of all others to which primitively the Book of Genesis was adjusted. In fact, there is much reason to suppose that this scheme had formed the basis or centre, perhaps, of an independent line of traditionary inheritances, differing in some respects from those preserved at Babylon. The account of Berosus, for instance, respecting the deluge is evidently Babylonian. The god Ilu, or El, supreme divinity of Babylon, assimilated to the Greek Chronos, appears in close connection with it. But the "Deluge Tablets" appear to me to contain an inde-

pendent tradition, in which the god Hea appears as chief personage.
Again, the Babylonian Noah was certainly assimilated to the zodiacal sign Capricorn, while the Noah of Genesis must be associated
with the sign Pisces. Finally, the Babylonian genealogy evidently
begins with the cosmical year itself, at the dawn of creation; while
the Mosaic, adjusted to the twelve stars of Phœnicia, properly
begins with history, with the first man. It will be seen from this,
and from the facts heretofore presented, that in the cosmical year
of Babylon Aries and Taurus properly represent the creation week
of Genesis. More than this it is impossible to say, considering the
obscurity in which all these matters are still involved.

We cannot conclude the present series of studies without another
brief allusion to the symbolic element that everywhere appears in
the cuneiform inscription from which our list of twelve stars has
been derived. If the Orientalists of a former period were in error
in making a too exclusive use of the principle of symbolism, the
majority of those of the present day have erred also in rejecting
entirely this principle. The tablet which has been studied, although so fragmentary, proves that such a thing as symbolism actually had an existence in high antiquity, and that no greater mistake
can be made than to ignore this fact in our studies of the ancient
systems. The geographical system of Sargon the ancient, like that
prevailing in nearly all parts of the world at that period, was thoroughly symbolic, and was inspired by religious conceptions. In
this scheme *Akkad*, denoting " highland," replaced the traditional
mountain, or " highland," reputed first abode of man. The four
countries surrounding it, located according to the cardinal regions,
• replaced the four great countries situated around man's primeval
home, also facing the four points of the compass. As this primitive
happy abode was the conceived centre of the universe, the divergent
point of all the cosmical divisions, so Akkad was conceived as such
centre, and in the same spirit all civilized countries were regarded.
They were centres of the world; each system was a cosmos, a new
creation, and the entire theory had been inherited from the mount
of paradise. Tradition, symbolism, cosmical and religious conceptions, constituted the basis of that system of which our cuneiform
tablet was undoubtedly a very ancient, as, indeed, it is a most remarkable exposition. It might be said almost to embody the entire
theory of the ancient civilizations. But we must now pass on to
other topics.

BOOK V.

-------•-------

CHAPTER XIII.

THE PROBLEM STATED AND ITS CHIEF POINTS ELIMINATED.

SEC. 136. I am aware that the attempt which is to be made in the present chapter, and in those immediately following, is an extremely hazardous one, and that those most familiar with the history of similar efforts heretofore will be likely to entertain the least confidence in a successful result. The conceived possibility of establishing a chronological date in high antiquity by means of the internal arrangements of the zodiacal system, in connection with the well known law of the precession of the equinoxes, was that which inspired the labors of M. Dupuis nearly a century ago, in his celebrated essay upon "The Origin of the Sphere," and a similar motive appears to have prompted the investigations of Dr. Gustave Schlegel upon the antiquity of the Chinese sphere, as given to the public in his recent and voluminous treatise on "Chinese Uranography." But it is now generally admitted that M. Dupuis' attempt was mainly a failure, and it is probable that the majority of scholars will refuse to place much confidence in Dr. Schlegel's results. Both these writers appear to be too much under the influence of preconceived opinions, and the fundamental principle involved in their method of research has been often called in question. M. Dupuis assumed that the Egyptians were the inventors of the zodiac, and thus construed everything with reference to the climate of Egypt. But Egyptologists have now shown that the zodiac, as transmitted to us by the Greeks, was not even in use in the Nile valley until a very late epoch. Dr. Schlegel attempts to show that

the Chinese were the inventors of the zodiacal system, and thus
bases all his calculations upon the latitude and climate of the Celestial Empire. But the investigations of Professor Whitney some
years since, not to mention here the labors of other scholars, have
rendered it extremely doubtful whether the Chinese can lay claim
to this invention. For myself, in view of all the facts, I believe it
most probable that the celestial sphere, including the zodiacal arrangement, had its beginnings in the period before the separation
of races from their common home on the high table-lands of Central Asia. Of this, however, in another connection. The method
adopted by M. Dupuis, and followed by Dr. Schlegel, is grounded
principally upon the supposition that the zodiac was the product of
the needs of agriculture, the constellations being designed by their
periodical risings and settings to announce the various seasons of
agricultural labors. Besides this, certain religious, civil, and social
customs and ceremonies were connected with the appearance of
these asterisms. In calculating the antiquity of the zodiac, then,
such positions in the heavens were assumed for these constellations,
that their appearance would correspond with the various labors of
the field, and the difference between such assumed positions and
the present actual positions of these constellations was then converted into time by means of the known rate of precession. Following this method, and adopting his calculations for the climate
and latitude of Egypt, M. Dupuis arrived at an antiquity of some
15,000 years for the origin of the sphere. Adopting the same
method substantially, and basing his calculations upon the latitude
and climate of China, Dr. Schlegel finds that the Chinese invented
the sphere some 18,500 years ago. But there is no certainty that
the constellations were originally designed to mark by their appearance the various seasons of agricultural labors. The ground principle of this method is a pure assumption, and it is for this reason
principally that scholars have refused to place confidence in these
deductions.

But notwithstanding the unsatisfactory nature of the results as
attained by these writers, the great importance of such an undertaking if it could be rendered successful, the immense service to
the learned world which a fixed chronological date in the prehistoric ages would be if once scientifically verified, must be regarded as a sufficient justification of still other and oft-repeated

attempts to this end, however hazardous they may appear to the generality of critics. For myself, therefore, I am willing to take the responsibility of another effort in this direction, and to incur the risk of another failure. But I have considerable confidence in a successful result, especially as my object is quite different from that pursued by the two writers referred to, and as my method of research is also different from theirs.

Primarily we have no desire to fix the date of the origin of the sphere. That which we seek to accomplish is in general this: by means of the zodiacal system, in connection with some of the extra-zodiacal constellations, in connection also with the law of the precession of the equinoctial points, to determine approximately the epoch to which the primeval traditions of mankind appertained. With respect to the method of research to be here adopted, it may be observed in general that we shall not seek to identify the people who invented the zodiac, nor the country in which this system originated. We shall have no reference to the agricultural labors, the customs, the local circumstances of any nation whatever. Our method will be, as principal aids, to interrogate the primitive traditions inherited alike by all the cultured nations of antiquity. The primary object of this interrogation will be to ascertain whether these traditions, many of them at least, *do not* INVOLVE AN ASTRONOMICAL REFERENCE *sufficiently precise, and so nearly agreeing one with the others, as actually* TO REVEAL THE STATE OF THE HEAVENS *at the period to which they all relate*, thus supplying the necessary data to constitute an astronomical problem that admits of a ready solution according to the ordinary rules. Those who are familiar with this class of studies will recognize at a glance that the method of procedure here outlined is essentially different from that heretofore adopted by any writer, and that it is wholly free from those objections already pointed out in connection with previous attempts of this kind. It is a very simple matter to ascertain, in the first place, whether the primitive traditions actually do involve an astronomical reference so definite and direct as to reveal the state of the heavens at the epoch to which they pertain. If we show that such is really the case, it will be a simple process; also, in the second place, to calculate the difference in time according to the known rate of precession between that epoch and the present era. The underlying principle is perfectly clear, although the crit-

ical analysis and proper arrangement of the mass of materials
appertaining to our general problem will be an arduous labor,
requiring no small degree of patience and perseverance. But the
foregoing will suffice respecting the chief object in view, and the
mode of attaining it. We proceed now to the consideration of
the more specific points involved in our problem.

Sec. 137. To render these investigations more intelligible to
those who have never made astronomy a special study, it is necessary to explain here briefly the internal arrangements of the
zodiacal system. First is the important distinction always to be
kept in mind between the twelve *constellations* and the twelve *signs*
of the zodiac. The two series bear the same names; that is, we
have the constellations Aries, Taurus, Gemini, etc., and the signs
Aries, Taurus, Gemini, etc.; but the constellations are one thing,
and the signs quite another. The constellations are the groups of
stars arranged in a complete circle round the heavens, which never
change their positions; and this circle marks the annual course of
the sun. The number of these asterisms is twelve, corresponding
to the twelve months of the year. The twelve signs of the zodiac
are supposed to mark the same spaces in the heavens as the constellations; but their positions are constantly changing; they have a
retrograde movement on the zodiac, due to the so-called precession
of the equinoxes. Since one series is fixed, while the other has a
slow retrograde movement, the signs are not always found to occupy
the same positions as do the constellations of the same name. At
the present era, all the signs have fallen back thirty degrees, or the
extent of one sign, from their correspondent constellations. Thus,
the sign Gemini is in the constellation Taurus, and the sign Taurus
is in the constellation Aries, etc. But about twenty-one hundred
years ago all the signs corresponded exactly in position to their respective constellations. According to the ascertained rate of precession, or of this retrograde motion of all the signs, it would take
about twenty-five thousand years for them to complete the entire
circle of the zodiac. It may be well to observe here that the vernal
equinox always occurs in the first degree of the sign Aries; the
autumnal equinox in the first degree of the sign Libra; the summer
solstice in the first degree of the sign Cancer; and the winter solstice in the first degree of the sign Capricorn. The retrograde movement of the signs always follows that of the equinoctial and solsti-

tial points, and in fact is caused by it. The results would be the same, if we conceive the signs as fixed and the constellations as moving forward on the zodiac, as to regard the constellations as fixed and the signs as moving backward on the zodiac. But the latter is the usual mode of notation among modern astronomers. These explanations, however, will be sufficient for the present.

We enter now upon the consideration of a very important element in our general problem, in relation to the following inquiry: At the period of the invention of the zodiacal system, did the constellations and the signs having the same name correspond to each other in postion? That is to say; was the constellation Aries in the sign Aries, the constellation Taurus in the sign Taurus, etc.? It will be seen that these suppositions are admissible in the absence of any proof to the contrary: 1st. All the constellations and signs may have corresponded in position at the date of the origin of the zodiac; or 2d. All the signs may have been behind their respective constellations, as is the case at the present period; or 3d. All the signs may have been ahead of the constellations on the zodiac; a supposition which we shall find to be the correct one, although it is opposed to the usual impression among practical astronomers. We quote below from a recently published text-book on astronomy: —

" When the first catalogues of the stars were constructed, the *signs* doubtless corresponded with their *constellations* in position; and we can therefore calculate the era when the earliest star charts were made. Thus, the *rate of precession* for one year ($50.24''$) is to *one year* as thirty degrees ($108,000''$) is to 2149.7 years. The zodiac was therefore constructed about two thousand years ago." [1]

This is a very easy method of settling the question in regard to the antiquity of the zodiacal system. But it is wholly chimerical. The signs of the zodiac are at present about thirty degrees, or one whole sign, in the rear of their respective constellations. Now the author quoted above *assumes* that when the zodiac was invented all the signs and constellations coincided in position; and since they are now separated only thirty degrees distant, the known rate of precession gives about 2100 years for the date of the origin of the system. The author's calculation would be perfectly legitimate and correct if his presupposition was correct, to the effect that this coincidence in position of the signs and their con-

[1] Brocklesby, *Elements of Astronomy*, p. 99.

stellations was really a primitive feature of the zodiac. But this
presupposition is erroneous, as we shall show ; and thus the calculation based upon it falls to the ground. We propose to substantiate the fact, independently, of the extreme antiquity of the zodiac,
amounting to many centuries, if not to thousands of years, prior to
the date 2100 years ago, fixed upon by the author just cited.
This being established, the inevitable result will follow that the
signs and constellations did not correspond in position primitively;
that, in fact, the signs were ahead of their respective constellations
on the zodiac, instead of being behind them, as at the present
day. The cuneiform inscriptions will afford us ample data on the
point before us. The following is Mr. George Smith's rendering
of an astronomical tablet discovered by him, with the substitution
of the degree and sign for the day and month, according to his own
suggestion : —

"From the first degree of the sign Pisces to the thirtieth degree
of the sign Taurus, the sun in the division (or season) of the great
goddess is fixed, and the time of showers and warmth. From the
first degree of the sign Gemini to the thirtieth degree of the sign
Leo, the sun in the division (or season) of Bell is fixed, and the
time of the crops and heat. From the first degree of the sign Virgo
to the thirtieth degree of the sign Scorpio, the sun in the division
(or season) of Anu is fixed, and the time of showers and warmth.
From the first degree of the sign Sagittarius to the thirtieth degree
of the sign Aquarius, the sun in the division (or season) of Hea is
fixed, and the time of cold. When on the first day of the month
Nisan, the star of stars and the moon are parallel, that year is right
(normal). When on the third day of the month Nisan, the star
of stars and the moon are parallel, that year is full (has thirteen
months)." [1]

The foregoing demonstrates sufficiently that at the date of the
tablet of which it is a translation the zodiac was in common use
among the Babylonians. The precise date of the tablet, however,
is a matter of conjecture. It must have belonged to the great collection of Assur-banipal at Nineveh, either as an original, or as an
Assyrian version of a much earlier Accadian tablet. Its date, therefore, cannot be assigned later than to the seventh or eighth century
B. C. This would be four or five centuries before the conjunction
of all the signs and constellations of the zodiac in the same posi-

_____
[1] *Assyr. Dic.*, pp. 404, 405, American Edition.

tions, about 2100 years ago. But we proceed to earlier dates. Alluding to the inscribed conical stones to which reference was made in a previous chapter, Rev. George Rawlinson remarks : —

" The accompanying representation, taken from a conical black stone in the British Museum, and belonging to the *twelfth* century before our era, is not, perhaps, strictly speaking, a zodiac, but it is almost certainly an arrangement of constellations according to the forms assigned them in Babylonian uranography. The Ram, the Bull, the Scorpion, the Serpent, the Dog, the Arrow, the Eagle or Vulture may all be detected on the stone in question, as may similar forms variously arranged on other similar monuments." [1]

I have already expressed the opinion that these figures were not intended for a complete zodiacal arrangement, but they prove the existence of the zodiac at this period, being over eight hundred years prior to the conjunction of all the signs and constellations of the zodiac to which we allude. Another inscription records the actual occurrence of the vernal equinox on the fifteenth day of the month Nisan, which would be in the fifteenth degree of the sign Aries.[2] Thus, this inscription goes to substantiate the existence of the zodiac at Babylon about 1000 years prior to the time when all the signs and constellations occupied the same positions. But we are able to demonstrate a still higher antiquity for the zodiac. The great book of astrology, dating from the times of Sargon the ancient, contains numerous texts in which the existence of the zodiac at that period is placed beyond question. M. Lenormant, as heretofore cited, assigns this monarch to the era between 1900 and 2000 years B. C. Hence, over 1600 years before the conjunction of the signs and constellations referred to took place, the zodiac was in ordinary use at Babylon.

SEC. 133. Having presented such evidence relative to the antiquity of this system as is most readily obtained from the monuments affording specific dates, we proceed now to show that the zodiac really appertained to the period of the earliest traditions of Babylon. In the eleventh chapter (Sec. 117), M. Lenormant and Dr. Movers were cited to the effect that the names *Alarus* and *Alaparus*, which head the list of antediluvian kings preserved by Berosus, are simple designations respectively of the zodiacal divi-

---

[1] *Five Monarchies*, ii. pp. 574, 575.
[2] 3d Rawl. Pl. 51, note 2.

sions *Aries* and *Taurus;* a fact that in M. Lenormant's opinion
proves the zodiacal character of the entire series of names. Now
the primitive character, traditionally speaking, of this antediluvian
genealogy will be readily admitted; yet its connection from the
first with the zodiac has been established by the two eminent critics
named above. To the data here presented should be added the following extract from the "Creation Tablets," as discovered and
translated by Mr. Smith: —

"Stars, their appearance (in figures) of animals he arranged.
To fix the year through the observation of their constellations,
twelve months (or signs) of stars in three rows he arranged, from
the day when the year commences unto the close."[1]

The precise date of the original tablets, of which those discovered by Mr. Smith were only Assyrian translations is unknown,
but this writer inclines to refer them to the period nearly 2000
years B. C. But the exact date of the documents is not for us the most
important point. They unquestionably record traditions, which
were absolutely the most primitive among the Cushites of Babylon, and they prove beyond doubt that the zodiac was associated
with those traditions. Indeed, I think it most probable, all things
considered, that the Cushites brought the zodiac with them to the
"land of Shinar," when they migrated from the sacred mount of
the east, the reputed father of countries and the first home of man.
We have shown in our first chapter that a Cushite civilization,
probably anterior to that of Babylon, primitively existed around
the headwaters of the Indus and Oxus, and it was proved in a subsequent chapter, that Mt. Meru of Hindu tradition was really the
true *Ararat*, a name corrupted from *Arya-rata*, originally applied
to Meru. In view of these facts, considerable significance attaches
to the language of Mr. Richard A. Proctor, in an article on "The
Origin of the Constellation-Figures," republished in Appleton's
"Popular Science Monthly" (Supplement) for November, 1877,
language which is as follows: —

"I think from 35° to 39° north (latitude) would be about the
most probable limits, and from 32° to 41° north the certain limits,
of the station of the first founders of solar zodiacal astronomy.
What their actual station may have been is not so easily established. Some think the region lay between the sources of the

---

[1] *Chald. Acct. Genesis*, p. 69.

Oxus (Amoor) and Indus; others think that the station of these astronomers was not far from Mount Ararat (in Armenia), a view to which I was led long ago by other considerations, discussed in the first appendix to my treatise on 'Saturn and its System.' At the epoch indicated (2170 B. C.) the first constellation of the zodiac was not, as now, the Fishes, nor, as when a fresh departure was made by Hipparchus, the Ram, but the Bull, a trace of which is found in Virgil's words, 'Candidus auratis aperit cum cornibus annum Taurus.'

"The Bull, then, was the spring sign, the Pleiades and ruddy Aldebaran joining their rays with the sun's at the time of the vernal equinox." [1]

Whether the considerations which led Mr. Proctor to associate the origin of the zodiac with the region of Ararat would apply as well to Mt. Meru, the primitive Ararat, I am not able to say. But we shall find much reason, as we proceed, to consider Meru as the earliest centre of astronomical science, as well as of civilization generally. In regard to the limits between which this writer would locate the first astronomical stations, it may be remarked that the situation of Babylon, although not exactly outside these boundaries, was still very near the extreme southern limit assigned. But the author's statement that *Taurus* was the first constellation of the zodiac at the period 2170 B. C., to which he is inclined to assign the date of the zodiac, has for us the especial value that it disproves the prevalent hypothesis which assumes the coincidence of all the signs and constellations in position at the time the zodiac was formed. For, if Taurus was ever the constellation in which the vernal equinox fell, then the *sign Aries* was at that period in this constellation. But we have one other fact to verify in this connection, namely, that the Babylonian zodiac was substantially the same as ours, transmitted to us by the Greeks, and for this purpose we quote M. Lenormant, as follows: —

"The nomenclature of the signs as they have been preserved to our day does not differ essentially from that which had been established by the astronomical priests of Chaldæa and Babylon. In fact, according to passages derived from astronomical monuments, and according to the zodiacal figures which are found upon a great number of monuments, especially upon the cylinders, we are able to establish to a great extent the series of signs as employed by

[1] Proctor in *Scientific Monthly* (Supplement) November, 1877. [This chapter was revised in 1878. S. M. W.]

the Chaldæans: 1st. The Ram, or the Ibex. 2d. The Taurus, sometimes represented with wings and with the human head. 3d. The Twins, expressed by two virile figures superimposed. 4th. The Crab, or the Lobster. 5th. The Lion, sometimes replaced by the group of a lion devouring a taurus. 6th. The Archeress (goddess *Ishtar*), mentioned by the texts, but of which no figure has as yet been recognized. 7th. The Pincers of the Scorpion. 8th. The Scorpion; it results from a tablet not yet edited, relative to the movements of the planet Venus, which I have had the occasion of studying in the British Museum, that they sometimes united this sign with the preceding, under the common name of the 'Scorpion,' which was then counted as a sign having double the extent of the others. 9th. The Arrow, which thus replaces, though rarely, Sagittarius holding the bow. 10th. The Goat, its body often terminating in the tail of a fish. 11th. The Water-bearer, this figure being more habitually reduced to that of a vase from which water is being poured. 12th. The Fish, or the Fishes. It was evidently for mythological reasons that these names and figures were assigned to the constellations of the zodiacal band; because we search in vain for any direct relation to the labors of agriculture and the phases of the seasons contemplated from this point of view. We know to what groundless conjectures the school of M. Dupuis resorted to find a relation of this kind, and how they were obliged to place the invention of the zodiac in an epoch fabulously remote, in order to arrive at the time when, thanks to the precession of the equinoxes, the presence of the sun in Taurus coincided with the season of the labors of the field," etc.[1]

SEC. 139. Various conclusions result from the data now before us. In the first place, there can be no longer question whether the Greeks were the inventors of the zodiac, as formerly maintained by M. Letronne, since its use among the Babylonians, at a period before the dawn even of Greek history, can hardly be doubted in view of the facts already presented. In the second place, it results from the statement of M. Lenormant above, which appears to be fully justified by his own researches, that the principle assumed by M. Dupuis, and recently by Dr. Schlegel, of a direct relation of the zodiacal constellations to agricultural operations has really no solid foundation. Mythological considerations were those, as M. Lenormant believes, which determined the names and figures of these asterisms. It appears to us that this statement is somewhat too broad and exclusive, for evidently symbolism had much to do in

<hr>

[1] *Premières Civilisations*, t. ii. pp. 68, 69.

the formation of this system. But that which is of the greatest consequence in connection with our general problem is the extreme antiquity of the zodiac, substantially in the form in which it has been handed down to us; an antiquity which appears to be most conclusively established by the various considerations already presented. It was associated with the earliest traditions of the Babylonians, even with those that related to the antediluvian dynasties, and to the creation of the world. In fact, as was pointed out in our criticism of the " Creation Tablets," the Babylonian account of the formation of the world, which these documents contain, actually includes the arrangement of the zodiac as constituting a distinct labor in the creative process. It must be regarded, therefore, as fully substantiated that the zodiac had been in existence some thousands of years when, about twenty-one centuries ago, it so happened that all the signs and constellations of the same name corresponded in position. This correspondence, then, could not have been an original feature of the system ; and certainly not, if at one time, as many reasons lead us to conclude, the vernal equinox fell in the constellation Taurus ; for in such case the sign Aries would be in this constellation, — in other words, all the signs would be thirty degrees, or one whole sign, in advance of their respective constellations. Thus, the first among the fundamental conditions or elements of our problem may be formulated as follows : —

At the period when the zodiac was invented, the positions of the signs and constellations were different, the signs being in advance of the constellations bearing the same name.

It will be recognized at once that the foregoing proposition, if it is to be admitted as firmly established, attaches to itself a high degree of importance. It is contrary to the usual impression among practical astronomers, and in fact it is contrary to that which we should naturally suppose to have been the case. Assuming it for our basis, the following conclusions may be fairly deduced from it :

1st. If the signs and constellations of the zodiac occupied different positions originally, then this was not the result of *accident*, but of *a specific, conscious design* on the part of the inventor of the system.

2d. If this difference in position was a primitive feature of the system, then the zodiac was strictly speaking *an invention;* it was not the result of a gradual, unconscious development, so to speak, from crude beginnings.

3d. If this difference in position was a primitive feature of the zodiac, then that difference must have been *greater than one sign merely*, or greater than that when the vernal equinox fell in the constellation Taurus.

4th. If there was originally a difference in position between the signs and constellations of the zodiac, then the *amount of that difference* must have been determined, not by accident, but by some *good and substantial reason*.

5th. When the zodiac was constructed, it doubtless represented the constellations in their *actual positions in the heavens*, while the location of the signs was a matter of choice, and was determined by some other considerations.

SEC. 140. In regard to the first deduction, as stated above, it would be natural to suppose that originally the signs and constellations occupied the same positions, it being due only to the law of precession that the signs had gradually fallen back into other positions. But we now find that such was not the fact; the signs were originally in advance of their constellations, and it is impossible to account for this circumstance except upon the ground of some especial design. The second deduction follows partly from the first. If there was conscious design in the arrangement, then the zodiac was properly an invention. In fact, if the system had grown up unconsciously from crude beginnings, it is almost certain that the signs and constellations would have occupied at first the same positions. With reference to our third deduction, it may be set down as wholly improbable, if there was a difference in position between the signs and constellations, that, either by accident or by design, it should be only one sign, or thirty degrees. Passing to the fourth deduction above, it is necessary to conclude, whatever difference was originally assumed between the signs and constellations, that it was not left to mere accident to determine its extent; some good and sufficient consideration in the mind of the inventor influenced his choice of location for the signs. The last deduction is almost self-evident. The zodiac must represent the constellations in their true positions in the heavens, or it would be worthless as the basis of a calendar. But the location of the signs was in a measure arbitrary, and might be influenced from various considerations.

It is necessary to introduce here for a brief consideration another inquiry. We have had much to say relative to the difference, both in character and in position, between the signs and the constellations ; but is it certain that the *signs, as distinguished from the constellations*, have really been in use from the time the zodiac was invented ? Before the era of Hipparchus, for instance, who is supposed to have introduced some improvements into the system, had the signs been distinctly recognized as different from the constellations ? Do we find among the Babylonians, among whom, as M. Lenormant holds, the zodiac originated, the distinct recognition of the signs as different from the constellations ? If such a distinction had not existed from the first, then the great importance of an original difference between the signs and constellations, and the various deductions which we have drawn from it, must be regarded as imaginary. The signs were not in existence from the first ; only the constellations constituted the zodiac. But the cuneiform texts afford sufficient proof, as it would seem, that among the Babylonians a distinction had always been made between the constellations and the signs. In the first place we have distinct names for the *months*, the *constellations*, and the *signs*, and these names are never employed interchangeably. The Accadian name for month is *Ab*, Assyrian *Arakh;* the Accadian for star, or constellation, is *Mul*, Assyrian *Cacab.* Finally, the Accadian for zodiacal sign is *Zu*, the Assyrian being the same. The moon-god is often termed *En-zu*, "lord of the zodiacal sign." At other times the numeral *30* is employed for writing the name of this divinity, in reference to the thirty degrees comprised in each sign, corresponding to the thirty days contained in each of the twelve months. The series of months, beginning with the Accadian *Bara*, Assyrian *Nisan*, corresponded exactly to the series of signs commencing with *Aries*. This arrangement was constant, or intended to be such, the first *day* of the month *Nisan* answering to the first *degree* of the sign *Aries*, and thus on to the end of the calendar.

Now, as the Babylonians had distinct *names* for the signs and constellations *Zu* and *Mul*, and even separate written *characters*, and as these names and characters were never substituted one for the other, it is safe to conclude that they had always made a distinction between the two series. Various other facts might be presented here tending to the same conclusion, but it is not necessary to give them in detail. The Babylonian astronomers noted the changes in the stellar world too constantly and too narrowly, not to have been aware that the constellations had an apparent advance movement relative to the signs bearing the same names. But this apparent advance movement of the constellations we now know to be due to an actual retrograde movement of the equinoxes, carrying all the signs with them. It is probable, in fact, that the Babylonians regarded the constellations as movable and the signs as fixed, since this accords better with the notion that the earth is fixed in the centre of the universe, the heavens revolving around it; a notion which seems to have prevailed generally for many ages.

SEC. 141. Returning, then, to our first general proposition, that the positions of the signs and constellations were different at the time the zodiac was invented, we seem to be wholly justified in reaffirming it here, and to proceed from it as an established fact, which we now do, and also from the various deductions already made from it, some of them being highly significant and important.

We have shown that there was a specific, conscious design on the part of those who invented the zodiac in giving to the signs different positions from those of the constellations, and we are now brought face to face with the inquiry : *What was that design?* My answer in brief is: *A zodiacal chronology.* It ought to be a very satisfactory proof of this, that the Babylonians actually employed the zodiac for this purpose. We have seen that they employed a great cosmical year, divided into twelve great cosmical months, corresponding to the twelve months of the ordinary year. This great cosmical year was supposed to have opened at the dawn of creation, and with the sign Aries. The ten antediluvian kings, Alorus, Alaparus, etc., were definitely associated with the signs; Alorus with Aries, Alaparus with Taurus, and thus down to the reign of the tenth king, Xisuthrus, the Babylonian Noah, who was associated with the tenth sign, or Capricorn. These facts have been fully substantiated by M. Lenormant and Dr. Movers, as heretofore cited. That these ten kings were believed to have succeeded each other chronologically, and that their reigns were associated with the signs beginning with Aries, does not admit of serious question. But this our present hypothesis supposes, that the inventors of the zodiac were aware of the fact which we designate as " the precession of the equinoctial points." It may be thought that this phenomenon could not have been observed at so early a period. The inventors of the zodiac, however, have sufficiently proved by the very nature of their invention that the fact of precession was familiar to them. The series of *signs*, as distinguished from the *constellations*, presupposes the knowledge of precession, because, if it were not for this retrograde movement of the equinoctial points, there could be no occasion for employing the signs as something different from the constellations. The two would always occupy the same positions and stand for the same thing. Thus, the very nature of the invention presupposes precession and the knowledge of it, as a *simple fact*, but not a knowledge of the exact *rate of precession*, which has required the superior advancement and facilities of the present day to determine.

Conceive, now, two circular, flat rings, each divided into twelve portions, and each of these into thirty degrees, making three hundred and sixty degrees in the aggregate. Let one of these rings represent the twelve constellations, the other the twelve signs. Consider one as fixed, the other as movable at the rate of one degree in about seventy-one years. Such would be a representation of the arrangements of the zodiacal band as it circles the entire heavens and marks the annual course of the sun. It will be seen that such an arrangement, in connection with the law of precession, would constitute a great *celestial clock*, keeping the time of the world throughout all ages. We believe this to have been the nature and design of the zodiac when it was invented, and that it was so consciously intended by those who originated the system. But all this does not conduct us nearer to the central point of our problem, and it will be better to leave for future developments the question of the design of the inventor in separating the signs and constellations in regard to their respective positions.

The great and fundamental inquiry before us now is this: *What was the distance primitively between the signs and constellations?* The answer to this question is no less than the solution of our problem. For whatever may have been that original difference in position, it has since been wholly traversed by the retrograde movement of the signs, so that even 2100 years ago the signs and constellations came together, and now the signs are one entire zodiacal division in rear of the constellations. If we knew the original distance between the two series, the rate of precession, which is known, would give us the date of the origin of the system. Or, if we knew the state of the heavens at that epoch, this would give us the distance at the time between the signs and constellations, and then again the rate of precession would enable us to calculate the antiquity of the zodiac. But these are the very points yet to be determined, and the next inquiry is: *By what means can they be determined?* If we speak indefinitely, and of probabilities, it may be said that the distance originally between the signs and constellations must have been more than thirty degrees, or the extent of one sign. This would have been the distance between them when the vernal equinox fell in the constellation Taurus, as is often held to have been once the case. The signs were then thirty degrees, or one whole sign, *in advance* of their constellations. But no good and sufficient reason can be given, nor be conceived even, why a difference of only one sign should have been assumed by the inventors of the system. If any difference was to be assumed, all the probabilities are that it would have been greater than this. But how much greater is the question, and this is that to which we now direct attention.

Sec. 142. The means by which we hope to fix the original distance between the signs and constellations have been already briefly explained. We propose to interrogate the primeval traditions of mankind, with the view to ascertain whether they do not involve an astronomical reference so definite as to reveal the state of the heavens at the period to which they pertain. Our first appeal will be to some important traditions of the Chinese relating to their sphere, which have been collected by Dr. Schlegel. This writer opens his first chapter as follows : —

" The most ancient division of the Chinese celestial sphere, which was at the same time the most natural, was that into four parts, corresponding to the four principal epochs of the year, springtime, summer, autumn, and winter. In these four parts they traced four great constellations, as the following passage demonstrates : ' At each of the four cardinal regions are found seven domiciles, or clusters of stars, which together form a single figure. Those of the east form the figure of a *Dragon*, and those of the west form the figure of a *Tiger ;* the heads of these figures are to the south, and their tails to the north. Those of the south form the figure of a *Bird*, and those of the north the figure of a *Tortoise ;* the heads of these figures are to the west, and their tails to the east.' The eastern part of the heavens was named the domicile of the Blue Dragon (*T'sang loung*); the northern part was named the domicile of the Black Warrior (*Hiouen Won*); the western part was named the domicile of the White Tiger (*Pé hou*); and the southern part was named the domicile of the Red Bird (*Tchou naio*). The first part corresponds to springtime, the second to winter, the third to autumn, and the fourth to summer, as we learn from the celebrated *Tchou-tsze*, who says: 'The *Dragon* corresponds to the medium of heat, the *Tiger* to the medium of cold, the *Bird* to the maximum of heat, and the *Tortoise* to the maximum of cold.'"[1]

That which is very remarkable in the foregoing statements is the fact that they represent the correspondence of the seasons to the four cardinal regions, exactly the opposite of their actual correspondence at the present era of the world. The Dragon, situated in the east, and which corresponds nearly to our Virgo, Libra, and Scorpio, is said to mark the period of springtime. But, in fact, in our day the east answers to autumn, and at the present epoch the Chinese Dragon marks the autumnal season. The White Tiger, situated in the west, answers nearly to our Pisces, Aries, and Taurus, and to-day it corresponds to springtime. But according to Chinese tradition, as stated above, it primitively marked the autumnal period. The Tortoise, or Black Warrior, is made up in part of our winter constellations, and actually corresponds to the winter season; yet it is located in the north, with which, properly, the summer should be associated. The Red Bird is composed of our summer constellations, and really marks the summer season, but it is situated in the south, with which we associate the winter. In reference to this marked discrepancy between the traditional and the present actual relation of these four great constellations to the four seasons, and after enumerating the constellations of our sphere which answer respectively to the Chinese Dragon, Tiger, etc., Dr. Schlegel observes: —

"But this distribution is all, in fact, inapplicable to any known historical epoch; for if we trace the constellations of the zodiac in a circle, we have for the eastern constellations, *Pisces, Aries,* and *Taurus;* for the western, *Scorpio, Libra,* and *Virgo;* for the southern, *Gemini, Cancer,* and *Leo;* and for the northern, *Sagittarius, Capricorn,* and *Aquarius.* Thus, the order in which they traced the ancient division of the Chinese sphere into four parts is wholly reversed."[2]

---

[1] *Uranographie Chinois*, 1st chapter.   [2] Ibid., ch. i. p. 3.

SEC. 143. Notwithstanding the apparent progress thus far made,
we are not yet able positively to affirm that those who invented the
great celestial clock set it at any particular hour of the world. If
so, it was necessary to assume two points upon the dial-plate, one
as fixed, a second as movable, and as having departed primitively
from the other. If we consider the zodiacal *signs* as fixed and the
constellations movable, then the solstitial sign for midwinter, being
that of Capricorn, constitutes our permanent factor, and it was in
this *sign* that the ancient traditions localized the birth of the world
and of man. This zodiacal division is then the fixed point on the
celestial dial-plate ; and we must seek among the *constellations* of
the zodiac for the movable one, which originally departed from
Capricorn. The very direct and seemingly conclusive proofs heretofore introduced (Secs. 118, 119), tending to connect the zodiacal
"Twins," or Gemini, with the paradisiacal man, the first human
pair in fact, naturally give rise to the suspicion here that this constellation realizes for us the required movable factor upon the dial-
plate of the stellar world. It may seem to the reader a very singular circumstance, and for this reason one of doubtful occurrence,
that a zodiacal constellation should be taken as a representative of
the first man, or of the first human pair ; and as this point is one
of such importance to our argument upon zodiacal chronology, it is
necessary to explain the manner in which such a circumstance
would very naturally occur. The facts affording the basis of the
explanation needed here, and which have been already fully verified,
are : 1st. That the sun's course, in its three phases as daily, annual,
and cosmical sun, was taken as a symbol of the life of man, the
course of the cosmical sun answering in this scheme to the history
of our race. 2d. That the antediluvian genealogy had been assimilated to the signs or constellations of the zodiac, beginning
with the first man, and ending with the hero of the deluge. It was
shown in the eleventh chapter that the Mosaic genealogy opened
with the sign Gemini, which represented thus the first man. We
thus understand why the traditionary first man was often identified
with the sun, and especially with the sun in Gemini.

SEC. 144. The evidences connecting the zodiacal Gemini, or
Twins, with the first man and woman, derived from the Hindu
legends relating to *Yama* and *Yami ;* from the Scandinavian traditions pertaining to *Askur* and *Embla*, especially in connection with

the Accadian *As-kar*, a name of Mercury signifying " the propitious summit," or " happy garden," and which answered to Gemini in the zodiacal arrangement, — these evidences, I say, embodied in the sections just referred to, ought to be regarded as sufficient, although other proofs will be developed as we proceed. Hence, we are authorized to assume that the constellation Gemini represented in primeval tradition : 1st. *The original progenitors of the human race.* 2d. *The lower hemisphere and the winter solstice.* Geographically speaking, we know from previous investigations that the birth of man took place in the upper hemisphere. We know, also, that nearly all the ancient cosmogonies centred in the sacred mount of paradise, geographically located in the upper hemisphere. Hence, the theory which assigned the origin of the world and of man to the lower hemisphere, and to the period of the winter solstice, must be taken in a purely astronomical sense. It will be seen, then, that we have here an important, fundamental condition of our problem ; and the only possible mode of satisfying it is to assume our present upper hemisphere, where Gemini is now found, as having been the lower hemisphere at the period to which these traditions pertained. In other terms, it is necessary to suppose that the *constellation Gemini*, assimilated to the first human pair, was then found in the *sign Capricorn*, marking the period of the winter solstice, and situated in the inferior heavens. This done, these traditions are at once made clear.

While treating upon the condition of our problem as stated in the last paragraph, I desire to place another one by the side of it, at the same time presenting incidentally other proofs of the original assimilation of Gemini to the first human pair. No characters appear more frequently in the hymns of the Rig-Veda than the *Asvins*, so termed, or " the horsemen ; " and Sanskrit scholars seem to take it for granted that they are to be identified with the Dioscuri of the classic mythology, that is to say, with Castor and Pollux, the zodiacal Gemini.[1] *Saranyú*, daughter of *Trashtar*, " the creator," is mother of two pairs of twins, *Yama* and *Yami*, as the older, and the *Asvins*, " the two horsemen," as the later. Professor Kuhn, as cited by Professor Müller, thus explains the legends centring in these various characters : —

[1] Vid. Whitney, *Orient. and Ling. Studies*, p. 38 ; Müller, *Chips*, ii. p. 91.

" *Trashtar*, the creator, prepares the wedding for his daughter *Saranyû*, i. e. the fleet, impetuous, dark storm-cloud, which in the beginning of all things soared in space. He gives to her as husband *Vivasvat* (*probably the sun*), the brilliant, the light of the celestial heights ; according to later views, which for the sake of other analogies I cannot share, the sun-god himself. Light and cloudy darkness beget two couples of twins : first, *Yama*, i. e. the twin brother, and *Yami*, the twin-sister ; secondly, the *Asvins*, the horsemen." [1]

It would be desirable, if possible, to locate these legends astronomically and geographically, and the Vishnu Purana helps us to do so in both senses. The version of these legends, as given in the Vishnu Purana (p. 266), informs us definitely that the *Asvins* were begotten " in the region of *Uttara-Kuru*." We are able to determine the situation of this region, both geographically and astronomically. In the passage cited from M. Renan (Sec. 66), on the location of the Gan-Eden of Genesis, Meru of the Hindus, etc., this writer observes : " There is the *Uttara-Kuru*, ' the country of happiness,' of which Megasthenes speaks." M. Lenormant confirms this view : " Among the Hindus, the men before, as well as those after the deluge, descended from Mt. Meru. It is there that we find the *Uttara-Kuru*, veritable terrestrial paradise, the traditions relative to which had been collected by Megasthenes." [2] But there existed an *Uttara-Kuru* celestial, as well as terrestrial. According to Persian conception, the sacred river was located on the summit of Albordj, or Meru, which penetrated the heavens in the region of the North Star. It was traditionally on the banks of this sacred river that the beautiful horses were engendered, evidently referring to the *Asvins* themselves. M. Lenormant adopts the language of M. Obry respecting the imperial gardens of Persia, called "paradises," as follows : " These terrestrial paradises represented, among the Persians, the celestial paradise of Ormazd, . . . planted upon Albordj, . . . it was on Mt. Meru, confounded with the *Uttara-Kuru of the firmament*." [3]

SEC. 145. Thus, the celestial *Uttara-Kuru* was in the region of Su-Meru, or the North Star, the terrestrial *Uttara-Kuru* being one with the earthly paradise. We have now another direct proof of

---

[1] Vid. *Lect. Sci. Language*, 2d series, p. 503.

[2] *Frag. de Bérose*, p. 304.

[3] Vid. Ibid., p. 219. Cf. Obry, *Du Berceau*, etc., p. 116.

the connection of Gemini with the first human pair and with the terrestrial paradise. The Hindu *Yama* and *Yami*, reputed first man and woman, being twins, are related to *Gem-ini*, by the etymology of their names, and this zodiacal division is represented by a man and woman in the Hindu sphere. The *Asvins*, the twin horsemen, usually identified with the Dioscuri, assimilated to Gemini in classic mythology and upon the Greek sphere, are now seen to have been begotten in the *Uttara-Kuru*, directly associated with the mount of paradise. Besides this, *Saranyû*, "the cloud," mother of the two pairs of twins, represents the primal chaos, and her father *Trashtar* is creator, and we have shown that all the ancient cosmogonies centred in the sacred mount, traditional first abode of man. The direct connection of Gemini, then, with the first human pair and their happy abode, considering all the facts developed in these studies, is a point upon which no more doubt can be entertained. But the " *Uttara-Kuru* of the firmament," astronomically speaking, is the region of the north celestial pole, the *Su-Meru* of the Hindus. Ample proofs will be presented in the next chapter that this celestial region is that of the source of the sacred river of paradise, and that it was upon its banks that the celestial horses were supposed to be engendered. It will be shown, also, that the *Asvins*, properly considered, are the *Centauri*, especially the zodiacal *Sagittarius*, directly opposite Gemini in the zodiacal arrangements. Lest this statement shall appear too hazardous in the mind of the critic, I offer here two considerations tending to substantiate it, reserving other proofs for another study.

In the old Accadian calendar we have for the name of the month answering to Sagittarius, *Ab Gan-gan-na*, "the month of many clouds."[1] According to classic mythology the Centaur, assimilated to Sagittarius, was born of *Ixion* and *Nephele*, the latter name signifying "a cloud."[2] As previously shown, the Phœnician star *Su-gi* appertained to Sagittarius. This word *Su-gi* signifies: 1st. The front part of a chariot; 2d. Rival, or rivalry, especially between horsemen with chariots.[3] All this corresponds perfectly with

[1] Vid. Norris, *Assyr. Dictionary*, i. p. 51.

[2] Smith, *Class. Dic.*, art. " Centaurus." Cf. Bernard, *Dic. Myth.*, art. " Centaurus."

[3] Sayce, *Trans. Bib. Arch. Society*, iii. p. 173, note 2. Cf. Delitzsch, *Assyr. Studien*, pp. 120, 121.

the character: (a) Of Centaurus; (b) Of the Dioscuri; (c) Of
the Asvins.[1] We are thus justified in assuming, provisionally at
least, the assimilation of the Asvins, whose mother was Saranyû,
" the cloud," to Sagittarius or the Centaur, as one of these twins,
the other being Pegasus, as will appear hereafter. The two pairs
of twins, then, born of the Hindu Vivasvat and Saranyû, "the
cloud," were for Yama and Yami, the zodiacal Gemini, and for the
Asvins, the celestial horses; these two couples being conceived in
some sense in opposition, just as Sagittarius and Gemini are in the
zodiacal band. To regard the Asvins as horses instead of horsemen is quite natural, as both their father and mother had assumed
the form of horses when they were begotten. Now, Yama and
Yami, as first man and woman, are thoroughly terrestrial characters, and appertain thus to the terrestrial Uttara-Kuru, or paradise;
thus, also, to the lower hemisphere and winter solstice as the traditionary place and period of the birth of man. But the Asvins, the
celestial horses, are strictly astronomical characters, appertaining
thus to the Uttara-Kuru celestial, that is to say, to the Su-Meru of
the Hindus, the region of the celestial pole. We have here another
essential condition of our problem. It is this, to harmonize astronomically these traditional and legendary characters and facts. But
we shall have sufficient space in the present chapter to point out
another condition of the general problem before us.

SEC. 146. M. Dupuis develops some remarkable facts relative to
the natal hour of the world as held by the ancient astronomers,
and as he cites his authorities constantly, they ought to be, simply
as facts, fully relied upon : —

" It does not suffice us to determine the positions of the planets
in the heavens at the moment of the departure of the spheres ; it is
necessary to know the position of the heaven itself relative to the
horizon, and consequently to the day, if we would find exactly the
position of the heavens at the instant when the first ray of light
illuminated the world." " This position is given us by Firmicus
and Macrobius. The last named observes : ' At the moment when
the day commenced which first illuminated the universe, and when
all the elements issuing from chaos arranged themselves under the
brilliant form which we admire in the heavens, — day which we may
with reason term the natal day of the world, — it is said that Aries,

---

[1] Smith and Bernard, art. " Centaurus," also " Dioscuri." Cf. Whitney, Orient.
and Ling. Studies, p. 38.

or the Ram, occupied the centre of the heavens.' 'At the horizon mounted Cancer carrying the crescent of the moon, followed immediately by Leo.' Firmicus places equally at the point called the Horoscope, as to the east, the middle of Cancer, at the moment when the heavens commenced to revolve. This astrological tradition upon the position of Cancer, at the instant when the march of Nature began, is confirmed by Eneas of Gaza, who says that the Hierophants of Egypt, among their opinions upon the origin and formation of the universe, made Cancer preside at the natal hour of the world. This accords perfectly with the sentiment of Porphyry, who assigns the commencement of the Egyptian year at the new moon of Cancer, at the rising of Sirius (dog-star), which rises always with this sign, and which presides at the birth of the world, the same as Regulus, which at Babylon rises with it. This agrees with the language of Solinus, on the occasion of the rising of Caniculus, that the Egyptian priests regard this moment as the natal hour of the world, that is to say, that they assign the commencement of the world and of all the revolutions to the beginning of their great year, or the Sothic period." Of Sirius the author adds: "It was a paranatellon of Cancer, says Servius, that is to say, as he himself explains this mode of expression, the principal star which always accompanies the rising of Cancer. Thus Cancer mounts the horizon at the moment of the birth of the world, and at the same time Sirius rises; both preside at the birth of all things, the one as sign, the other as paranatellon. Behold, then, the state of the heavens well determined, fixing in a manner the most precise the position of the sphere at the moment when the revolutions commenced." [1]

A striking confirmation of these traditionary notices connecting the birth of the cosmos definitely with Cancer and the rising of Sirius, astronomical events associated likewise in Egypt with the periodical deluge or overflow of the Nile, is to be found in connection with the Aryan *Trashtar*, father of *Saranyû*, "the cloud." He appears, as we have seen, in the character of "creator," and according to the Persian legends he causes a deluge for the destruction of the wicked.* At the same time he is identified with Sirius as a star, and assigned to the fourth month of the calendar, answering to the sign Cancer.[2] Another confirmation of this tradition is to be found

[1] *Origin*, etc., iii. pp. 175, 176.

[2] Carré, *L'Ancien Orient*, ii. p. 342. Cf. Benfey, *Monatsnamen*, pp. 54–57, 94, 95; Lenormant, *Frag. de Bérose*, p. 277. *Trashtar*, the creator, is said to give his daughter in marriage to *Vicasvat*, the sun. This giving the daughter in marriage was an ordinary locution in China for the *heliacal rising of a star*; thus Dr.

in the facts: 1st. That the Egyptian Thoth, or Mercury, presided
over the commencement of the calendar, which coincided nearly
with Cancer, or the summer solstice.   2d. That the same Thoth,
as infernal Mercury, appears in the character of creator, and as
such is especially associated with the moon, which, according to
Macrobius, "accompanied the rising of Cancer at the natal hour
of the world." [1]

The fact that the zodiac as known to us was employed in Egypt
only at a late date does not essentially affect the foregoing statements, since all appears to have hinged upon the rising of the dog-
star, or Sirius, which in later periods was known to correspond to
Cancer.   For the rest, the facts noted concerning Trashtar as creator and as Sirius; concerning *Saranyû* as daughter of Trashtar,
as personifying the primal chaos; finally, concerning Thoth as creator, and as presiding over the opening of the year, all tend to confirm the astrological tradition stated, as being not merely Egyptian,
but widely known among other peoples, including the Aryans of
Asia.   But Cancer, in the minds of the authors cited by M. Dupuis,
was understood for the *summer solstice*, and the attendant heliacal
rising of Sirius; this supposition, however, is in direct conflict with
the traditions equally reliable, ancient, and wide-spread, which
assigned the natal hour of the world and the birth of man, not to
the summer, but definitely to the *winter solstice*.   We know even,
from the data produced in the present chapter, that the Egyptians
themselves, from the remotest period, had associated these events
with the region Sutensinen, and with the period of the winter solstice.   Thus, we have here another, and quite important, condition
of our general problem.   It is, to reconcile these two contradictory
traditions, equally ancient as well as perfectly well authenticated.
M. Dupuis has correctly seen, in my view, that the summer solstice
was not that primitively intended in either tradition.

Schlegel says : " The rising of a star in the morning was termed, its *expansion*,
and the rising in the evening its *contraction*.   The heliacal rising of a star was
equally termed, ' to give his daughter in marriage,' as the sun seemed to conduct
the young star to the nuptials" (*Uranographie*, p. 21).   This obviously offers
some explanation of *Trashtar* as creator, on one hand, and as associated with
Sirius on the other, giving his daughter in marriage at the heliacal rising of this
star, the sun being in Cancer, marking the period of the birth of the cosmos.

[1] Vid. Rawl. *Herod.*, ii. pp. 237–240.   Cf. Brugsch, *Nouvelles Recherches*, pp.
14, 15, 22 ; also, De Rougé, *Nomes de l'Egypte*, p. 25.

SEC. 147.   It is a very significant circumstance that a palpable contradiction, to be compared with the one just developed, is found to exist between the Chinese *traditions* of a primitive character, on one hand, and the later Chinese *uranography*, on the other.   It has been shown (Sec. 142) that four great constellations, each comprising seven lunar mansions, had been originally assumed by the Chinese as marking the four seasons of the year, and as corresponding to the four cardinal regions.   The constellation of the Blue Dragon was put for the east; that of the White Tiger for the west; the Red Bird corresponded to the south; and the Tortoise, or Black Warrior, to the north.   For the east and west, the lunar houses, according to their modern notation, agree perfectly in position with the ancient order of the four great constellations above named.   But it is a remarkable fact that the lunar asterisms of the Black Warrior, put for the north, are now found in those zodiacal divisions appertaining to the lower hemisphere, as Sagittarius, Capricorn, Aquarius, etc.; while the lunar mansions composing the Red Bird, primitively put for the south, are to-day located in those signs appertaining to the upper hemisphere, as Gemini, Cancer, etc.[1]   The singular discrepancy thus presented in the Chinese sphere, which is indeed quite a mystery, constitutes the chief basis of Dr. Schlegel's calculations upon the antiquity of the sphere among the Chinese; an antiquity which, according to his interpretation, amounts to about 18,500 years from the present time.   My opinion is that the author's construction of the discrepancy referred to is fundamentally erroneous, and that his resultant chronology is too great by many thousands of years, but of this hereafter.   We have in the data here produced an additional element and condition of our general problem, and our space forbids the development of any others in the present· chapter.   We can only recapitulate the various facts now before us, including those brought out in previous chapters, having a like reference to our problem, as constituting really essential conditions of it.

SEC. 148.   1st. In the seventh chapter it was fully established that the primitive traditions relating to the terrestrial paradise had been, at an unknown period, localized astronomically in the lower hemisphere, and connected with the Greek Hades, likewise assigned astronomically to the same region.   But in the fifth chapter it had

---

[1] Schlegel, *Uranographie*, pp. 1, 2 ; cf. pp. 171, 172, etc.

been shown that the lower hemisphere, astronomically speaking, was
put symbolically for the earth, according to the principles embodied
in the zodiacal temple; but in the sixth chapter, this earth was
found, to have been really the traditional earth known to the first
men; that is to say, the terrestrial paradise. All this is perfectly
consistent; but when it is shown, as has been done, that the constellation Gemini had been assimilated to the first man and woman,
we encounter a difficulty in the fact that Gemini is now in the
upper hemisphere, and really marks the period of the summer solstice. If we assume, however, that owing to the law of precession
the inclination of the earth's axis to the ecliptic was just the opposite of what it is to-day, at the period to which the traditions of
paradise appertained, we should then find the constellation Gemini
in the lower hemisphere zodiacally, and all the discrepancies otherwise appearing would be at once removed.

2d. The fact has been verified that, according to primitive tradition inherited by distantly separated races, the creation of the
world and the birth of humanity took place *in* the lower hemisphere, and *at* the period of the winter solstice. In all the cosmogonies, the organization of the cosmos and the birth of man are
placed in immediate connection; and it has been shown heretofore
that the traditions relating to both centred in the paradisiacal mountain, the birthplace of man, located geographically in the upper
hemisphere. Assuming the same original inclination of the earth's
axis to the ecliptic as before, we shall find a consistent explanation
of these traditions. 1st. These notions must be interpreted astronomically; since geographically man was created in the upper
hemisphere. 2d. The lower hemisphere astronomically was assimilated to the terrestrial paradise, where man was actually created. 3d. On the present assumption relative to the earth's axis,
the constellation Gemini, which was put for the first human pair,
was actually in the lower hemisphere, and marked the period of
the solstice of winter, instead of that of summer, as is at present
the case.

3d. According to Hindu legends, especially as contained in the
Vishnu Purana, the *Asvins*, or "horsemen," were begotten in the
Uttara-Kuru, "the country of happiness;" that is to say, in the
region of the mount of paradise. As the *Asvins* are purely astronomical characters, we must assume here the "*Uttara-Kuru* of the

firmament;" in other terms, the *Su-Meru,* or the north celestial pole. But in point of fact, the *Asvins* must be assimilated to the Centauri, especially to Sagittarius, which is in the lower hemisphere. Again, *Yami* and *Yami,* having the same parentage as the *Asvins,* are traditionally put for the first man and woman; are thus eminently terrestrial in character, and to be associated with the terrestrial *Uttara-Kuru,* or the terrestrial paradise. At the same time they are assimilated zodiacally to Gemini, which is now in the upper hemisphere. It will be seen at once that the essential conditions of these legends require a complete reversal of the present positions of the two constellations assimilated to the two pairs of twins. Gemini should be in the lower hemisphere, symbol of the terrestrial paradise, and Sagittarius in the upper hemisphere, corresponding to the celestial *Uttara-Kuru.* The assumption of the primitive inclination of the earth's axis, as already proposed, would reconcile all these discrepancies. Sagittarius would then mark the period of the summer solstice, while Gemini would mark that of the winter solstice.

4th. We have shown that according to an ancient doctrine held by the Egyptian priests, and apparently prevalent in various quarters of the old world, the natal hour of the universe was marked by the heliacal rising of Sirius, or the dog-star, with the sun in Cancer; that is to say, at the summer solstice as supposed. Now this astronomical theme is not only in direct conflict with other Egyptian traditions, equally primitive and authentic, but it contradicts the Chinese doctrines, which in every other particular accord perfectly with the Egyptian, and which assign the birth of the cosmos from the conflict of light and darkness to the period of the winter solstice, marked by Capricorn. As correctly apprehended by M. Dupuis, the original doctrine had reference to the winter solstice, and thence to the period when Cancer (considering the signs movable) marked that period; and thus again our supposition relative to the inclination of the earth's axis at the epoch referred to constitutes the basis of explanation.

5th. The simple fact that the great Chinese constellation termed the Tortoise, or Black Warrior, was put for the north leads us naturally to suppose that the lunar asterisms comprising it would be found in the constellations of the upper hemisphere; but in fact they are now found in those appertaining to the lower hemisphere.

The same in regard to the great constellation of the Red Bird, primitively put for the south ; the lunar houses of which are not found in the zodiacal divisions of the lower hemisphere, but in those of the upper. For the great constellations of the east and west no such anomalous circumstances appear. Dr. Schlegel's proposed method of explanation of these discrepancies results, as before stated, in a chronology of 18,500 years for the Chinese sphere. But I think it is demonstrable that his conclusions, considering the data, are far from correct. It appears to me that the assumed reversion of the inclination of the earth's axis, which offers so ready a solution of all the other contradictions that have been stated, affords the only true ground of explanation in the present instance.

SEC. 149. Properly, another very essential condition of our problem, relating to the primitive pole star, might be included with those already presented. But as I have wished to develop this entire subject to the reader's mind in the precise order in which it gradually revealed itself to my own mind, the point referred to will be taken up in another connection, where the facts relating to it will take the form of a confirmation of what has gone before. At present, the *constellation* Gemini is in the *sign* Cancer, and both mark the period of the summer solstice ; while the *constellation* Sagittarius is in the *sign* Capricorn, both marking the period of the winter solstice. Our assumed primitive inclination of the earth's axis, in order to fulfill the various conditions of the general problem before us, supposes that at the period referred to the constellation Gemini was in the sign Capricorn, answering to the winter solstice, while the constellation Sagittarius was in the sign Cancer, corresponding to the summer solstice. In other words, I have assumed that the solstitial points, and of course the equinoctial, have fallen back just half the circle, or six entire divisions of the zodiac, since the period from which the primitive traditions of mankind were first derived ; and it is on this assumption, as a constant principle applicable alike to each of the conditions of the problem, that I have sought to reconcile the contradictions involved in them. Upon ordinary questions pertaining to high antiquity, the force of evidence derived from the ready explanations afforded by means of the principle assumed would be regarded as entirely sufficient in proof of the correctness of that principle. But in a question so very important as the one before us, I do not think it is sufficient, and shall not ask the reader to receive it as such. We are, however, fully justified in a *provisional* assumption that our principle is correct. It has been inductively established thus far, and it remains to test it according to the same inductive method.

As the reader is probably aware, all the zodiacal *constellations* are to-day one entire division in advance of the *signs* having the same name; otherwise expressed, all the signs are one division of the zodiac behind their respective constellations. This difference involves a period of about 2100 years ago for the time when all the signs and constellations coincided in position. Since our theory supposes an advance movement of the constellations, or a retrograde one of the signs, to the extent of *six* entire divisions, the result would be that there was a relative difference of *five* divisions of the zodiac between the signs and the constellations at the period to which the primitive traditions of mankind related. This would involve an antiquity for that period of about 12,500 years from the present time. It was at that hour of the world, if my views are correct, for which the hands on the dial-plate of the heavens were originally adjusted. So far as concerns the data now before us, my theory, even if admitted as correct, does not absolutely suppose the existence of a completed zodiacal arrangement at the epoch indicated. It presumes simply the existence of certain constellations, traditionally associated with certain events and with certain epochs of the year, especially the solstitial periods. The relative positions of the same constellations to the same annual phases of nature to-day suffice to determine the length of the period approximatively between the two astronomical eras. But the zodiacal arrangement, when perfected, bears obvious reference to those very traditions whose antiquity we seek to determine. For the rest, we must seek further light.

CHAPTER XIV.

PRIMITIVE ADJUSTMENT OF THE ZODIACAL SYSTEM.

SEC. 150. The astronomical, or properly the astro-chronological theme which is to constitute the subject of the present study the reader finds in our fourth plate. It represents the relative positions of the zodiacal signs and constellations which it is necessary to assume, in order to satisfy the various conditions of our problem, as these were brought out in the last chapter. In the theme presented I have preferred to consider the *signs as fixed*, and the *constellations as movable*, since such appears to have been the custom at Babylon, according to the recorded observations of the vernal equinox already noticed, and for the reason that such method agrees better with the apparent aspects of nature. This, however, is a matter purely optional, the result being the same in either method of notation. In the plate, the inner circle is taken to represent the signs, — Aries being in the west, at the vernal equinox, Libra in the east, to mark the autumnal equinox, the reader being supposed to *face the south* in examining the plate. The second circle is put for the constellations, their relative positions to the signs being that which, according to my theory, truly represent their actual positions at the period to which the primitive traditions appertained. The third circle contains the names of the twelve stars of Phœnicia, which I supposed to have agreed rather with the constellations than with the signs. Finally, the antediluvian genealogy is placed in the fourth or outer circle, according to the adjustment of the same to these Phœnician asterisms. Our main interest, of course, centres in the two inner circles. As before intimated, it is assumed that the relative positions of the constellations in the second circle truly represent the aspects of the heavens in relation to the signs and to the earth at the epoch from which the primeval traditions were derived. The *constellation* Gemini

PLATE IV.

N.

Enoch. Jared.
Methuselah. Sugi. Nammakh.
Qagsidi. Capricorn. Sagittarius. Scorpio. Nibeanu. Mahalaleel.
Lamech. Bir. Leo. Cancer. Gemini. Libra. Cainan.
Aquarius. Virgo. Taurus. Aries. Ninmakh.
Virgo.
E. Libra. Astro-Chronological Theme. Virgo. T.
Pisces. Leo. Barta bbaqal geel.
Noah. Ungal. 1st Inside Circle = Zod. Signs. Pisces. Enos.
Scorpio. 2nd Circle = Zod. Constellations. Aquarius.
Aries. 3rd Circle = Stars of Phœnicia. Cancer. Sir.
La ba. Sagittarius. Capricorn. Seth.
Creation. Taurus. Gemini. 4th Circle = Antediluvian Genealogy.
Allap. Askar
Week. Adam.

S.

was then in the *sign* Capricorn, marking the period of the winter solstice. The constellation Virgo was in the sign Aries, answering to the vernal equinox; Sagittarius was in the sign Cancer, corresponding to the summer solstice; while Pisces was in Libra, marking the period of the autumnal equinox. Thus, the aspects of the heavens relative to the earth were precisely the opposite of what they are at the present day. Such, then, is the astro-chronological theme which we have provisionally assumed with a view to submit it to the most rigid tests possible.

A few additional remarks are submitted here in explanation of the ground principle constituting really our method of research. The various traditions and legends pertaining to the first ages of humanity *are the facts for whose existence we have to account.* The two chief elements comprising these facts are certain geographical and astronomical references, recollections of which had been inherited alike by peoples the most widely separated in antiquity. The aspects of nature on one hand and of the heavens upon the other which were most familiar to the first men, forming in fact the world in which their sensuous, intellectual, and moral beings had their birth and development, constituted the normal and producing cause giving rise to the conceptions transmitted to their posterity, these being the traditions and legends to which we refer. We cannot account for their origin and transmission upon any other natural and rational principles than the surroundings of primeval humanity. Geographically we have determined beyond question what were the general aspects most familiar to the first men. Such was the work of the sixth chapter. This element is thus a known factor in our main problems. That which is unknown is the astronomical element. What we have to do is, therefore, to assume such a phase of the heavens, in relation to the geographical element, as renders it possible to explain therefrom the origin and transmission of those ideas which have been passed in review, and this upon natural principle. We very well know that, in the present relation of the heavens to the earth, it would be impossible to reconcile the two chief factors, the geographical and astronomical, forming the basis of primeval tradition; impossible to account for the rise, development, and transmission of these conceptions, except upon arbitrary grounds or mere conjecture. The question for us now is: Does the theme presented in the fourth plate satisfy all the conditions of our problem? Does this scheme enable us to say, Such was the state of the heavens in relation to the earth which gave rise naturally to the primitive conceptions transmitted to after ages, as well as to the various doctrines growing out of them? It will be seen from these remarks that the method proposed is the only legitimate one in such an investigation as this now pending. The course also to be pursued in the present chapter will be clearly apprehended, although our inquiries will not be confined to the simple conditions of the problem, as these were established in the last chapter. In fact, it is proposed to include quite a number of points in the present study beside the conditions named, all having a direct bearing, however, upon the main question before us.

Sec. 151. Enlarging the sphere of the present inquiries, as just intimated, we recur here to a matter previously brought out, namely, the location of the terrestrial paradise astronomically in the inferior heavens. The fact of such location was shown in the seventh chapter, and no doubt of it can be entertained. It agrees perfectly with that other wide-spread doctrine, established in the last chapter, which assigned the birth of man to the lower hemisphere and to the period of the solstice of winter. We know from previous investigations that the birth of man was in the upper hemisphere geographically, and this fact compels us to see in this doctrine a distinct astronomical reference. It so happens that the very constellation to which the first human pair had been assimilated, being the zodiacal Twins or Gemini, marks to-day the period of the summer solstice. But we take the tradition referred to in earnest, giving it an astronomical interpretation, since it will not in any sense bear a geographical one. Thus, in our plate, we have so adjusted all the constellations to the signs that Gemini shall be found in the lower hemisphere, in the sign Capricorn, in fact, which answers to the period of the winter solstice. This, then, as we suppose, was the state of the heavens in relation to the earth which was one of the chief causes of the transfer of the terrestrial paradise zodiacally to the lower hemisphere, and which contributed to the rise of that doctrine, so prevalent in antiquity, that the birth of man took place in this hemisphere, and precisely at the epoch of the winter solstice. The birth of man certainly took place in the terrestrial paradise, and as it has been shown that this was located in the inferior heavens, the two dogmas agree perfectly, confirming each

other, and our plate exhibits the necessary astronomical arrangement to coincide with them. Assuming this position of the heavens at the period referred to, and considering the assimilation of
Gemini to the first human pair, we see at once that these ancient
ideas would naturally take their rise under such a state of circumstances, or from the outward phenomena then most familiar to the
sensuous nature of man.

But we should bring into our present view another important
fact, which was amply verified in the fifth chapter, namely, that
according to the symbolism embodied in the primitive sphere, especially in the zodiacal temple, representing the cosmos itself, the
lower hemisphere astronomically was put for the earth, particularly
the terrestrial paradise, while the upper hemisphere represented the
heavens, more definitely the celestial paradise. This symbolism
also perfectly agrees with the doctrines just noted as to the location of paradise in the inferior heavens, and in regard to the birth of
man in the same region, precisely at the period of midwinter. The
symbolic arrangement here described constitutes, in fact, another
condition of our general problem, and the plate exhibits the exact
state of the heavens relative to the earth to coincide with all, and
to account for the rise of such conceptions. The lower hemisphere
astronomically, to answer all these conditions, has to be taken, 1st.
For the earth; 2d. For the terrestrial paradise; 3d. For the place
of the birth of man: while the constellation Gemini must represent, 1st. The original progenitors of mankind; 2d. The period of
the winter solstice. Our plate fulfills every one of these conditions,
and shows the precise state of the heavens relative to the earth
which would naturally give rise to these notions. It would be utterly
impossible, I think, by any other zodiacal arrangement to explain
these traditionary ideas, except by resort to pure hypothesis and
arbitrary speculation.

The causes, then, which operated in after ages to transfer the
terrestrial paradise to the lower hemisphere astronomically were
for the most part perfectly natural, and our zodiacal theme explains
to us precisely their nature. The constellation to which the first
human pair had been assimilated was then found in this region of
the heavens, and it marked the period of the winter solstice. In
process of time, and as due to the phenomenon of precession of the
equinoxes, these aspects of nature wholly changed, the inferior por-

tion of the zodiac became the superior, as we find it in the present
period. But during this change tradition remained constant and
faithful to its origin. It was the traditionary idea which still per-
sisted in localizing the terrestrial paradise and all the cosmical
legends in the inferior heavens, and the construction of the sphere
in later epochs was made in many respects to correspond.

SEC. 152. In the ninth chapter (See. 100), we have cited M. De
Vogüé upon the twofold cause of all things, heaven and earth,
father and mother, light and darkness; these operating uniformly
according to the law of *union in opposition*. This principle applied
universally, as held by the ancients, and the author just named
thus alludes to its connection with the zodiac : " These twelve signs
in their turn are apportioned between the two principles, accordingly as they are considered male or female, and the same is true
of the thirty-six principal constellations which preside, some over
the celestial and others over the subterranean (really the terrestrial) world. This entire celestial army is animated and active. . . .
The reciprocal action of all these things, their combinations and
antagonisms, produce all the phenomena of the sensible world, since
nature is composed of contraries, and harmony is born of the reaction of contraries. We might almost add, From the *identity of contraries*, since this celebrated formulary constitutes actually the basis
of the entire system." We have seen that, according to the Chinese and Egyptian traditions, the birth of the world and of man
took place as the result of a peculiar union in opposition between
heaven and earth, light and darkness, which occurred precisely at
the period of the solstice of winter. The same notions prevailed at
Babylon, as embodied in the cosmical legends relative to the battle
between Bel and Omarka, male and female principles; a battle
that becomes a conjugal relation in the theogony, as M. Lenormant has correctly observed. This principle of union in opposition, then, from which the birth of the cosmos and of man took
place, precisely at the solstice of winter, was an ancient, a widely
prevalent, and a fundamental doctrine. It formed the basis, in fact,
of the religious philosophy of antiquity. Now, in order to realize
this primitive and traditionary dogma assigning all to the inferior
heavens and to the winter solstice, it has been necessary to bring -
the constellation Gemini into the sign Capricorn, which marks this
solstitial period, and thus to adjust the entire zodiacal system

according to this scale. The peculiar combinations which a slight examination of our fourth plate reveals, as the result of such adjustment, afford a striking illustration of the principle of *union in opposition*, and most fully explain the notion of the birth of the world and of man from the remarkable concurrence of such circumstances.

Confining ourselves to the two inner circles of the plate, the reader observes in the lower portion the *constellation* Gemini connected with the *sign* Sagittarius, by the line *a, b*. Gemini, assimilated to the first human pair, is placed here in the *sign* Capricorn, which marks the winter solstice, in order to realize the traditionary idea of the birth of the world and of man at that period. Gemini and Sagittarius are directly opposed to each other in the ordinary zodiacal arrangement, being also in opposite hemispheres. Here the opposition is between the *constellation* Gemini and *sign* Sagittarius. But from the line *a, b*, in the lower portion, cast the eye to the line *c, d*, directly above it in the uppermost part of the circle, where we have now the *sign* Gemini opposed to the *constellation* Sagittarius. By inspection it will be seen that this law of the combination of opposites prevails throughout the entire circle of asterisms. We proceed, then, to point out another one. Note the opposition of the sign Sagittarius, marked *a* below to the sign Gemini marked *d* above. This is a normal feature of the zodiac. But see the same feature reduplicated between the *constellations* Gemini, marked *b*, and Sagittarius, marked *c*. This law also prevails throughout the whole circle. Finally, we have to note the opposition between the *constellation* Gemini, *b*, and the *sign* Gemini, *d*, directly above it; on the other hand, between the *sign* Sagittarius, *a*, and the *constellation* Sagittarius, *c*, also directly over it. The same law obtains through the entire circle. Here is a threefold law of the combination of opposites which prevails in every quarter of the system, as represented in our plate. It is either an *unparalleled accidental circumstance*, or a *most ingeniously contrived arrangement*. The writer of these pages certainly never contrived it, for he did not discover it till after the last chapter had been written, which determined the assumed adjustment of the zodiacal system with a view to answer the conditions of our problem. Compare now the tradition of the birth of the world and of man from the union in opposition of the two princi-

ples, supposed to take place at the winter solstice, with the threefold law of opposition which results from the assumed zodiacal
arrangement dictated by the terms of that tradition. The direct
relation between the two is extraordinary; and it is difficult to
avoid the conclusion that our plate actually represents the state of
circumstances which gave rise to this ancient and traditionary
dogma which prevailed so widely in antiquity.

1st. We have seen that the sphere, especially the zodiacal temple,
was a symbol of the cosmos. 2d. That the fundamental doctrine of
all the cosmogonies was that of *union in opposition*, like male and
female, heaven and earth. 3d. The constellations on one hand and
the signs on the other were the most appropriate symbols of the
twofold principle from which all originated. 4th. The threefold law
of opposition between the constellations and signs as shown in the
fourth plate offers, then, the most striking explanation of, and literally the key to, that condition of things in which all these notions
originated. We see what was the precise state of the sphere, of
the zodiacal temple, when first taken as the symbol of the cosmos,
and as the basis of the ancient cosmogonies. All considered, it is
impossible to resist the conviction that we are here in the presence
of a great fact, and no fortuitous circumstance. It is easy to perceive here *why* the conditions of the problem, as involved in the
primitive traditions, forced us to assume the present astronomical
theme: it was the identical state of the heavens in relation to the
earth which gave rise to those traditions and the conditions involved
in them.

Sec. 153. I desire to trace briefly here the relation of the zodiacal scheme before us to the fundamental doctrines of the "celestial
earth," or "terrestrial heaven," as developed in the three chapters
devoted to this subject. The primitive connection of the celestial
earth with the mount of paradise, from whence the ideas involved
in it had been traditionally derived by the nations of antiquity, was
fully established in the studies referred to. At the base the notion
was that of a *heaven + earth ;* or an earth divided off in the manner
of the heavens, according to the cardinal regions, and placed in
direct relation to a certain celestial space, of which the terrestrial
was esteemed a reproduction, or express imitation. All things on
the earth were held to be copies of things celestial, from which they
were supposed to be derived as their original. As before remarked,

actually the primitive celestial earth was the paradisiacal mountain itself, the first abode of man. But it was shown in the seventh chapter that both had been transferred astronomically to the lower hemisphere in subsequent tradition, an explanation of which fact was submitted in previous chapters, and particularly in the very last section. Again, with the first human pair the constellation Gemini had been identified; and the birth of this original cosmos had been assigned to the period of the winter solstice. If, now, we compare these notions of a "celestial earth," together with the other data connected with it, with the astronomical theme offered in our plate, it will be impossible not to recognize at once the various and striking analogies existing between the two orders of ideas. Their identity in origin can hardly be doubted. It must have been this zodiacal scheme, and no other, that gave rise alike to the fundamental doctrine of union in opposition, and to the notion that all things on earth were reproductions of heavenly originals. The "celestial earth," in fact, was only the completed cosmos, both being produced from the operations of one and the same law; namely, that so ably exposed to us by M. De Vogüé, and so perfectly embodied in the zodiacal arrangement before us. The two principles, male and female, heaven and earth, light and darkness, the upper and lower hemisphere, — these two principles, I say, whose inherent and fundamental union in opposition was the ground thought of all, are perfectly symbolized in the constellations and signs, under the peculiar method of grouping them exhibited in our plate. No contrivance could more significantly represent to the eye those notions which constituted the basis of the cosmical theories of antiquity. But we return now to a consideration of the main conditions of our problem, some of which have been noticed already in the present study.

All that has to be done to realize the Egyptian tradition relative to the natal hour of the world, at the same time to reconcile the two versions of it as presented in the last chapter, is simply to change our method of notation, considering the signs as movable, instead of the constellations; and then to assume the same amount of precession, as done in the scheme represented in the fourth plate. In such case, we transfer the sign Cancer to the position of Capricorn, marking the winter solstice; and the sign Capricorn to the position of Cancer, marking the period of the summer solstice.

We thus harmonize the traditions derived from the classic authors, respecting the natal hour of the universe, with those of the monuments, which localized the organization of the world in the region of Sutensinen, when Osiris, symbol of the life of humanity, is manifested to the light. It was obviously to the period of the winter solstice that both traditions referred; and M. Dupuis was perfectly correct in adopting this view. Thus understood, the Egyptian and Chinese traditions become almost identical. This leads us to notice briefly the marked discrepancy existing between the traditions of the Chinese and their uranography, to which reference was made in the last study.

SEC. 154. The four great constellations of the Chinese sphere, each composed of seven lunar mansions, each put for one of the four cardinal regions, are very properly regarded by Dr. Schlegel as primitive; as appertaining to the earliest epochs. The constellation of the Red-Bird, as has been seen, was located traditionally in the south; and it is remarkable that it answers to our Gemini, Cancer, etc., at present in the upper hemisphere. On the other hand, the Tortoise, otherwise termed the Black-Warrior, was traditionally assigned to the north; while it is made up of our Sagittarius, Aquarius, etc., which appertain to the lower hemisphere. In accordance with these facts, it seems that Chinese astronomers associate the south with summer, although the sun at this season is in the north; and likewise, that they associate the north with winter, notwithstanding the sun is then in the south. But there is much reason to believe that this habit of thought was not primitive; that it originated rather in the anomalous connection of asterisms appertaining to the upper hemisphere with the great constellation of the south, or the Red-Bird; and so in reference to the Black-Warrior put for the south, really composed of asterisms, many of them belonging to the inferior heavens. It is worthy of especial note here that the Chinese regard the Red-Bird put for the south as a Phœnix; and this circumstance adds force to the suggestion which would identify the Chinese Phœnix with the Bennen, or Phœnix, located by the Egyptians in the mystical region of Sutensinen, in whose great nest the sun was supposed to renew itself at the period of midwinter. If such identification be admitted, then the Red-Bird located in the south answered likewise, primitively, to the sun in the south, or in the lower hemisphere. The difficulty then would be found in the fact that this asterism is made up of our Gemini, Cancer, etc., which at present are found in the upper hemisphere. On the whole, I believe the position of the heavens as exhibited in our fourth plate offers the true explanation of these contradictions. The asterism termed the Red-Bird had its origin at the period when the constellation Gemini was found in the sign Capricorn, marking thus the period of the winter solstice.

We turn attention briefly to the great constellation of the north, the Tortoise, composed of our Sagittarius, Aquarius, etc. Assigning this asterism to the same period as before for its origin, we should find Sagittarius, Aquarius, etc., in the upper hemisphere, as seen in the plate. I think it more than probable that we should recognize here some primitive connection of ideas with the cosmical Tortoise, conceived by the Chinese as having upon its back the image of the eight celestial regions and of the seven stars of the chariot, the legend relating to which was produced in a previous chapter. But more important for us here is the fact, as abundantly established by Dr. Schlegel, that the constellation of the east, or the Blue-Dragon, was held traditionally to represent the *vernal equinox*.[1] It will be readily perceived that this tradition supposes a complete reversion of all the seasons as compared with the present age of the world; and yet our plate exhibits an astronomical theme which implies just such a state of things. Thus, it must be admitted, I think, that the hypothesis assumed by us, relative to the primitive position of the constellations of the zodiac, affords a ready solution of the difficulties referred to as pertaining to Chinese uranography.

SEC. 155. We proceed now to consider the Hindu legends relating to the two pairs of twins, offspring of Vivasvat and Saranyû. The fact has been often stated in previous studies that the mount of paradise was supposed to unite the heavens and earth. Sometimes this sacred locality was conceived as an immense conical hill; but at others, as a vast column. The summit, in either case, was held to penetrate the heavens in the region of Su-Meru, the celestial pole; and this region was taken for the celestial paradise, united thus to that of men by means of the sacred mount. It was fully shown in the last chapter that both pairs of twins were directly associated with this mountain. Yama and Yami, the first pair, were the reputed first man and woman; and their connection with

[1] Vid. *Uranographie Chinois*, pp. 56–58.

Gemini, or the "Twins," is placed beyond doubt. As for the second pair, the Asvins, the Vishnu Purana proves that they were begotten in the region of Uttara-Kuru, evidently the Uttara-Kuru celestial, since they are purely astronomical characters. This celestial region was that penetrated by the summit of the sacred mount, the Meru of the Hindus, Albordj of the Persians. With these facts in mind, it is important to note this text from the Zend-Avesta: " I invoke, I celebrate the divine summit, source of waters, and the water given by Mazda," to which M. Carré adds the note: "This source is the Arduissur (or *Avanda*) at the summit of Albordj, from whence issue all the waters that flow upon the earth."[1] To these extracts, we must add still another, cited by Professor Benfey from an ancient author as a comment upon the foregoing text from the Avesta: "The divine Burga (*Albordj*), Ized of women, whose nature is water; it is the source of generations. It is the navel of waters, because upon it is the source of the river Avanda, from which *the beautiful horses were generated*."[2] Thus, the Asvins are begotten in the celestial Uttara-Kuru, identical with the source of waters, from which the beautiful horses were generated. Connect with these data the fact well known that the very name *Pegasus*, applied to the celestial mare, has the sense of "source of waters," since she was supposed to have been born near the source of Oceanus.[3]

The foregoing statements sufficiently justify us in identifying the flying horse, Pegasus, as one of the Asvins; and the proofs introduced in the last chapter tend equally to connect Sagittarius, or the Centaurus, with the other. The objection that these are *horses*, and not *horsemen*, as the term *Asvins* denotes, can have but little force; for the Asvins are begotten by Vivasvat and Saranyû, both having assumed the form of horses (Vish. Puran., p. 266). If, now, we examine the astronomical theme exhibited in the fourth plate, it will not be difficult to recognize its perfect accord with all the data now established relating to the two pairs of twins. The constellation Gemini, assimilated to Yama and Yami as first man and woman, is found in the lower hemisphere, where we are to locate the terrestrial paradise, or Uttara-Kuru. Sagittarius is found in the upper hemisphere in that zodiacal position which best answers

1 *L'Ancien Orient*, t. ii. p. 341.
2 *Monatsnamen*, pp. 208, 211.
3 Smith's *Class. Dic.*, art. "Pegasus."

to the Uttara-Kuru celestial. In point of fact, Sagittarius is located upon the borders of the milky way, which was really the sacred river from which the beautiful horses were born; but this is a point to be verified in the next chapter. We merely assume it now, in order to complete our explanation of these legends.

It is proper to remark here that all Sanskrit scholars concede the difficulties to be encountered in a satisfactory interpretation of the legends relating to the Asvins. For the most part, their character has been viewed with reference simply to the diurnal aspects of nature. They evidently relate to the sun, as evinced by their frequent connection with chariots, that is, the chariots of the sun. Now the sun-god really involves three phases of character, namely, the daily, annual, and cosmical; and the notion of *opposition*, so fundamental in that of both pairs of twins, actually underlies all these phases. There is the opposition between day and night, summer and winter, and so of the primordial night from which the first light is born. In all, also, there is the notion of union; hence union in opposition. Finally, as regards the Asvins, their reference to the diurnal phase of nature is generally admitted; Professor Kuhn has shown that they have a cosmical import, since their mother is the dark "storm-cloud" representing chaos; and for the annual phase, I believe the facts developed in these researches will be regarded as conclusive. It seems to me that the fourth plate accurately represents the state of the heavens when all these legends had their birth.

SEC. 156. Thus far in the present study we have been mainly occupied with the attempt to show that the astronomical theme before us actually fulfills in a most satisfactory manner all the conditions of the chronological problem which were established in the last chapter. But in addition to this, we have shown the existence of several remarkable features, as involved in our assumed zodiacal arrangement, that seemed to explain the origin of some of the fundamental doctrines pertaining to the cosmical and religious philosophy of antiquity. We now turn attention to matters less general in their bearing, but which will afford not less striking proofs, tending to the support of our general hypothesis, all connected with the zodiacal scheme before us.

In the tablets of the twelve stars of Phœnicia, we found the star *Al-lap*, or the Taurus, put for the country of the east, and the star

Lab-a, or Aries, for the country of the west. These points of the
compass being directly opposite, it was necessary, in order to realize such an arrangement, to conceive the Taurus as located in Libra,
formerly the Pincers of Scorpio. This called to mind the artistic
representations of the sacrifice of the taurus, in which a scorpion is
uniformly seen attacking the generative organs of the victim. The
first circumstance gave rise to the suspicion that there was a mystical allusion, on the part of the scribe, to the cherubim placed at
the east of the Garden of Eden ; especially as the sacred mount was
located zodiacally in Libra. In the chapter devoted to the subject
of these symbolic figures, the arrangement of the scribe and the
representations upon the art monuments contributed essentially to
the discovery of the fundamental notion involved in the cherubim.
But this direct association of the Taurus with Libra seemed quite
arbitrary, and wholly the result of a symbolical conception. There
was not the slightest suspicion at the time that any such relation
had ever actually existed. However, in our attempt to ascertain
the astronomical period to which the primitive traditions of the
world pertained, including those relating to the Garden of Eden,
we have been led to assume the zodiacal scheme now being studied
as accurately representing the state of the heavens at that epoch.
It must be admitted, I think, as somewhat remarkable that we find
here the constellation Libra in the sign Taurus, indicating thus a
direct, primitive association of these two zodiacal divisions. It will
be seen at once that this fact affords a complete explanation, as well
as an important confirmation, of our theory respecting the fundamental import of the cherubic animals, as developed in Sec. 130.
On the other hand, the data there presented serve powerfully to
support the hypothesis that our fourth plate actually represents the
astronomical condition of things from which the traditions of Eden
must be dated. In a word, the two propositions explain and support each other.

But another coincidence, not less striking than the one just discovered, is the fact that, according to our plate, the constellation
Scorpio was found in the sign Gemini at the period to which we suppose the primeval traditions of the world pertain. Gemini represents, of course, the first man and woman according to the abundant
proofs heretofore presented. As regards Scorpio, recall here the
ideas found centring in the two signs Libra and Scorpio in the chap-

ter on the Cherubim. Thus, it is during the sun's passage through Scorpio, according to Egyptian tradition, that Typhon kills Osiris, or the god with the head of an ox. To this must be added the fact, not mentioned in the twelfth chapter, of the definite association of Dan with the same sign in Jacob's last vision, thus: " Dan shall be a serpent by the way, an adder in the path, that biteth the horse's heels" (Gen. xlix. 17). The word for " serpent " here is the Hebrew *Na-khash*, the same as applied to the serpent of Eden ; and it is well known that the serpent was often substituted for the scorpion in the zodiac. Now, according to our astronomical theme, the constellation Scorpio, or serpent, was found in the sign Gemini at the period of the origin of the primitive traditions. Finally, we note the appearance in the sign Gemini of the name of the patriarch, *Jared*, " the descent," which was interpreted in the twelfth chapter, of the *fall of man ;* and we have now a very conclusive proof of the correctness of this view. Thus, the identical conceptions relative to the fall of man which were seen, in the twelfth chapter, to centre in the signs Libra and Scorpio are now found repeated with equal clearness in the signs Taurus and Gemini. It is obvious that they originated at the astronomical period, when the *constellations* Libra and Scorpio were in the *signs* Taurus and Gemini ; and that ages after, when the astronomical period had passed out of mind, the same ideas continued to be associated with the signs Libra and Scorpio taken alone ; and to aid in perpetuating this association of ideas, the sacred mountain and the cherubim were mystically represented in Libra, and the name of Dan, as biting serpent, connected with Scorpio. Our fourth plate, then, offers a complete explanation of all these singular conceptions, both as to their origin and import; and its actual scientific value in this respect begins now to vindicate itself. But we pass to the consideration of still another coincidence, equally remarkable in character.

SEC. 157. Immediately following Taurus and Gemini, in the order of the signs, is that of *Cancer.* In the Asiatic calendars, the dying sun-god was associated usually with this zodiacal division. Thus, the Hebrew name of the month answering to Cancer is *Tammuz*, and the Assyrian name is *Dâ-zu*, both being well known titles of the solar deity, who suffers a violent death, and is afterwards raised to life. As previously stated in these researches, in the Arabic zodiac, the lion's paw stretches out over the space occu-

pied by Cancer, forming thus a distinct lunar mansion called the
" Paw." But the dying sun-god was also Hercules, whose association with the cherubim, located zodiacally in Libra, and assimilation
to the promised seed of the woman, were points fully established in
the twelfth chapter. We are prepared now to explain the enigmatic
allusions in Jacob's prophecy relative to Dan: " A serpent by the
way, an adder in the path, that *biteth the horse's heels.*" Not only
is there an obvious reference here to the serpent of Eden (*Na-
khash*), but to the expression, " Thou shalt bruise his *heel*," referring to the promised seed. The same word for " heel " occurs in
both texts. In one case, the direct allusion is to Hercules, assimilated to the seed of the woman ; and in the other, to Sagittarius, the
celestial horse, proved by the representation in some zodiacs of a
serpent biting the heels of Sagittarius. It is evident that in both
instances the reference is to the same primitive conceptions. 1st.
Our zodiacal scheme shows the constellation Sagittarius in the sign
Cancer, this last being uniformly connected with the dying sun-
god, and thence with Hercules. 2d. That which confirms all is the
Assyrian name of the month *Kisi-livu*, corresponding to Sagittarius,
a name in which Rev. A. H. Sayce recognizes the term *Kisil*,
" giant," that is, Hercules. Add to these data the ancient legend
according to which the giant receives a sting in the heel from the
scorpion, from the effects of which he dies. Thus we reproduce the
entire circle of ideas relative to the fall and redemption of man
which were previously found centring in the signs Libra and
Scorpio, as having been primitively connected with the three zodiacal divisions, Taurus, Gemini, and Cancer. The astronomical
period when these ideas were first given a zodiacal expression
could have been no other than that exhibited in our fourth plate.
But long after this epoch had been forgotten, the same notions
were perpetuated in the signs Libra, Scorpio, and Sagittarius. In
this case, however, it was necessary. in order to complete the representation, to locate the sacred mountain in Libra ; to picture there
the great mother and her child, etc., all which demonstrates the
existence of conscious design and scrupulous care in preserving the
astronomical record of traditions dating from the beginning of
human history.

   It will be remembered that in the twelfth chapter the god *Hea*
was identified with Hercules as true fish-god on one hand, and with

the man-bulls or Kirubi, really one with the Biblical cherubim, on the other hand; these symbolic animals being located zodiacally in Libra, or the sign of the Pincers. It was quite unaccountable to find Hercules closely connected at different times with three distinct zodiacal divisions: 1st. With Pisces as the true fish-god; 2d. With Taurus as *Alap Shamas*, "taurus of the sun;" 3d. With Libra as localization of the sacred mount and of the cherubim. Our fourth plate affords a satisfactory explanation of these apparent anomalies. We see that primitively the *constellation* Pisces occupied the *sign* Libra, and that then, also, the *constellation* Libra was found in the *sign* Taurus. Thus we readily account for the intimate association of Hercules with the three asterisms named. It must be, I think, that a zodiacal scheme which offers so many valuable explanations, pointing so significantly to the origin of so many traditionary ideas, attaches to itself some scientific character. But to give here an additional illustration, we recur to the name *Centaur*, applied to Sagittarius. It is derived from two Greek words: *Kenteō*, "to prick, stab, goad, sting;" and *Taurus*, "the Taurus or Bull;" hence the name *Ken-taurus* or *Centaur*. For the first element, then, we have the fact that Sagittarius was often represented on the Babylonian sphere by an *arrow* simply, the instrument for pricking, stabbing, and killing. For the second element, compare the fourth plate, where we find the constellation Taurus in the sign Sagittarius, frequently marked by the "arrow" alone. Nothing can be plainer than the fact that our zodiacal theme represents the state of the heavens, which gave rise to the name *Kent-taurus*. But it is evident that the first element *Kent* is much older than the Greek verb *Kenteō*.

SEC. 158. In the chapter showing the adjustment of the antediluvian genealogies to the zodiac, the assimilation of Jared to Scorpio, and of Enoch to Sagittarius, owing to the want of sufficient data at the time, was left only partially explained. In the present study, we have developed already an additional significance in the connection of Jared with Scorpio. Whether as associated with the sign Scorpio following Libra, or with the constellation Scorpio found in the sign Gemini, we see now that the meaning of *Jared*, "the descent," or "he who descends," is very significant. In both instances, as already explained, it involves very plainly the notion of the fall, of the descent or expulsion from the paradisiacal mount into the region of darkness and death. In relation to the name *Enoch*, connected with Sagittarius, it has the sense of " the initiated," from a verb meaning " to choke, to straiten, to make narrow," etc. Our plate shows this name accompanying the constellation Sagittarius, in the sign Cancer. We are here evidently in close contact with the mysteries appertaining to the dying sun-god. Here is the " lion's paw," calling to mind the Biblical expression applied to Christ: " The Lion of the tribe of Juda," who alone could open the book of seven seals (Rev. v. 1–5). It was in the sign Cancer that the Egyptians added the five intercalary days, and it was during these five days that the Osiriaf divinities were born. The mysteries of Osiris, as Sir G. Wilkinson has informed us, formed the central point in the Egyptian religion. Osiris' death and resurrection constituted the subject of those mysteries. Finally, the years of Enoch's earth life, 365, are interpreted by many writers of the 365 days of the year. including the five intercalary days. The theory is, that at first the year contained only 360 days, equal to the number of degrees of the zodiac, or 30 for each sign ; that afterwards it was found necessary to add five more, in order to complete the year. These were the five days in which the Osirian deities had their birth. As before said; they were added in the sign Cancer; and the 360 + 5 years of Enoch's life may refer to this fact. Another circle of ideas centres here. Sagittarius was one of the Asvins, whose connection with the chariot of the sun has been pointed out. The Roman chariot races were, as M. Dupuis states, in imitation of the sun's annual course. The name of the Phœnician star *Sugi*, also found in the sign Cancer, signifies "front part of a chariot," and "rivalry," especially of chariots. Now in the solar race the termination was in Cancer. But it was found that the chariots did not arrive "on time," that they were falling back at the rate of five days in each year. This accounts for the legends respecting the Centaur, one with Sagittarius, as being an instructor in the chase. He taught medicine also, and this agrees with Enoch's character, as identified with Thoth or Hermes. But it is unnecessary to enlarge upon these matters.

It will be urged that these data, in addition to those heretofore introduced, render it difficult to attach any historical character to the Mosaic genealogy before the flood. This may be so ; but it does not alter the nature of the data themselves, nor modify their force as grounds of scientific conclusions. For myself I still believe that this genealogy was in some sense historical, but as for positive proofs to this effect I have not as yet discovered them. Indeed, the evidences thus far go to justify a different conclusion. However, the zodiacal scheme exhibited in our plate has not yet been exhausted of its meaning, and I am confident that a long time will elapse before it has yielded up its full import. Thus, while stating the facts as they have appeared to my mind, I do not consider it safe to affirm as yet that no historical value can be attached to the genealogy in question.

SEC. 159. I have long suspected that the Chinese dragon, put for the east, was primitively identical with the dragon to which Damascius alludes, having the four faces of a man, a bull, a lion, and a serpent. According to Chinese tradition this symbolic animal had the head of a camel, the eyes of a serpent, the belly of a frog, the scales of a fish, the claws of an eagle, the feet of a tiger, and the ears of a taurus.[1] In this description the mention of the serpent and eagle, substituted for each other in the cherubim, of the tiger, which well answers to the lion of the cherubic animals, and of the taurus, although these forms are associated with various others, according to mythological fancy, seems to justify the supposition of an original connection with the dragon of Damascius, which personified the Hercules and Chronos of Western Asia. In both instances the symbolic figure relates to the sun in its annual course, so frequently taken, as we have seen, as a symbol of the life of man. But that which must be regarded as conclusive here is the conceived death and resurrection of the dragon of the Chinese, according to Dr. Schlegel, as follows: " But the death and resurrection of the dragon took place at the same epochs as of the sun, namely, at the two equinoxes. Thus, we must regard the dragon as a symbol of the sun."[2] The sun, however, in this case, and as is now apparent, represented in its course the life of man, especially the promised seed of the woman. Evidently the dragon, both of China and of Western Asia, was identical with that chosen for their *emblem* by those primitive " corporations," that ancient order of priest-kings, constituting the subject of the third chapter, and we are now able fully to comprehend the traditionary significance of this emblem.

---

[1] Schlegel, *Uranographie Chinois*, pp. 49, 50.

[2] Ibid., p. 53.

In all cases the reference was to the cherubim placed at the east of
the Garden of Eden. Recollections of these symbolic figures, and
even of their import, had been carried into every quarter of the
Asiatic world by the peoples primitively diverging from the sacred
mount of paradise, that "mount of the assembly" around whose
summit turned the chariot of the seven stars, associated with the
earliest traditions of all the ancient nations.

But if the dragon of China must be identified with that of Western Asia, and if both are to be taken as symbols of the sun-god,
especially under the form of Hercules, this last must be assimilated
likewise to Saturn, the time-god, the Chronos of the Greeks. As
shown in our study upon the cherubim, noted especially in the last
section of it (135). Hercules is intimately associated with three different zodiacal signs, with Pisces as fish-god, under the name of
Hea, with Taurus as *Alap Shamas*, "taurus of the sun," and finally
with Libra, where we find located the cherubim, symbol of Hercules. Our assumed zodiacal scheme explains all this, for, as before stated, we find the constellation Pisces in the sign Libra, and
the constellation Libra in the sign Taurus. On one hand, the primitive situation of the constellation Libra, formerly the Pincers of
the Scorpion, in the sign Taurus, perfectly illustrates the origin of
the fundamental notion involved in the cherubim, as heretofore
explained. On the other hand, we see how Hea, as fish-god, as
taurus of the sun, and also as Hercules, could be so closely connected with Libra, for the constellation Pisces was primitively in
this sign. But Hercules is time-god, and this suggests the notion
of chronology, in this case a *zodiacal chronology*. Hea unites in
himself two remarkable characters: 1st. As *Auv-Kinuv*, Existent
Being, who impregnates the primeval waters of chaos. In this
character he answers well to that of *Adar*, or *Nin-ib*, "lord of generation." 2d. Hea is god of the deluge, as seen in the "Deluge
Tablets." Add to this the fact that the diluvian mountain was
uniformly identified with the mount of paradise in ancient tradition. This very mountain, as we have shown, was located zodiacally in Libra. Hea's connection with the sacred mount thus situated calls to mind the Egyptian god *Tum* associated with the solar
mountain placed in the same zodiacal division. "Thou enterest
into thy palace," says a scribe in praise of an Egyptian king, and
as previously cited by us, "as Tum into the solar mountain." The

Egyptian Tum was the sun of the lower hemisphere, the same as Hea, identified with *Nin* or *Adar*. Compare again with the great mother, represented in the Egyptian solar mountain, the cuneiform text heretofore rendered by us, "The sublime mountain of Tum = the goddess Ishtar." Thus do the earliest religious conceptions of different peoples of antiquity perfectly accord, and thus does our zodiacal scheme furnish the key to their origin and import.

With respect now to a zodiacal chronology: the sacred mount located in the sign Libra with which Hea, in his double character as *Auv-Kinuv* and as God of the deluge, is connected, — this sacred mountain, I say, must represent two distinct epochs, that of creation and of the deluge. If it was the epoch of creation alone, the location of the solar mount should be in the sign Capricorn, put for the winter solstice, at which period the birth of the world and of man was supposed to occur. But the attempt has been made, in the arrangement before us, to represent the two eras of creation and of the deluge. Hea is connected with both, hence the constellation Pisces, the fish of Hea, is found in the sign Libra. The indications are that, while the creation epoch is assigned to the winter solstice, that of the deluge answers to the spring season, and this accords with the tablet translated by Mr. Smith (Sec. 137), which assigns the commencement of spring to the first degree of Pisces. The chronological difference, then, between the two eras is that represented by the sun's annual course from the period of the winter solstice into the first degree of Pisces. All this, however, only by way of suggestion to future investigators. To me, at present, the matter of the deluge epoch is very doubtful. The era of creation alone, of the birth of man, is that positively indicated by our zodiacal arrangement.

The investigations of the present chapter have served to demonstrate, it seems to me, that the astronomical theme presented in the fourth plate has nothing fanciful for its basis; but that it must be regarded as possessing an actual scientific value. We proceed, then, to the final confirmation of our chronological hypothesis in the chapter following.[1]

---

[1] A passage in the *Litany of the Sun*, as translated by M. Naville, appears to me to afford a direct proof, and this of an entirely independent character, that Gemini, or the Twins, as a constellation, was originally found in the sign Capricorn. As was shown in the thirteenth chapter, and also in previous studies, the

sun-god was supposed to die of old age, and to renew itself under the form of a child, precisely at the winter solstice, being then in Capricorn. The Egyptians represented this aged sun by the figure of a man bent down with years, supporting himself with a staff. Now the *Litany of the Sun* hinges to a great extent upon the annual course of the solar orb, setting forth the various forms which the sun-god assumes during its journey. The passage to which I allude is as follows (*Litany*, p. 64) : —

"The great Senior (eldest, first-born, or aged) who resides in the Empyrean, Chepri who becomes the two infants, the image of the bodies of the two infants."

The determinative of the word which M. Naville renders by the French *Aine*, "elder, first-born," etc., is an aged man supported by a staff as before described; and the passage regards the sun-god as having assumed this form, which refers, of course, to the winter solstice. The translator offers the following by way of comment : —

"This is evidently the Aged One, the great Senior, he who has achieved his existence ; this is one whose life approaches the end ; for he has also the name of Chepri, which, as we have seen in Nikennu 2, is the Searabæus who folds its wings, who reposes in the Empyrean, and who is re-born as his own son. It is the same idea expressed here under a little different form ; the great Senior, the Aged One reappears, is re-born under the form of two infants." "The re-birth is, for the rest, the cause of the duality; we have the proof in a passage of the *Book of the Dead*, relative to the re-birth of Osiris: 'Osiris arrives at Mendès, where he finds Ra, and, behold, they embrace each other; and he becomes a spirit under the form of *two twins.*' These two twins, according to the *Book of the Dead*, are the two different forms of Horus, or Osiris and Ra, or yet Shu and Tefnut ; these two last divinities represent the male and female principles. This duality, these two principles, are, then, not the result of the primordial creation, but rather of a re-birth, a second creative act." (Id.)

I must differ from M. Naville respecting the opinion expressed in the last sentence. The daily and annual courses of the sun were taken as types of the primordial sun; and Osiris involved in his character these three phases. Hence, as the *Litany* itself is a cosmogony, the primordial creation must be that here intended. Although familiar with the fact that the sun, conceived as an aged man, was supposed to die at the winter solstice, and to be re-born under the form of a child, I was not aware that the sun was ever conceived as renewing itself at this period under the form of two twins, till made acquainted with this important truth through the researches of M. Naville. The bearing of this fact upon our theory of the original adjustment of the zodiac will be readily perceived : —

1st. This zodiacal scheme is supposed to exhibit the actual state of the heavens at the period to which the ancient cosmogonies pertained ; and the *Litany of the Sun* is itself preëminently a cosmogony.

2d. According to the ancient cosmical traditions, the creation of the world and the birth of man took place in the lower hemisphere, and precisely at the winter solstice, when the old sun dies and renews itself in infancy.

3d. Having seen that the constellation Gemini, or the Twins, represented traditionally the first human pair, and thus, according to the cosmogonies, the winter solstice also, we have placed this constellation in the sign Capricorn, and adjusted our zodiacal scheme upon this basis.

4th. We find now that, in the Egyptian cosmogony embraced in the *Litany*, inscribed upon the walls of the royal tombs at Thebes, the old sun, when it died at the winter solstice, was supposed to renew itself under the form of two twins; sometimes represented as two males, like Castor and Pollux of the Greeks, at others as a male and female, like Yama and Yami of the Hindus.

Call to mind that the sun in its course was taken as a symbol of the life of humanity, and that the birth of the primordial sun to the light was a type of the birth of humanity itself, and we have a complete conception of the force and bearings of the data which M. Naville has made available to our purpose. It is obvious, I think, that the passage of the *Litany* here produced appertains to the epoch when the zodiacal twins represented the period of the winter solstice.

CHAPTER XV.

SEC. 160. By the primitive pole of the heavens we mean that answering to the period from whence the primeval traditions of mankind were derived, especially those with which we have been principally occupied in the present treatise. But before entering directly upon the subject here proposed, it is necessary to offer some explanations with the view to a clear apprehension of it.

If the reader notices the wabbling motion of a top as its rotating force diminishes and it begins to fall over, a perfectly clear idea will be gained of that movement of the earth in the plane of the ecliptic, which causes the phenomenon termed the "precession of the equinoxes." It is not the annual motion around the sun, nor the daily revolution on its axis, both of which are geographically speaking from west to east, but a very slow movement of the earth's axis itself from east to west, describing circles in the celestial spaces opposite the terrestrial poles, about 47° in diameter, or equal to the northern and southern declination of the sun during its annual course. For the two poles of the earth, here supposed to be projected into the heavens, to describe the circles alluded to requires an immense period of time, equal to about twenty-five thousand years. Confining ourselves to the northern circle, since little was known of the south pole in high antiquity, it results from the movement just described that the polar star is not always the same. This depends wholly upon the point in the circle referred to, toward which the earth's axis for the time being is directed. If we were to suppose a point opposite to that of the present direction, the celestial pole would be about 47° distant from its position to-day, and the inclination of the earth's axis to the plane of the ecliptic would be the opposite to its present inclination. The position of the earth in its orbit, that now results in summer north of the

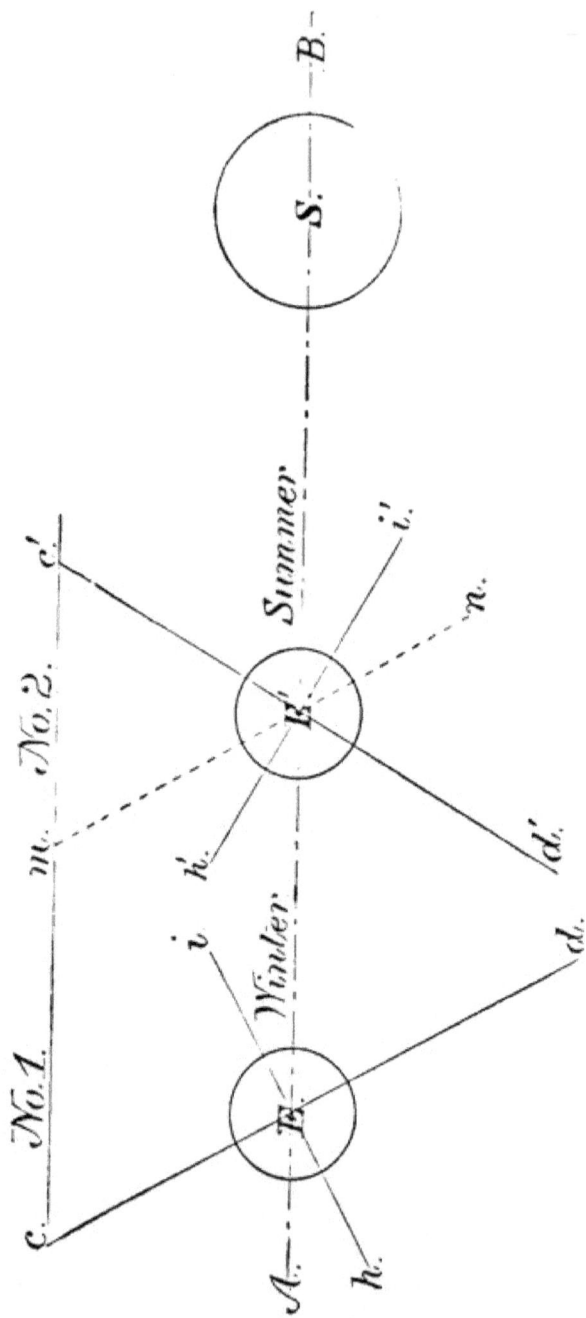

PLATE V.

B.

S.

Summer

No. 2.

c'.

i'.

n.

E'.

h'.

d'.

No. 1.

m.

i

Winter

d.

c.

E.

A.

h.

equator, would then produce winter in the same terrestrial region, and a similar reversion of the equinoxes would likewise take place. The summer solstice would fall in Capricorn and the winter solstice in Cancer, the vernal equinox in Libra and the autumnal in Aries. The reader finds all these matters illustrated in the fifth plate, to which we now refer.

The line A, B represents the plane of the ecliptic in which is S, the sun, and E, E', the earth, exhibiting the supposed primitive and the present actual inclination of its axis. We will take No. 1 for the former, No. 2 for the latter, with the sun in Cancer. Then c, d shows the inclination of the axis in its primitive state, and c', d' its opposite obliquity at the present day. The lines h, i, and h', i', represent the two opposed inclinations of the earth's equator to the ecliptic. Properly there should be but one figure of the earth shown here, in which case the dotted line m, n would take the place of c, d, and m, c' would then truly represent the diameter of the polar circle. But the two figures enable us to present the subject more palpably to the eye, if the fact is distinctly understood that E, E' are put for one and the same position of the earth in its orbit, but showing its two opposed conditions at the different epochs. Examine now position No. 2, representing the present period : the sun is in the sign Cancer, the time of year is the summer solstice. As the line h', i' represents the earth's equator, it will be seen that the sun appears geographically in the north where it is mid-summer. The constellation Gemini is now in the sign Cancer, and thus marks the period of the summer solstice as it occurs in our times. Observe, then, the condition of things as represented by position No. 1. The earth is in exactly the same point of its orbit as before, the sun is in Cancer, the constellation Gemini ; but the time of the year as regards the seasons is wholly reversed. The sun appears geographically in the extreme south, and it is winter, the exact period of the winter solstice, the place and time to which all the primeval traditions assigned the birth of the world and of man. It is necessary only to add here that the point c' represents the position of the present pole star, while c, more correctly m, indicates the relative position of the primitive celestial pole about 47° distant.

SEC. 161. The conditions of our problem as developed in Chapter XIII. presuppose precisely that opposite inclination of the earth's

axis, and the reversion of all the seasons, compared with the present era of the world, which is illustrated in position No. 1 of the
plate. In the fourth plate, forming the subject of our last study,
I have preferred for several reasons to consider the signs as fixed
and the constellations as movable. The constellation Gemini, assimilated to the first human pair, marks to-day the summer solstice,
exactly the opposite period of the year to which the birth of man
and the world were assigned in tradition. The sun, also, whose
course symbolizes the life of humanity, is at its highest exaltation
in Gemini to-day, whereas to conform to tradition it. ought to be
in the period of infancy at mid-winter. The sign Capricorn represents this period, and hence I have adjusted the constellations in
such manner in the fourth plate that Gemini, traditionally associated with the first human pair, should fall in the sign Capricorn.
This operation supposes that the constellations have moved forward
on the zodiac (or the signs fallen back) just half a circle since the
period of the birth of man, according to the earliest and most authentic traditions. Now half the circle zodiacally presupposes a
like amount in that described in the northern heavens by the slow
movement of the earth's axis already described. The primitive
pole star, therefore, was in the point of this circle exactly opposite
to the present pole star, or about 47° distant if we regard the
diameter, 180° if we have respect to the circumference, which would
represent the so-called "right ascension" of all the fixed stars since
the period whose chronology we seek to determine.

If the primitive celestial pole was at a point directly opposite the
present pole star in the circle referred to, or 47° distant, it is very
easy to determine its position on the globe as well as in the heavens.
It was in the *constellation Lyra*, near the *star Vega*, sometimes
called *Lyra* also, and this I believe to have been the actual pole
star at the period to which the primeval traditions of mankind pertained. It was the original *Su-Meru* of the Hindus, *Shemal* of the
Haranite Sabians, the *Su-mi-lu* of the cuneiform texts. It was precisely this central region which was then penetrated by the summit
of the traditional mount of paradise, and it was thus the supposed
celestial paradise united to the terrestrial by the sacred mount.
The terrestrial paradise was taken for the centre of the earth, while
the celestial was regarded as the centre of the heavens, the one
being conceived as directly over the other, and these two points

constituted the supposed axis of rotation of the entire universe. What we now seek is the original centre of the heavens, since, as due to the earth's wabble in the plane of the ecliptic, this centre is constantly changing; in other terms the pole star is not always the same. We have assumed the star *Vega* in the constellation *Lyra* as that original centre, and as the primitive *Su-Meru*. Hence it was there that the summit of Meru penetrated the sky, giving rise to the Sanskrit phrase *Svarga-bhoumi*, " the celestial earth," as already stated to us by M. Obry. It is a remarkable fact, as affirmed by this writer in a passage soon to be produced, that the terrestrial paradise corresponded, as supposed, to the celestial paradise in almost every particular. One was the image of the other, and the resemblance extended even to the details. It is this traditional correspondence between the two that will constitute the *underlying principle* of nearly all the facts to be presented in the present chapter. In this identical region of celestial space, now assumed by us as having been the primitive pole of the heavens, we are to find, as it were written in the sky, another version of the Mosaic account of Eden, both geographically and historically considered. This will constitute the severe additional test to which we desire to submit the zodiacal theme studied in the last chapter, and which was dictated by the conditions of our problem upon zodiacal chronology as developed in Chapter XIII. The reader is thus advised of the course to be pursued in the present study.

Sec. 162. As a general introduction to the facts upon the consideration of which we are now to enter, I produce the extracts from M. Obry that follow : —

" It is certain that the Persians and Hindus often confounded their terrestrial paradise with the celestial paradise of the great divinities; the same as Ezekiel and St. John after him confound the renewed Jerusalem with the celestial Jerusalem, constructed in the form of a square, and situated upon a high mountain." " But in going back to the Mosaic account of Eden, we readily see that the analogies traced by the doctors of mythology repose often upon posterior fictions. The truth is, so far as concerns the topography, that the ancients have generally made the heaven in imitation of the earth, and the infernal region in imitation of heaven. Thus, the celestial Eden and after this the infernal Eden were formed successively upon the model of the terrestrial Eden, with all its principal accessories. The one was placed in the superior hemisphere, either at the north pole, sojourn of the gods and of the just, or in the eastern quarter of the sphere, where the sun, at its rising, renews the light, heat, and life. The other, in its turn, was relegated at first, and for some ages, to the bowels of the earth; afterwards, when the spherical form of the heavens was known, either at the south pole, abode of dæmons and reprobates, or in the west, where the great star sets in darkness, cold, and death." "We conceive, then, that the four rivers and animals of paradise had been assigned successively to the superior and inferior hemispheres. This was all the more natural to the Aryans of India and Persia, since their sacred mount was thought to embrace and to unite the three worlds in such manner that the divine source of waters that supplied them (Ganges or Arduissur) was conceived to divide itself into four rivers in heaven and in the infernal region, as well as upon the earth. In all cases, it is evident that the people who have located these four rivers, either in heaven or in the infernal region, or in both, have given them names derived from those of the terrestrial paradise." "If at a later period, the priests have made these four favorite rivers descend from the milky way, from the north pole, or from the zodiacal band, at the four points of the intersections of the colures, this has been only to render them more sacred in the eyes of believers." [1]

M. Obry has called attention in the foregoing passages to many important facts, some of them confirming positions taken by us in other parts of our work, and others affording evidence applicable to the subject of the present chapter. But the author has not in all cases been able to view these facts, as I am persuaded, in their true light. As respects the three Edens, the terrestrial, and then the celestial and infernal, successively copied from it, the principles established in the seventh chapter, relative to the transfer of the terrestrial paradise into the Greek Hades, offer a better point of view for the facts noticed by our author. As to the topography of the celestial Eden, placed either at the north pole or in the east, and of the infernal Eden, located at the south pole or in the west, it all depended upon the particular phase of the sun's course assumed for basis. The north and south related to the annual course, and the east and west to the daily revolution. The spherical form of the heavens was a fact primitively known, doubtless, since they take no other form to the eye, and the revolution of the seven stars of the chariot around the pole star must have taught the first men the

---

[1] *Du Berceau*, etc., pp. 183-186.

notion of the rotation of the sphere on its axis. As for the cardinal points, they were necessarily first determined from the movements of the heavenly bodies, and the terrestrial paradise was uniformly, in tradition, conceived with special reference to them. That the celestial Eden was only an imaginary reproduction of the terrestrial, at a later epoch, is a point very doubtful. The earliest traditions seem to recognize the existence of both ; and they appear to be equally primitive. M. Obry was not aware, at the time he wrote, of the remarkable coincidences in a geographical sense between the celestial and terrestrial Edens, which are to be developed in the present study; and much less, if we speak in the historical sense. He alludes to the celestial source of the four rivers as being sometimes conceived as the north pole, at others, the milky way, and still again, as the four points of the zodiac cut by the colures. The fact is that all these sources coincided almost perfectly, or at least harmonized with each other, when the celestial pole was in Lyra, near the star Vega; for the pole then bordered directly upon the milky way, and the latter was cut directly by the colures.

SEC. 163. In the same region with the Lyre, we find the constellation of the Eagle, or *Aquila*. The two are situated nearly on opposite sides of the milky way, the Eagle being a little more to the south and east. Between these, and floating on the celestial river, identified with the milky way, is the constellation *Sagittarius*, or the Arrow. I mention these facts, in view of certain statements by Dr. Schlegel relative to the Chinese sphere, which are to follow. Upon the asterism called *Tsien-Tai*, "the terrace of filtrations," corresponding to β, δ, γ, ι, of the constellation Lyra, this author remarks: —

" They excavate two or more reservoirs, one above the other, in the form of terraces, or of stages, which can communicate together only when one reservoir has filled and overflows into the one below. They conduct into the upper basin a natural current of water. In the lowermost reservoir is an arrow that rises with the water, so adjusted and marked that it notes the periods of time. They obtain thus a measure of time quite regular and sure, because the period consumed by filling one reservoir, its overflow into another, and that consumed by the arrow in marking the successive divisions, being compared with the movement of the sun or of the stars, suffices for all purposes of marking the periods of time among a primitive people." " Thus the astrologues inform us that

the asterism *Tsien-Tai* is a terrace from which the waters flow, and
that it presides over the horologes of water on earth. The 'Exegesis of the Celestial Sovereigns' says that this asterism presides
over both the sun-dials and the horologes of water ; and it adds that
they verify the epochs of the seasons, their precession or retrograde,
by observations of the celestial bodies, in order to determine the
changes and movements of the signs.   The harmonious sound made
by the falling water, and the noisy cadence accompanying it, as it
descends slowly from one reservoir into another, gives the idea of
nature's music, and it is for this reason that the astrologues make
the asterism itself preside over music." [1]

The author then describes the *Lien-Tao*, "the route of the chariots," corresponding to η, θ, and three other stars of Lyra : —

"The characters composing *Lien* are those representing two men
and a chariot; thus a chariot drawn by two men." "This chariot
served not only for the imperial promenades, but also for the purpose of war, when horses were attached to it. As such, the
'Route of the chariots' represents the path of the galloping horses
of the emperor; and when it was invisible they presaged that the
imperial routes were to be occupied by warriors." [2]

The same writer refers to another asterism situated in Lyra,
called *Tchi-Nin*, "the female weaver;" also to one in the lower
hemisphere, *Nin-Sin*, "domicile of the virgin," marking the period
of the winter solstice. Still another group, called *Kien-Nin*, "conductor of the Taurus," being a, β, γ of the Eagle, formerly associated with the solstice of winter, is described. This last asterism
was otherwise named *Ho-Ku*, "drum of the river;" the allusion
to the "river" having reference to the milky way, regarded as the
celestial river by the Chinese.[3] Returning now to the "domicile of
the virgin," Dr. Schlegel observes : —

"This virgin was placed there to recall the epoch of marriages,
which took place near the time of the winter solstice, for the
reason that the mysterious marriage of nature was assigned to the
same season. At the solstice of winter, say the ancient Chinese,
the two principles, male and female, *Yin* and *Yang*, are united
amorously, heaven and earth have commerce together, and abandon
themselves to each other. Of this marriage is born the new light
(also man), that is to say, at the solstice of winter the principles of
heat and light are revealed. The displacement caused by the pre-

[1] *Uranographie Chinois*, pp. 188, 189.   [2] Ibid., p. 192.
[3] Vid. ibid., pp. 184, 196, 197, etc.

cession of the equinoxes having brought the asterism of the Female
Weaver to the position formerly occupied by the Virgin, it was the
former that became the emblem of the marriage of Nature at the
winter solstice ; and as it culminated at the same time as the Conductor of the Taurus, the two asterisms were considered as symbols
of the *male* and *female* principles of nature." "These two asterisms are separated by the milky way, which rolls a double stream
between them. That the Conductor of the Taurus might join the
Female Weaver, it was necessary to cross this river. Thus, they
imagined that at one time during the year, or at the instant of the
winter solstice, a bridge was thrown over the milky way, which
enabled the Female Weaver to join the Conductor of the Taurus,
for the purpose of consummating their marriage." "There are no
constellations upon which the Chinese have founded so many fables
and legends as upon those of the Conductor of the Taurus and the
Female Weaver." [1]

SEC. 164. It is necessary now to study the different bearings of
the facts relating to the Chinese asterisms, and the various notions
associated with them, for the knowledge of which we have been
indebted to Dr. Schlegel.

1st. Are the two personified asterisms located respectively in the
constellations Lyra and Aquila, taken as symbols of the two principles, *male* and *female*, *light* and *darkness*, *heaven* and *earth*, whose
annual marriage, precisely at the period of the winter solstice,
represents that of nature, giving birth to the cosmos and to man ?
It is quite obvious that these ideas were definitely connected with
the Chinese cosmogony, and especially with the tradition heretofore
alluded to, which placed the beginning of the revolutions, of the
world, and the birth of man, at the instant of the winter solstice.
This tradition, as will be recollected, constituted one of the conditions of our problems, as developed in the thirteenth chapter, which
dictated the zodiacal scheme presented in the last chapter, and according to which we have been obliged to assume Lyra as the
primitive celestial pole. It is very remarkable that we should find
localized in this identical region those very traditionary ideas constituting the conditions of the problem to which we refer. No evidence more direct and forcible could well be conceived, to the effect
that Lyra was actually the polar region at the period from which
those ideas had been inherited. But it is quite certain also that

[1] *Uranographie Chinois*, pp. 494, 495.

these personified asterisms represented the first man and woman created at this period. Of this, however, hereafter.

2d. In what more significant region was it possible to locate the celestial horologe of water, supplied from the celestial river, to mark the very beginning of time, when the revolutions of the spheres first commenced?   Dr. Schlegel thinks that the Chinese were the first to conceive the idea of marking the periods of time, by means of an arrow floated on the surface of a stream, or in the method of horologes.   Probably the notion had its birth before the Chinese were a distinct nation.   Be this as it may, we find the Arrow, *Sagitta*, floating upon the river of the celestial paradise, exactly between Aquila and Lyra.   But the terraced hill or mountain, serving the purpose of the horologe of water, recalls the fact that the pyramidal temples, constructed in stages, were artificial reproductions of the mount of paradise, thus conceived as a mountain of degrees.   It was upon its summit penetrating the heavens that the Persians located the celestial river, the waters given by Mazdla, and it was upon the same summit that the Hindus conceived the divine Ganges to pour its waters, which flowed down from thence through the three worlds, being gathered into a single reservoir in each region.   Are not all these traditionary ideas connected with the celestial horologe supplied from the heavenly river that rolls its silvery tide between the constellations Aquila and Lyra?   If so, such was the location of the primitive Su-Meru.

3d. Here we find the route of the chariots. This recalls the legend of the Asvins, the Vedic "horsemen," who were born in the "*Uttara-Kuru* of the firmament."   They ride together "in chariots, all the parts of which are in threes."   The celestial horses were born on the banks of the sacred river, and this river, identified with the milky way, actually cuts the zodiacal band in one instance at Sagittarius, in the other at Gemini, two asterisms which I have assimilated to the two pairs of twins born of *Saranyû*, the "cosmical storm-cloud," daughter of Trashtar, "the creator." Pegasus was born near the source of waters, likewise, and the head of the winged horse lies less than 30° from the milky way in the vicinity of *Sagitta*.

SEC. 165. It is well known that the constellation Lyra was represented anciently under three different forms, namely, that of a *Tortoise*, of a *Vulture*, and finally of a *Lyre*. To each of these forms certain important ideas were attached, which I wish to present; and first of the *cosmical Tortoise*, according to Chinese tradition. Dr. Schlegel, to whom again we appeal for information, relates the facts as follows: —

" In all the cosmogonies, the solstice of winter was considered as the epoch of the creation, the various peoples believing that the world had been created at that instant. From this notion is derived the Chinese fiction, that in the times of *Thao-tang*, that is, *Yao*, there was in the state of Laos a divine Tortoise a thousand years old, and more than three feet square. Upon its back there were characters in the form of tetrads, which recounted all the events that had occurred since the separation of chaos; for this reason it was termed by the emperor the ' chronology of the Tortoise.' " [1]

Although the subjoined passage has been in part presented before, it ought to appear here entire in connection with the foregoing: —

" 'To the west of the mountain *Yuen Kiao* is the lake of the stars, which is a thousand Chinese *Li* in length. In this lake is a divine Tortoise, having eight feet and six eyes. Upon its back it carries the images of the seven stars of the chariot, of the sun, of the moon, and of the eight celestial regions. Upon the lower side of its shell is the image of the five summits and of the four canals.' That is to say, upon the back of this animal is traced the celestial map, and upon its lower side the terrestrial, as said in the Book of Rites: ' Above, it is round, imitating the heavens; below, it is square, imitating the earth.' " [2]

The reference of this legend to the mount of paradise, and thus the definite association with this mountain of the Chinese cosmogony, were points fully established in the fifth chapter (Sec. 61). Nor can there be any less doubt that the other legend, relating to the " chronology of the Tortoise," had primitive reference to the same locality. In such case it is necessary to identify the constellation Lyra, otherwise termed the Tortoise, with that upon whose back the chronology of the world, since the separation of chaos and the birth of man, was supposed to be recorded. Admit that this asterism constituted the celestial pole at the period to which these legendary conceptions pertain, and we have a complete explanation of their origin, and it would be almost impossible to account for their origin upon any other hypothesis. In fact we have very clear and direct proof of this hypothesis, if we compare these two legends

[1] *Uranographie Chinois*, pp. 64, 65.　　　[2] Ibid., p. 61.

of the Tortoise with the two personages also assimilated to Lyra and Aquila, as symbols of the two principles, male and female, heaven and earth, whose marriage at the period of the winter solstice results in the birth of creation and of man. To my mind, then, it is very difficult to avoid the conclusion derived from these facts that Lyra was actually the celestial pole at the epoch of the origin of these primeval and traditionary notions, pertaining alike to the cosmogony and the first abode of humanity.

The origin and name of the Lyre, as connected with this constellation, is thus related by M. Dupuis: —

"This constellation passed for being the Lyre which Mercury made from the back of a Tortoise, and which he gave to Orpheus." "They say that the Nile, after its overflow, having retired to its bed, left upon the sand a Tortoise, which had fallen to putrefaction except its nerves (and shell), which being touched by Mercury put forth sounds. Mercury, in imitation of what he had done with this shell, constructed a musical instrument in this form." "They say that at the first the number of the strings was seven, equal to that of the Pleiades, of which Maïa, his mother, was one." [1]

Note in the above account that the Lyre is made from the shell of the Tortoise, evidently one with the cosmical and chronological Tortoise of the Chinese traditions. As for the form and name of the Vulture, it is too well known to need authentication. Our celestial globes frequently represent the Lyre as being held in the talons of the Vulture. The two names respectively, *Vultur cadens* and *Vultur volans*, were often given to the Lyre and Eagle by the Latins. The celestial Lyre was supposed to symbolize the harmony of the sphere, and the same idea seems implied in the fact that the Chinese asterism of the "Horologe of water" presided over music. The connection of the Tortoise with the cosmos, and with chronology, of Mercury, as cosmical deity, with the Lyre made from the shell of the same Tortoise, symbolizing the harmony of creation, seems strongly to favor the supposition that, quite extensively in antiquity, this very constellation was definitely associated with the period when creation's harmonies first pealed forth.

SEC. 166. We come now to the identification of the constellations Aquila and Lyra as the *first man and woman*. In reference to the Eagle we have the following legendary history as related by Dr. William Smith: —

---

[1] *Origin*, etc., iii. pt. 2d, p. 132.

" *Merops* (Μέροψ). 1. King of the Island of Cos, husband of the nymph Ethemea, and father of Eumelus. His wife was killed by Diana b. cause she had neglected to worship that goddess. *Merops*, in order to rejoin his wife, wished to make away with himself, but Juno changed him into an Eagle, whom she placed among the stars. 2. King of the Æthiopians, by whose wife, Clymene, Helios became the father of Phaëthon." [1]

Thus, *Merops* is to be identified with the constellation of the Eagle, or Aquila. But to the above extract should be added the following : " *Meropes* (Μέροπες), an ancient name of the inhabitants of the Island of Cos, from an early king, *Merops*" (Lidd. and Scott, Gr. Lex., sub Μέροπες). It remains now to show that the name *Merops* is only another form of the word *Meru*, applied to the first men issuing from the sacred mount, that is to say, the *Meropes*. In reference to the locality of this original human abode, M. Lenormant observes : —

" The culminating points are the Belurtag and the vast plateau of Pamir, so well calculated to sustain a primitive population yet in a pastoral state, and of which the name under its first form was *Upa-Meru*, ' the country under the Meru,' or, perhaps, *Upa-Mira*, ' the country near the lake,' which itself had been the motive of the appellation *Meru*. There are again certain traditions of the Greeks that force themselves upon our notice, particularly that involved in the sacred phraseology *Meropes anthrōpoi* (Μέροπες ἄνθρωποι), which can mean only ' *the men issued from Meru.*' " " Among the Hindus the men before the deluge, as well as those after the deluge, descend from *Meru*. It is there that we find the *Uttara-Kuru*, veritable terrestrial paradise, the traditions relative to which were collected by Megasthenes. It is there, also, that we are conducted by the paradisiacal myth of the *Meropes* among the Greeks, that is, the ' men of *Meru*,' a myth which had been transported to Greece and localized in the island of Cos." [2]

As the interpretation here given to the Greek phrase *Meropes anthrōpoi* may seem to many classical students unauthorized, it is well to note the fact that it had been suggested long before by no less a critic than M. Renan, thus: " Compare the paradisiacal myth of the *Meropes* among the Greeks, and the expression *Meropes anthrōpoi*, ' the men issued from *Meru*.' " [3] It is certainly a strik-

---

[1] *Class. Dic.*, art. " Merops."

[2] *Frag. de Bérose*, pp. 303, 304. Cf. Homer, *Iliad*, A. v. 250, B. v. 285, G. v. 402.

[3] *De l'Orig. du Langage*, p. 228, note 2.

ing confirmation of the interpretation thus supported, that the same
name *Merops* is connected with that of Æthiopia, a country particularly mentioned by Moses in his geographical description of Eden,
this being of course the Asiatic and not the African Æthiopia.
Finally, the name *Cos* given to the island ruled by *Merops* can be
no other than *Cush*, by which the Cushite race was so generally
known. *Merops* is, then, only another name of the first man, applied to the primeval abode of humanity under the form *Meru*, like
Asgard from *Askur*, among the Scandinavians, and *As-kar*, "propitious summit," among the inhabitants of the Euphrates valley.
It is very significant for us here that the great cosmical or *world-
tree* of the Scandinavian traditions had its roots in the earth, its
branches reaching the heavens, while on its top was perched an
eagle surveying all things below.   The *Garuda*, or Eagle, of Aryan
mythology, also is assigned to the highest summit in *Gorotman*,
"paradise," where he wages war upon the serpents. Thus, it is impossible to avoid the conclusion from these data that the constellation of Aquila was definitely associated with Meru and with the
first man.   It was the original Su-Meru, or closely connected with
that region.

    The literal sense of the Greek phrase *meropes anthrōpoi* is "ar-
ticulately speaking men," and Mr. Buckley so renders it uniformly
in his version of the Iliad.   Thus *merops* (μέροψ) signifies literally
"dividing the voice;" that is, speaking, endowed with speech, this
being the characteristic of man, as distinguished from animals.
On the other hand, Professor Curtius agrees with Professor Fick,
in deriving *merops* from *marp*, "to seize, to gripe," thence, "he who
seizes, holds, gripes," from which the notion of *griffon*, fundamental idea of *cherub* in "cherubim."   Thence comes again the
notion of touching, "he who touches, feels, understands,"[1] etc.
The *Garuda*, or Eagle of the Hindus, was one of the griffons, and
thus all goes to show that Doctors Gesenius, Delitzsch, M. Renan,
and others were correct in assimilating the *idea* of the "cherubim"
to the Aryan *Gruphes*, etc., although the *etymology* of the term is
purely Semitic, as proved in our twelfth chapter (Sec. 130).   Thus,
Merops must be assimilated variously; 1st. To the constellation
Aquila; 2d. To the paradisiacal man issued from Mt. Meru; 3d.
Especially to the man of Meru as endowed with speech and under-

---

[1] Grundzüge, p. 456.

standing. Being so directly associated with the traditions centring in the sacred mount, this asterism must be taken for that region very nearly answering to the primitive Su-Meru, or celestial pole, supposed to be penetrated by the summit of that mountain.

SEC. 167. We turn attention now briefly to the constellation Lyra, under the form of the Vulture, assimilated to the great mother, evidently in this case the first woman placed at the side of the first man. M. De Rogüé has the following upon the Egyptian hieroglyph of the Vulture: "The Vulture was the symbol of the *mother*. The (phonetic) value *Ma* comes from the word *Ma-t*, 'mother.'"[1] Sir G. Wilkinson observes that "*Maut*, the 'mother' of all, or the maternal principle, probably the *Mot* (*Mud*) of Sancho-niathon, appears to be sometimes a character of *Buto*, primeval darkness, from which sprang light."[2] Professor Romieu furnishes the proof that in the Egyptian system of Decans, the hieroglyph of the Vulture actually designated the constellation Lyra, other-wise represented as a Vulture, thus: "I believe myself . . . to be in the truth in identifying the asterism *Ma-t* with the Greek con-stellation of the Lyre."[3] It is probable that some connection exists between *Maïa*, name of the mother of Mercury, associated with the Lyre, constructed of the back of a Tortoise, and the Egyptian *Ma*, from *Ma-t*, "mother." The direct analogy of conception also between *Mot*, "Mud," as mother of all, and the habits of the tor-toise, living in muddy places, is worthy of note here. Finally, to these evidences are to be added the Chinese legends respecting the "Conductor of the Taurus" and the "Female weaver."

The primitive reference, then, of the two asterisms Aquila and Lyra to the first human pair is a matter too plain to admit of serious question, in view of the facts now before us. If the primeval tradi-tions respecting paradise centred geographically in Mt. Meru, then the same traditions centred astronomically in these two constella-tions; and since the summit of this mountain was supposed to pen-etrate the heavens precisely at the celestial pole, these asterisms constituted the *Su-Meru* of the period, to which the traditions themselves pertained. But Lyra, or the Tortoise, was rather the true pole, and this explains the ancient idea, represented even upon the art monuments, that the heavens rested upon the back of a

---

[1] *Chrest. Egyptienne*, 1st pt., p. 61, m. 7. [2] Rawl., *Herod.*, ii. p. 241.
[3] *Sur un Décan*, etc., p. 39.

tortoise.   M. Lajard produces the figure of an Etruscan candelabra, represented as being held in the hands of a female, clothed in a garment of the constellated heavens, and standing upon the back of a tortoise.[1]   M. Dupuis very properly insisted, I think, that the obvious meaning of this legend was, that the *axis* of the heavens rested on the back of the Tortoise, or, in other terms, that this constellation was actually the celestial pole, where the legend took its rise.   It is utterly impossible, in fact, to give any other meaning to it, except the most puerile, and at the same time the most improbable, namely, that the ancients really supposed the heavens to rest on the back of a huge, living tortoise.

We direct attention now to certain constellations, located in the immediate vicinity of Lyra and Aquila, obviously involving some distinct references, astronomically speaking, to the history of the first human pair as detailed in the Mosaic record.   No point seems better established in Hebrew lexicography than the fact that the word *Na-khash* (נחש), designating the serpent of Eden, denotes astronomically the constellation *Draco*, or the "Great Serpent," coiled around the present celestial pole.   As represented upon my globe, the head of Draco is not over 6° from that of the Vulture, identical with Lyra; the Vulture, as we have seen, being put for the great mother, evidently the mother of Eden. The nearer proximity of Draco to the asterism put for the great mother than to that assimilated to the great father accords perfectly with the Mosaic history of the fall, which represents the woman as being first deceived.   To proceed, however, it is said in relation to the serpent: "I will put enmity between thee and the woman, and between thy seed and her seed; it shall *bruise thy head*, and thou shalt bruise his heel" (Gen. iii. 15).   That the Asiatic Hercules represents the seed of the woman has been already well established.   Turning then to the constellation Hercules, we find his *heel* less than 2° from the *head* of Draco, in the *act of stamping upon it!*   The combinations here established are certainly very remarkable; and it would be difficult to conceive of anything more striking, more direct and conclusive.   The Mosaic record itself is hardly more intelligible, as pertaining to these matters, than the symbolical representations here presented to view.   That which makes everything plain and palpable is the fact, now placed be-

[1] *Culte de Venus*, Pl. III. A.

yond question, that Aquila and Lyra really represent the paradisiacal man and woman, or the great father and mother. Such being the case, the close proximity of the *head* of Draco and of the *heel* of Hercules, in the very act of bruising it, shows for itself a direct relation to the Biblical account of the fall and of the promise. This, then, is the third time that we have found these events recorded in the heavens. The vital relation of these three discoveries to each other is quite important to be considered. The first representation is connected with the signs Libra and Scorpio. To all appearance, it has no relation to any other. But the zodiacal theme assumed in the fourth plate reveals the existence of a second one, which more fully explains the origin and import of the first. Finally, on assuming a celestial pole answering to the theme of the fourth plate, we make the discovery of the third record connected with it. Thus each one in some sense conditions the others, and the three combined prove the existence of an express arrangement, of a scientific principle underlying them all. In other terms, they demonstrate the truth of our chronological hypothesis, and that the star Vega in Lyra was the pole star at the period to which the primitive traditions of the world pertained.

SEC. 168. I desire to recall here certain facts previously verified, and to present some new data, the united force of which will be found to be very conclusive. We have seen that the cuneiform *Ak*, "to make, to build, to create," was a symbolic designation of the Babylonian Mercury, who had his sanctuary in the chapel of a cubical form, constituting the eighth stage of the tower of Borsippa, which represented consequently the eighth celestial region, answering to Su-Meru or the pole star. The connection of Mercury with the heavenly Lyre, constructed with seven cords in honor of his mother Maïa, has been also established. It has been shown, moreover, that Mercury was a Cabirus; and the term *Ak*, by which he was denoted, was identified with the Aryan radical *Ak*, "to penetrate, to pierce," as the rocky summit pierces the sky; from the same radical were derived the various forms *Akman*, "a stone," also "heaven;" *Aktan*, "eight," and the Greek *Akman*, "heaven," likewise "anvil," the same term being the name of a Cabirus. Again, the reader is familiar with the fact that the Phœnician *Eshmun*, whose name signifies "the eighth," was considered the eighth Cabirus, and expressly associated with the eighth celestial region, answering thus to that of the pole star; and from all the facts it is evident that Mercury should be connected with the same region.   Now the mother of Eshmun, as shown by Dr. Movers, was traditionally the pole star; and according to the same author Eshmun was identical with the Egyptian *Thoth*, which is only another name for Mercury.[1]   Hence, as the two personages are one, Eshmun and Mercury must have the same mother, who is also to be identified with the celestial pole.   This cannot be the present pole star, as Dr. Movers seems to think, for these two Cabiriac personages were cosmical divinities; a fact that enforces us to assume the primitive pole star as their mother.   Now Maïa, the mother of Mercury, although regarded as one of the Pleiades, must be in some way connected with the celestial pole; for her father was Atlas, and the axis of the heavens was thought to rest between his shoulders; at other times, however, Atlas was conceived as a mountain, on whose summit the heavens and all the stars were supported.   This reminds us of the Tortoise, on whose back the axis of the heavens rested.   It is obvious that all these data should lead us to assume the Lyre as being the pole star, which was the mother of Eshmun as well as of Mercury.   This is confirmed by the fact that Lyra was otherwise represented as a vulture, symbol of the great mother. Finally, the character of Atlas, father of Maïa, when he takes the form of mountain god, answers only to that of Meru, on whose summit the heavens rest, and especially the axis of the heavens.

Thus, when it is said that the mother of Eshmun was the pole star, it must be admitted that no other asterism fulfills the conditions of this statement so well as Lyra; and when it is said that the mother of Mercury was Maïa, her father being Atlas, a mountain god, it is difficult to understand any other mountain than Meru, or any asterism but the constellation of the Lyre, constructed by Mercury in her honor.   Our object, however, in the present section has not been merely to confirm the hypothesis, already so well supported, respecting the primitive celestial pole.   We have aimed to show, also, that the eighth Cabirus was directly associated with this polar region, and consequently with this constellation.   It was precisely here, as the key-stone of the celestial vault, the heavenly temple, that was placed the dressed stone *Akman*, symbol of heaven, as well as of the Cabirus bearing this name.

[1] *Die Phœnizier*, i. pp. 527, 530.

SEC. 169. But it is time to bring the investigations of the present chapter to a close. The characteristic features, geographically and historically, of the *Gan-Eden* of Genesis we have found inscribed in that particular celestial region around and centring in the constellation Lyra; and it is impossible to doubt, in view of the facts before us, that this region constituted the original celestial paradise, abode of the divine hierarchy, corresponding to the terrestrial paradise, the home of primeval humanity. Just as all the traditions and legends centre geographically and historically in the one, so the same traditions and legends centre astronomically in the other. The analogy, or better, the identity, between the two orders of conceptions is everywhere complete; and the coincidence thus presented to the mind has no parallel in the history of antiquarian researches. We establish, then, beyond question, the precise era astronomically, and thence historically, of the paradisiacal man, that era from which the most primitive traditions of the world were ultimately derived. Lyra was the original *Su-Meru* of the Hindus, *Shemal* of the Haranites, *Su-mi-lu* of the cuneiform texts. It was in the particular region of Su-Meru that the summit of the sacred mount penetrated the heavens; and since, according to all the traditions, the region thus penetrated was exactly the celestial pole, such was the constellation just named. The chronology which results from this, as the practical astronomer will perceive at once, is, in round numbers, 12,500 years from the present time. Each of the three chapters devoted to this subject results independently in this date; they supplement and confirm each other, therefore, affording a threefold basis for our hypothesis, which, in my estimation, entitles it to be considered a scientific fact.

It will be now apparent to the mind of every reader, not only that the zodiacal system and the entire sphere, as primitively constructed, were a most ingeniously devised scheme, but that they were a practical embodiment of the sacred tradition and science which formed really the basis of the ancient religions and the civilizations growing out of them. The date of the origin of those ideas constituting this sacred lore, as will be readily perceived, is itself definitely fixed by the very zodiacal arrangement which is seen to have been designed to incorporate them. Nor is there anything factitious or merely mythical in the results that have been attained. The geographical locality of the *Har-Moad*, the " moun-

tain of the assembly," has been definitely ascertained; the primitive traditions centring in it from almost every quarter of the old
world have been successively authenticated; and in these traditions
have been detected certain uniform astronomical references that
furnish the key to the original condition of the heavens in relation
to the earth, from which our chronology has been deduced. But
this is not all; we know now by whom that marvelous system embodied in the primitive sphere was wrought out: by whom the
sacred science, originating at a period forgotten by the world, had
been transmitted to after ages, and incorporated in the religious and
civil institutions of antiquity; by whom those grand ideas pertaining to the cosmos, the temple, and to man had been taught to mankind; by whom that primeval revelation, more enduring than parch-
ments, than columns of brick or of stone, or even the massive
pyramids themselves, — we know, I say, by whom this revelation
had been embodied in those celestial hieroglyphs, more ancient than
the paleographical systems of Egypt, of China, or of Chaldæa.
The dressed stone that had been raised to the centre of the heavenly vault, symbol of the completion of the great world temple,
when the "morning stars sang together, and all the sons of God
shouted for joy," — this stone, I say, had a mark upon it; and it is
impossible to mistake its import, or the hand that placed it there.
We distinctly recognize, in fact, in these wonderful achievements
now present to our minds the work of the Cabiri, of the temple-
craft. These were the *brothers par excellence*, and they hail us to-
day from the heights of the *Har-Moad*, and from across the vista of
twelve thousand years! Theirs was the noble civilization of the
far distant era, which had been planted, not around the pyramids
of Sakkara nor of Borsippa, but upon the shining slopes of that
propitious mountain of which these were but artificial reproductions. Beautiful as the heavens, where rolled the chariot of the
Aryas, or issued the celestial river from beneath the throne of
God, was this golden age of humanity. But a frightful calamity
had befallen the human family, and the light of that happy civilization had gone out in darkness. If it had been forgotten by the
world; if its glory had faded out from the recollections of even the
oldest races known to history, like Merops, its representative, it had
been translated and placed among the stars. There we have read
the record of its loss and its ruin; there, too, of its last hope — in
the promised seed of the woman!

CONCLUDING REMARKS.

CHAPTER XVI.

A SUMMARY OF RESULTS.

SEC. 170. The point of view from which we are now able to contemplate the ancient world affords advantages such as have been rarely accorded to modern investigators. It is not from the summits of the pyramids of the Nile valley, nor from those of the mounds scattered upon the banks of the Euphrates and Tigris, that we look down to-day upon those scenes where the thrilling events of so many epochs have transpired; but it is intellectually from the heights of that sacred mountain which overlooks all Asia, of which the pyramids themselves were but artificial reproductions, and in which the earliest recollections of all the races centred, that we are now able to trace out their divergent paths, strewed with the relics of more than a hundred centuries. Nor is it, again, the desolate waste of the overflowing Nile that stretches out before us, nor yet the annual floods burying beneath their surface the fertile plains of Babylon and of Chaldæa, whose noisy tumult strikes the ear. On that vast mountain pile, the great watershed of the Asiatic world, where the cosmogonies were first revealed and the diluvian annals first recorded, we seem to listen to the surging billows of the deluge itself; of that watery chaos, in fact, from whose thick darkness the Voice went forth: "Let there be light! and there was light." So far as the sacred books of antiquity afford any indications, so far as the earliest known traditions offer any reliable hints, we are here at the beginning of history, and the ground beneath us is that first trod by the foot of man. If other historical epochs anterior to this are to be supposed, or if other portions of the earth's surface had been previously inhabited by men, the evidences to this effect must be drawn from sources entirely different from those to which we have referred. The proofs have been derived from every quarter of the old world, pointing unmistakably to the high table-lands of Central Asia as the original centre of the populations of our globe.

Such, then, is the standpoint from which we now contemplate the historic development of mankind. That which adds immeasurably to its advantage is our ability to attach to it a definite chronological value, an antiquity of 12,500 years. The so-called geological evidences of the antiquity of man, believed by many writers to prove his existence on the earth at an epoch tenfold more ancient than that here indicated, give rise to a problem purposely avoided by us heretofore, for the reason that we were in no sense prepared to discuss it. From the position which we now occupy the case is quite different. Can it be affirmed that these geological data enable us to go back of Mount Meru, and to prove the existence of man at a period prior to the date which the primeval traditions centring in this very locality have been made so clearly to establish? So far as concerns the cultured races of the old world we have proved their common origin from the great Asiatic Olympus, and this by means of their own traditions. Do these geological data show the existence of barbarous races still more ancient than the *Meropes anthrōpoi*, " the men issued from Meru"? For myself, I should hesitate to affirm it. In the first place, it must be considered that at the epoch verified by us the obliquity of the earth's axis was exactly the opposite to its present inclination. The order of the seasons was then completely reversed. The winter season in the northern hemisphere, where, as all goes to show, man first appeared, occurred when the earth was at its maximum distance from the sun, while during the summer it was at its least distance. The effect of such conditions must have been to create drift deposits north of the equator at a rate far more rapid than has occurred during the last six thousand years. It is the estimated rate of these deposits which constitutes the chief element of calculation for man's antiquity, based upon geological evidences. It is obvious, therefore, from the foregoing considerations, that the standard heretofore adopted is in no sense reliable. For several thousand years after the birth of man, according to our data, the changes upon the earth's surface, especially in high latitudes, took place with a rapidity to which nothing can be compared in more recent periods. But even at the present rate of deposits, some of the best authorities undertake to account for all known facts upon the basis of comparatively short periods of time. M. F. Chabas, in a recent and thoroughly critical work, dealing with these geological evidences in a manner such as but few writers are able to do, believes himself able to account for all the facts within a period of ten thousand years, admitting the necessity, however, of a moderate extension of it under certain conditions.[1] Finally, it should be observed, so far as concerns geology, that the evidences are not all in, that too many elements of uncertainty exist, not to render. it premature to assume fixed conclusions. Future discoveries may change wholly the conditions of the problem.

Such, then, are the reasons chiefly why we hesitate to go back of Mount Meru, and to assume on the strength of geological data a period of human existence anterior to the paradisiacal man. If such an epoch can be fairly proved, we shall feel free to accept it, but it will be obvious that the facts, as developed in our present researches, have very materially changed the complexion of this entire problem. Thus, in the light of the present we reaffirm our standpoint, chronologically and geographically, at the beginning of human history.

SEC. 171. The question which regards the *primitive condition of man* has assumed an importance in our day fully equal to that respecting his antiquity. The evidences having a direct bearing upon this subject, accumulated in different parts of our treatise, will be readily recalled by the reader. It is not too much to say that they demonstrate the existence of a noble civilization, centring in the sacred mount of the northeast, the prevalence of an exceedingly high order of conceptions at the period to which the earliest recollections of the old world appertained. It is evident that the fundamental ideas constituting the theoretical basis of this civilization were identical with those of the cosmos itself.

---

[1] *Etudes sur l'Antiquité historique*, pp. 9, 10. M. Chabas studied these geological indications of man's antiquity, so far as the more important localities in France are concerned, on the very ground where the relics of antiquity were originally discovered. He gives it as his opinion that the distribution of these relics was such that no successive chronological eras could be fairly deduced from them.

This was, so to speak, the first heaven and earth, whose model was the particular, material heaven and earth in which humanity had its birth. All the institutions of society, in fact, were constructed after the same model, and a profound symbolism presided over the whole. Whether as a cosmogony, a religious, political, or social theory, it is obvious that the sexual principle was regarded as the most sacred, the most important of all great truths. The world and man, the very constitution of things, depended upon it.

The proofs of the existence of a high civilization centring in the traditional mount of paradise, and at the immensely remote epoch heretofore established, arrange themselves, for the most part, under the following general heads : —

1st. All the great civilizations known to ancient history have been traced genealogically to one common centre, namely, Mt. Meru of the Hindus, Alhordj of the Persians, Gan-Eden of the Hebrews, Kharsak-Kurra of the cuneiform texts, to which we might add the Asgard of the Scandinavians, the Olympus of the classic nations, and even the solar mountain of the Egyptians and the five summits of the Chinese. Call to mind here the fact of the common inheritance of the doctrines embodied in the "celestial earth" or "terrestrial heaven," of the symbolical system of geography ; the theory of territorial divisions, of land surveying, of the orientation of sacred edifices, and the rules for founding towns, cities, and states.

2d. All the ancient cosmogonies had a common origin, and they have been traced in the main to the same central region inhabited by the first men. Not only do they involve uniformly the same fundamental ideas, but the most striking resemblances often occur in the details. That which is the most remarkable of all is the fact, developed in the thirteenth chapter, of a common astronomical reference so precise and definite as to enable us actually to determine the period to which they pertained ; and this has been done in the three chapters on zodiacal chronology. Nor is it alone the date of the origin of these systems which is by this means ascertained, but the locality from which they were primitively derived is made equally certain by the same process. The discovery of the celestial pole at the epoch referred to proved beyond question that it was the original celestial paradise, traditionally united by means of the sacred mount to the terrestrial. The particular allusions in the

Chinese legend of the Tortoise, as well as the Greek myth relating to *Merops*, a term evidently connected with *Meropes* and *Meru* in origin, have each a like tendency to fix the locality where the cosmogonies had their birth.

3d. Perhaps the most conclusive proofs of the existence of a high civilization, of the prevalence of an elevated order of conceptions, at the period and in the geographical region here assumed, are to be found in connection with the ancient sphere. The important fact must be admitted that a well-defined and ingeniously contrived relation existed between the internal arrangements of the sphere and the primeval traditions of mankind, fundamentally, with respect to the cosmogony and the history of the paradisiacal man. In fact, the evidences have constantly accumulated during these researches, that the sacred science and tradition of the old world, inherited alike by all the cultured nations, had been symbolically embodied in the primitive sphere. It is necessary to go farther even, and to admit that the first prophecy ever uttered to man was recorded in the heavens; for the promise relative to the seed of the woman may be read to-day, as plainly inscribed on the celestial sphere as in the third chapter of Genesis.

The tradition of the "golden age," then, was not a myth. The old doctrine of a subsequent decadence, of a sad degeneracy of the human race, from an original state of happiness and purity, undoubtedly embodied a great, but lamentable truth. Our modern philosophies of history, which begin with the primeval man as a savage, evidently need a new introduction. Those writers who would derive the origin of religion and civilization from a condition of savagism should go back of Mt. Meru to do it, and not content themselves with citing the customs of existing barbarous tribes. No; the primeval man was not a savage. He was born of the Heaven-Father and the Earth-Mother. He was the beautiful, pure image of both. Sweet Nature caressed him on her generous lap; she told him her secrets without asking, for she fondly trusted that he would not betray her. Heaven itself conversed with him; and the constellations taught him the music of the spheres. There was nothing that he did not love, and there was nothing that did not love him. All things whispered to him what they were and why they were. The sun and moon were his companions,—almost a brother and sister. To the primeval man, Nature was conscious;

and her consciousness was a part of his own. Eternal Mind was present to him, in all that he beheld, in all that he felt. The golden gates of the senses were continually thronged with tender sympathies, with loving messages, from the great world about him. Such was creation's first-born child, with whom the Holy One himself came down to dwell.

But there came that cruel hour when man fell! Nature was ashamed! and drew the veil over her face. Man, too, was ashamed, and sewed fig-leaves together to hide his nakedness. But God was angry; and He cursed the ground that had witnessed an act, a calamity, so terrible. The betrayer also met his doom; and his everlasting sentence was written in the sky. Thus, the light of that beautiful civilization flickered for a while, like a candle in its socket, and then went out.

SEC. 172.   But before the flames had died down on the primitive altars, a faithful band had kindled their torches, that they might conduct the race through the long night, and finally renew the fires in other times and in other climes. If the evidences adduced in the third chapter did not fully establish the fact, then I think the frequent additions of proof in later studies have served to demonstrate that there existed an ancient order of priest-kings, having its origin in the very dawn of history, through whom the sacred tradition and science had been transmitted to subsequent ages. The striking uniformity in the several versions of the primitive doctrines as inherited by different nations so widely separated, and at a period so early as to preclude the idea of their being derived by one from the other — a uniformity so great that we have been able to detect a precise astronomical feature common to all — this surprising analogy, I say, cannot be accounted for on the principle of ordinary transmission of ideas from age to age, especially in the absence of written documents scrupulously preserved. Nor are these exact resemblances discoverable only in the cosmogonies; they crop out in many a legend or custom where we are least prepared to find them. To illustrate, recall the Chinese legend of the Tortoise, having the images of the seven stars of the chariot, of the eight celestial regions, and of the five summits on its shell; a triple reference to the sacred mount of paradise, which admits of no other interpretation. But from China we go now to Rome, where we find in the location of the axis of the Pantheon another reference

to the seven stars, and to the eighth celestial region, considered as
the seat of the gods, especially of Jupiter; a singular proof of the
exactness with which the primeval traditions had been preserved
by these two nationalities so distantly removed from each other.
It seems to me impossible that such accuracy should be maintained
through ages even, and by different races, except by the vigilant
care of a class of personages, regularly organized, and specially
charged to preserve the ancient doctrines in their purity. Eneas,
whose Cabiriac character is quite well established, is supposed to
have brought the sacred science from Troy to Rome. The seats of
the Cabiriac worship were the most primitive of any, both in Egypt
and Babylonia. It is hardly to be doubted, I think, that members
of the same mystic order laid the foundations of the Chinese empire. When the old civilization centring in Mt. Meru was broken
down, it is evident that those ancient priest-kings, whose symbol
was the Dragon, one with the Biblical cherubim, conducted the
great migrations diverging from the original focus of populations
into the different quarters of the world; into China, Egypt, Babylonia, etc. ; carrying with them the primeval doctrines which
served the theoretical basis of subsequent foundations in these
various countries. The fact tends powerfully to support this
hypothesis that, uniformly in antiquity, the Cabiri were the reputed
founders of the ancient civilizations and kingdoms. That even the
Chinese empire should be included in this category is quite clearly
proved from the employment of the written character *Tsing*, denoting the constellation of the Dioscuri, who were certainly Cabiriac
deities, as otherwise the symbol of territorial divisions, and of the
rules for founding the state.[1]

[1] M. G. Maspero has translated an Egyptian text, quite difficult in many portions, which seems to me to indicate the prevalence in the Nile valley of those
peculiar ideas and customs which we have associated with the Cabiri, or temple-
craft, and at a period prior to the exodus of the Israelites, according to Dr. Birch
and others ; that is to say, in the reign of Seti I , of the nineteenth dynasty.
The text contains a panegyric upon this monarch on the occasion of the building
of a certain temple ; and the language of the address, in agreement with prevailing practice, is put by the scribe into the mouths of the gods themselves. The
following extract is taken from the introductory part, and relates to the founding
and consecration of the edifice : —

" It is I who have founded it with *Sokar*, says *Ptah*. I have determined with a
line the circumference of its walls; while my mouth pronounced the great formulas, *Thoth* assisted at the ceremonies with his *sacred books ;* — has consolidated

It is, then, quite apparent to whom we are indebted for the transmission of those grand ideas which had constituted the theory of a
brilliant civilization, whose antiquity was so great that its memory

the walls of the temple ; *Ptah To tunen* has measured the ground, *Tum* was also
present (*a*). The stake which I had in hand was of gold ; I struck upon it with
the hammer (*b*). As for thee (*Seti*), thou wast with me as geometrician ; thy
two hands held the spade, to the end of establishing the angles of the edifice, according to the cardinal regions of heaven (*c*). The formulas of the conservators
were pronounced, the ceremonies of the protectors were performed by *Neith* and
Selk. Completed with labors, that ought to ensure them to eternity, the walls of
the temple come to be nine,'' etc. (*d*). (*Bibliothèque de l'école des hautes études.
Genre épistolaire*, etc., pp. 91, 92.)

(*a*) The function of the god *Tum*, sun of the lower hemisphere, in these ceremonies is difficult to determine, the text being quite obscure in this place.

(*b*) For this sentence, M. Maspero refers otherwise to Dr. Brugsch's rendering,
which is to be preferred, probably : "The hammer in my hand was of gold ; I
struck with it upon the mallet." Even this version makes but poor sense ; and
for "mallet" it is obvious we are to understand a block of stone, like the corner-
stone, which, in the ceremonies, is adjusted by blows from the golden hammer, or
gavel.

(*c*) Here, again, Dr. Brugsch's rendering conveys a better idea, if it is not
more exact also : "Thou wast with me in thy function of geometrician ; thy
hands carried the measuring-instrument, to establish in an exact manner the
angles (of the temple), according to the cardinal regions."

(*d*) Finally, the concluding sentence of the extract is very obscure, and M.
Maspero offers another, which is, however, hardly more intelligible. The number
"nine" is in some way associated with the walls of the structure. For myself,
I conjecture a reference here to the *ninth* Cabirus, as the father of the eight
Cabiri is sometimes included with them ; or possibly we should understand the
plot of nine squares, as exhibited upon the cosmical seal, heretofore explained,
this again referring to the nine earths, or principal divisions, in the symbolic system of geography.

What, then, are the indications that the foregoing extract contains direct allusions to the Cabiri, or ancient temple-craft ? 1st. *Ptah* was himself the chief of
the Cabiriac divinities ; and his worship was primitive in the valley of the Nile.
2d. *Thoth*, likewise, the Egyptian Mercury, and reputed author of the sacred
books, was a Cabiriac deity ; and the city of *Sesun*, "eight," was the principal
seat of his worship. The numeral "eight" applied to *Thoth*, and it connects him
with *Eshmun* of the Phœnicians. 3d. The "golden hammer," in the hand of
Ptah, chief Cabiriac personage, was evidently the insignia of authority, like the
gavel, and its use in these ceremonies reminds us of the custom of officially adjusting the corner-stones of temples. The instrument for establishing the angles
of the edifice, according to the cardinal points, was obviously another implement
appertaining to the craft. 4th. Some mystical sense appears to have centred in
the number "nine ; " and this, too, could relate only to the Cabiriac mysteries.
Thus, I am strongly of the opinion that we have here a reference, dating from the
period as early as the time of Moses, to the ideas and customs of the ancient

even had been but faintly preserved at what is usually termed the opening of the historical period. We are fully justified in attributing the invention of letters and the authorship at least of many of the sacred books of antiquity to the same class of personages. All this goes to show that the existence of authentic records appertaining to the very first periods of history, which had been preserved with scrupulous care from age to age, is a matter upon which no serious doubts need be entertained, considering all the facts now before us. That the same doctrines and the same events which had been symbolically represented upon the sphere should be embodied also in written documents is naturally to be supposed. In many instances, without doubt, both records must be attributed to the same authorship. In such cases the analogies existing between the two, when properly understood, would be very striking, of which some remarkable instances have been developed as regards the Mosaic records. The great antiquity and perfect reliability of these records, therefore, has been thus most satisfactorily demonstrated.

Like everything human, it is doubtless true that the ancient order of priest-kings, otherwise termed the Cabiri, suffered a gradual degeneracy and corruption, although it is probable there were some rare and noble exceptions to the general rule. It was through the fidelity and devotion of these few, under the guidance of a Divine Providence, that the wisdom of the past was still preserved, serving the basis of new epochs and new dispensations. Thus, the sacred fires, first kindled on the heights of the *Har-Moad*, were never

temple-craft. But the facts here developed call to mind a remark of M. Mariette-Bey : —

"*Ptah* is thus the *Demiurgus*, the creator ; he is the Egyptian Vulcan, the first of the Cabiri ; the Demiurge armed, as M. Guigniaut has remarked, with the CREATIVE HAMMER." *La Mère D'Apis*, pp. 39, 40.

Thus, the mystical *hammer* in the hands of *Ptah*, which appears in the ceremonies of founding the temples as symbol of authority, reappears as symbol of the creator ; and it appertains to *Ptah* in both instances. We show here, 1st. That the Hammer, Hebrew *Pattisch*, like the cuneiform *Pa-te-shi* which gave the name to one of the Cabiri, appertained especially to this Egyptian divinity, reputed father of the Cabiri ; 2d. That there was a definite connection between the *temple-craft*, the founders of sacred edifices, and the *cosmical deities ;* and thus between the *temple* itself and the *cosmos*. For the rest, it is quite probable that *Thor*, the Scandinavian god whose symbol was the *hammer,* ought to be classed among the Cabiriac divinities.

wholly extinguished on the earth. They were successively renewed
on the holy "highlands" of the past, and their light has been reflected through all the ages. We see, therefore, that history has
not been wholly a blind struggle of human forces, with no God to
shape their ends. A divine priesthood, charged with the preservation of the truth and to effect the final redemption of the world,
has been contemporaneous with the entire life of humanity. God
has organized himself in humanity from the beginning, and it is
through this organization that He has presided at the opening of
all the glorious epochs of the world.

SEC. 173. No question has given birth in modern times to a more
extensive literature, to a great extent controversial, than that which
regards the problem of "Genesis and geology." Since my first
tolerable acquaintance with the doctrines of the ancient cosmogonies, it has seemed to me that the standpoints respectively of
Genesis and geology from which to view the creation of the world
and of man involved actually but very little common to both, and
thus that the necessary elements did not exist for a satisfactory
solution of this problem. Any attempt to construe the first and
second chapters of Genesis — for the Garden of Eden is but the completed cosmos — upon the principles of modern geology, especially
in connection with the "nebular hypothesis," must necessarily result, in my view, in really forced if not fanciful constructions of
both theories, and absolutely in a downright injustice to the Mosaic
system. To illustrate the hopeless disparity of views between the
two schemes, it is well known that all the ancient cosmogonies
assume for the "beginning" of things a primordial watery chaos.
But modern science demonstrates the original condition of all matter to have been that of a luminous ether or gas. Now we know,
on one hand, that a watery chaos never *preceded*, in the order of
creation, this luminous, gaseous state of all material substances, and
we know, therefore, on the other hand, that the ocean of waters,
which after the lapse of immense periods actually did envelop the
earth, was not in any sense the true "beginning," according to
scientific apprehension. To assume, in agreement with the theological interpretation, that the watery chaos of Genesis was created
ex nihilo at the "beginning" only shuts the door more effectually
against a reconciliation with the modern theory, for this theory can
never admit for a moment that, *since* the earth was covered with

an ocean of waters, all matter has at any time assumed the gaseous state. It knows that this condition of the earth was not the "beginning." Hence, what is affirmed as the primordial state in all the cosmogonies can never be admitted as such by science.

That the ancient cosmogonies involved the notion of a creation in the modern sense is probably true; but I think it was more a *philosophy* than a *history* of the creative process, this process being conceived fundamentally as one of *division*. I am confident, moreover, that the entire scheme in its higher and more habitual interpretation had a religious, or better, a politico-religious import. It constituted, in fact, the theoretical basis of the ancient civilizations, each of which was a new heaven and earth, modeled after the primitive heaven and earth. Creation was conceived as a temple, and the temple as an image of creation; but the state was only an expansion of the idea of the temple, and thus preëminently a world or cosmos. The foundation of the world was a type of all other foundations, and all other beginnings had for their type the beginning of time. As before observed, paradise itself was the completed cosmos, and this was the primitive "celestial earth." Accordingly, all the ancient kingdoms were celestial earths, from which comes the fundamental notion of "the kingdom of heaven." In this sense chaos may be interpreted of human society, before the introduction of the principle of *order* denoted by the term *cosmos*.

But I think the highest interpretation given to the cosmogony by the sacred writers themselves was that of a *divine dispensation*, of the establishment of an open communication between heaven and earth, the divine and human, a notion involved in the expression *Har-Moad*, "mountain of the assembly," or "mount of the divine presence," of which the *Beth-Moad*, or Hebrew tabernacle, was in every sense a reproduction. The sacred mount of the northeast was supposed to unite the heavens and earth, and the blending of its summit with the sky itself gave rise to the phrase *Svargabhoumi*, "celestial earth." The pyramid in stages, like the tower of Borsippa, was an express imitation of the traditional mount, which was thus regarded as a mountain of degrees or stages. The Egyptian hieroglyph of such a pyramid signifies "a ladder," and M. Lenormant has shown us that "Jacob's ladder" was only another mode of reference to the paradisiacal mountain. According to Scandinavian mythology, the bridge of *Bifrost* unites the heaven and earth. At the destruction of the world the final ruin of everything is when the giants jump upon the bridge of Bifrost, breaking it down. But afterwards, a new heaven and earth appear. The idea of a new heaven and earth, then, is very explicit, and not at all poetical ; it relates to the reopening of communication between God and humanity after it has been broken down by the increasing violence of wickedness among men, in other words, when the giants have jumped upon the bridge Bifrost, and all has gone to ruin. The Scandinavian mythology, derived from the ancient Asgard, has preserved perfectly the notion attached to the primitive *Har-Moad*. As long since shown by Dr. Faber, the doctrine of "a succession of similar worlds" extensively prevailed in antiquity, and it obviously arose from the idea of successive reconstructions or reorganizations of the heaven and earth on the principles of the primitive cosmos, thus replacing the ladder of communication between God and man.

There was many a time in antiquity that a grand civilization crumbled into ruins, overpowered by the incursions of barbarian hordes, or dragged down by popular violence, when passion ruled the hour and all virtue had decayed. It was only when modern civilization had been firmly established, intrenched behind so many frowning fortresses girding the earth, and maintained by the noble philanthropy of so many nations, that the ultimate redemption of humanity from savagism was perfectly assured. It was in such dark periods of history, when all was lost, when every holy interest of man had been betrayed, that another ladder was let down from the sky, on which the angels descended to inaugurate a new era upon the earth ; that another bridge was thrown across the deep chasm gullied out by the angry waters, or that an ark was built, and the helmsman bidden " to steer to the gods." [1]

SEC. 174. No more important truth has been developed by these investigations than the fact of the assimilation of the terrestrial paradise to the lower hemisphere of heaven, and subsequently to the Greek *Hades*, or region of the dead. The proofs embodied in

---

[1] It is stated by Berosus that Xisuthrus, when warned of the deluge and commanded to save himself in a ship, inquired : " Whither he should sail ? " and was told, " To the gods, with a prayer that it might fare well with mankind." The seat of the gods was the summit of the paradisiacal mount, identified also with that of the deluge. Ancient mariners were wont to steer their vessels by the group of seven stars that, like the rudder of the great world ship, swings around the pole.

the seventh chapter render it impossible, as it seems to me, to avoid the conclusions here stated. In the legend of "Ishtar's descent into Hades," one of the designations of the infernal abode is the Semitic phrase *Beth-Hedi*, which is ordinarily employed in the Hebrew Scriptures to denote the tabernacle, or place of assembly of the Israelitish congregation; while in the Syriac version of the New Testament, the last element of this phrase is put for the church, or Christian congregation. Mr. Talbot supposes that *Hedi* and *Hades* were originally the same word. Now it has been shown that, in the books of Moses, a direct relation was conceived to exist between the *Beth-Hedi*, otherwise termed the *Beth-Moad* and the *Har-Moad*, or mount of paradise. One was the "house of the assembly," the other the "mount of the assembly," the latter being the primitive type of the former. Mt. Sion, on which the temple was erected, only replaced the *Har-Moad* of primeval tradition. Thus, that which replaced the *Har-Moad* among the Hebrews was identified with *Hades*, in the legend of Ishtar and among the Babylonians. Again, the "celestial earth," under the form of a floating island, is seen in the vision of Timarchus borne on the bosom of the Stygian abyss, or river of *Hades*. The "celestial earth," however, was originally one with the terrestrial paradise. Finally, the Stygian river itself is situated astronomically in the lower hemisphere, taking its rise in the constellation Libra, directing its course from thence towards the south pole. This fact proves that both Paradise and Hades were located in the inferior heavens. It is into this region of darkness that the sun, chosen representative of humanity, makes its descent at night or at the period of midwinter. We have the most palpable proof, in this symbolism connected with the orb of day, of the definite location of *Hades*, like the Egyptian *Ament*, in the inferior hemisphere.[1] There exists, then,

[1] Besides the term *Ament*, denoting the abode of the setting sun, the region of the dead, the Egyptians employed another expression whose reading is unsettled, although that of *Tuaut* has been often adopted. It corresponded to the abode of the blessed, like the Greek Elysium, or Paradise of the later Jews. Egyptologists are in considerable doubt as regards its particular location. Thus, M. E. Naville remarks: "M. Pierret . . . names this region the 'region of the gates, the inferior hemisphere;' MM. Lauth and Brugsch translate it by 'deep; the subterranean world;' MM. Chabas and Lefebure by 'the inferior heaven;' M. Reinisch . . . by 'splendor, abode of glory.' By comparing the various examples of its use in these texts, *Tuaut* appears clearly indicated as being placed

the conclusive evidence: 1st. That the sacred mount of paradise was an actual geographical and historical fact, being located on the earth's surface; 2d. That it was associated at a subsequent period with the Greek *Hades*, or the infernal abodes; 3d. That both these were located astronomically in the lower hemisphere. It is for us to seek some explanation of these strange and contradictory notions.

From the point of view afforded by modern science, we know that the inferior heavens have really nothing to do with the abode of man after death. If the Greek Hades was located in this celestial space, it had just as little to do with the actual sojourn of the

in opposition to the *superior heaven*" (*Litanie du Soleil*, etc., p. 21). Quite inconsistently, as it seems to me, M. Naville finally elects to render *Tuaut* by the term *Empyrean*, that is, "the highest heaven." He observes again, "I have called *Empyrean* the region designated by a word which is as difficult to translate as the Greek *Hades* ('Αιδης) or the Hebrew *Sheol* (שאול): this is the sojourn of the dead, the place where the defunct ardently desire to arrive, and which is opposed to heaven; from whence it has been most frequently called the *inferior heaven*. This region . . . is probably situated at the centre of the earth; although this is not positively affirmed" (ibid., p. 126). The region in question appertains definitely to the nocturnal sun, to Osiris, and it is placed in opposition to the superior heaven, and is evidently located in the *Ament*, just as Elysium and Paradise were situated in Hades. All goes to show that the usual view, which locates *Tuaut* astronomically in the inferior heaven, is perfectly correct, and these texts translated by M. Naville, dating from the period of Seti I., afford another conclusive proof of our views, as set forth in the seventh chapter. That the Egyptian *Ament*, or region of the dead, was located astronomically in the lower hemisphere, abode of the nocturnal sun, admits of no doubt; in fact, M. Naville identifies expressly the two regions, as in this phrase, "At the same time he (*Ra*) illumes the *Ament*, the inferior heaven" (p. 30). Again, the eleventh and twelfth *hikennus* of the *Litanie* itself confirm this point : —

"He who descends into the spheres of Ament, his figure is that of *Tum*.

"He who descends into the mysteries of Anubis, his figure is that of *Chepra* (*Atmu*)."

The god *Tum* is notably the sun of the lower hemisphere, and M. Naville shows that *Chepra* is identical with him. Of *Anubis*, the author remarks: "Anubis, the dark jackal, the guardian of the coffins, is the emblem of the (*western*) horizon and of the inferior hemisphere. Jablonski has demonstrated this by numerous examples derived from Greek and Latin authors" (vid. pp. 31, 32). Thus the location of the Egyptian *Ament* in the lower hemisphere, where *Ra*, the sun, under the form of *Tum* or *Chepra*, descends at night, is a fully demonstrated fact; nor is it less certain that the Greeks derived their notions concerning *Hades* principally from those of the Egyptians respecting *Ament*, so that the same astronomical locality for both must be admitted as perfectly substantiated.

dead. Nevertheless, it is necessary to admit the reality of a future life, and thus of the abode of those inheriting it. There can be but little doubt that the mount of paradise itself was considered as such abode during the first ages of the world, although its actual location on earth has been so fully verified. In early Aryan tradition, as is well known, the mansions of the blessed were thought to be situated on the slopes of Meru or Albordj. The *Har-Moad* of Isaiah appears to have been regarded in a similar light. Hence, the sacred mount of paradise was definitely associated, in some sense, with the life of man after death. This fact is of capital importance, as it offers an explanation of the connection of Paradise with Hades. Both were regarded as the sojourn of the dead. There was, however, this difference between them : with the first men, the dear departed were believed to have gone up the sacred mountain, and to dwell on its shining slopes. In later conception, centring in Hades, all was transferred to the under world, the inferior heavens, the region of darkness. From whence arose these differences in conceiving the locality of the abodes of men after death? We must remember that, according to the symbolic mode of representing the cosmos by the two halves of the sphere, the inferior heavens were put for the earth, and especially the particular earth identified with the terrestrial paradise. Hence, the sacred mount, as well as Hades, conceived as equally the abode of the departed, was located astronomically in the lower hemisphere. But this celestial region was only a symbol put for the earth, on which the sacred mount had its literal location. The symbolic principle had been neglected and finally forgotten; and as a consequence the lower heavens assumed the character of a separate region, a world of darkness. Paradise and Hades, which had been assimilated to this region *symbolically*, were now conceived, by an entire misapprehension, to be *literally* situated there.

In primeval tradition, as we are now prepared to affirm, the actual dwelling-place of the departed was on the *Har-Moad*, "the mountain of the assembly." In this sense *Hades*, the Semitic *Beth-Hedi*, one with the *Beth-Moad*, was originally connected with it, and both were symbolically referred to the inferior heavens. It was only when the real motive of this last assimilation had been forgotten that all was transferred literally to the under world, the region of darkness, — a sad perversion of primitive doctrines, that

has resulted in clothing the future life of man in impenetrable
gloom. We are now able to comprehend how the sacred writers
were equally correct in assimilating the terrestrial paradise to the
state of the dead, to *Hades*, and at the same time to the *Beth-Hedi*,
considered as a reproduction of the sacred mount of tradition. It
was on the holy mountain of degrees, which, like Jacob's ladder,
united the heavens to the earth, that primeval tradition had located
the beautiful mansions of the just made perfect.

SEC. 175. Another very important fact contemplated from the
point of view which we now occupy relates to the uniform conception in high antiquity that the divine powers had their seat in the
extreme northern heavens, or precisely in the celestial region penetrated by the summit of the paradisiacal mountain. Among the
Romans, according to Dr. William Smith, the gods were supposed
to be seated in the north; and this is confirmed by the location of
the axis of the Pantheon only five degrees from the pole star; the
reference being to the seven stars of the chariot; but especially to
the eighth region in relation to them, regarded as the throne of
Jupiter. The Phœnician Eshmun, the eighth Cabirus, as shown
by Dr. Movers, must be associated likewise with this highest, central point of the heavens. The god Shemal of the Haranite mysteries, a divine name equal to the Hebrew Semal, which appears
originally to have been common to the Semitic races, was connected directly with the north celestial pole. The Persian Ormazd
had his abode on the divine summit, the Albordj, source of the
sacred river Avanda. The Aryans of India located the paradise of
the gods on the heights of Meru, or in the region termed by them
the Su-Meru. The Kharsak-Kurra of the cuneiform texts, or the
"Bit-Kharris of the east," where, according to the passage heretofore cited, the great divinities and their spouses reign forever,
is identified by M. Lenormant with the rotating centre of the
superior heavens. As the same writer has observed, the last term
of hope of the Babylonian and Assyrian kings was to be conducted
at death to this mystical region of the skies, there to dwell with
the immortal gods. The prophet Isaiah makes use of this very
fact, in his bitter sarcasm upon the fall of the Babylonian monarch: " For thou hast said in thine heart, I will ascend into heaven,
I will exalt my throne above the stars of God (*El*); I will sit also
upon the mount of the congregation (*Har-Moad*), in the sides of

the north ; I will ascend above the heights of the clouds; I will be like the Most High (*Elyon*). Yet thou shalt be brought down to hell (*Sheol*), to the sides of the pit " (xiv. 13–15). The "stars of *El* " were previously identified with the seven stars of the chariot, whose rolling motion around the pole gave rise to the notion of strength, as involved in the term *El*, "the Strong One." The celestial space above this group of stars was that appropriated to *Elyon*, "the Most High," and it answered to the eighth region, to which also the Phœnician *Eshmun* was assimilated. Thus, the blasphemous king of Babylon had said in his heart that he would exalt his throne above the seven stars of *El ;* that he would sit upon the very summit of the *Har-Moad ;* that he would ascend even into the eighth celestial region, and be like *Elyon*, "the Highest."

It is perfectly clear from the prophet's language that he intends to associate the two divine names, *El* and *Elyon*, not only with the *Har-Moad* situated geographically in the north, but with the particular and central region of the northern heavens, supposed to be penetrated by the summit of this sacred mountain ; that is to say, with the same region precisely which all the Asiatic nations regarded as the seat of the divine powers. Such being the case, it is necessary to admit also that this inspired writer intends to indorse, and does so expressly, the primitive doctrine which located the divine abode in this quarter of the heavens. I think, moreover, that the sacred writers generally countenance this idea. Heaven is often put for the Deity, and especially for the dwelling of God, to which corresponded the Holy of Holies of the tabernacle, as shown by Dr. Bähr. But it is quite certain, from the facts developed in these researches, that this is not the heaven in general, but the particular and traditional heaven, specially connected with the *Har-Moad*, identified with the mount of paradise, and of which the *Beth-Moad* was an image. The cubical space constituting the Holy of Holies, like that forming the eighth stage of the pyramidal temples, and which was in an especial sense the divine abode, obviously represented that particular heaven, and no other, which was united to the terrestrial paradise by means of the sacred mount. Thus, the two divine names, *El* and *Elyon*, were expressly connected with this celestial region, and the Holy of Holies represented it. Besides this, we have the "gate of heaven,"

the "door of heaven," the "gate of El," and the "house of El,"
— expressions whose primary reference to the central region of
heaven, connected with the *Har-Moad*, admits of but little doubt.
If we speak, therefore, of the heavenly dwelling, of the divine
abode *par excellence*, it can be no other than that to which
uniform tradition pointed, the definite reference to which was embodied in the most ancient temple structures, the notion of which
was fundamental in all the Asiatic religions. It was on the slopes
of the sacred mountain that the mansions of the blessed were
situated. It was its summit on which the divine powers were conceived to dwell, and they dwell there to-day. Theirs are the hands
that firmly grasp the celestial rudder steering the ship of the
world.

SEC. 176. The two divine names, *El* and *Elyon*, so frequently
occurring in the Hebrew Scriptures, are well known to have appertained to the earliest periods of Semitic development. As in the
texts cited from Isaiah, they are often employed separately, — the
first signifying "the Strong One," importing a character similar
to that of a Hercules; the second having the sense of "the Highest," or "the Most High." But in the majority of instances, perhaps, *Elyon* is employed as a qualifying word to *El*, in the phrase
El-elyon. "El the highest," usually rendered "the Most High
God." The proofs were presented in a previous chapter to the
effect that the term *Elyon* involves the notion of "an ascent," as
of a series of stages, a staircase, or ladder, hence also a chamber,
the chamber of heaven, etc. In all cases, this word imports "the
highest," from whence its peculiar significance in relation to the
seven stars of El, referring to the eighth celestial region, to which
the Phœnician Eshmun corresponded as the eighth Cabirus. The
Phœnician and Hebrew languages are substantially the same, but
as a distinct nationality the Phœnicians were far more ancient.
In the remains of Phœnician literature, *El* and *Elyon* both occur
as divine names, particularly in the cosmogony, which is doubtless
equal in antiquity to the Mosaic and Babylonian systems. According to the "Fragments of Sanchoniatho," *Elyon*, called *Elioun*,
the *Hypsistus* of the Greeks, plays a very important part in the
Phœnician cosmogony; and by a comparison of the facts as stated
by this author with those already before us. it will be seen that
they reflect much light upon each other. The language of Sanchoniatho is as follows : —

" But from *Sydyc* (the just) descended the Dioscuri, or Cabiri, or Corybantes, or Samothraces; these first built a ship." " Contemporary with these was one *Elioun* (*Elyon*), called Hypsistus (the most high); and a female named Beruth." " By these was begotten *Epigeus* or *Autochthon*, whom they afterwards called Ouranus (heaven)." " He had a sister of the same parents, who was called *Ge* (earth)." " Hypsistus, the father of these, having been killed in a conflict with wild beasts, was consecrated, and his children offered libations and sacrifices unto him." [1]

All these personages are treated by the author as purely human characters; but it is well understood by critics that they are properly cosmical potencies. The seven sons of Sydyc, " the just," including the Dioscuri, or Twins, zodiacal Gemini, are Cabiri, to which another son, born subsequently, Eshmun, or " the eighth," is to be added. As for *Elyon*, called Hypsistus, he is the father of heaven and earth, or of *Ouranus* and *Ge*. It is important to compare with these data the facts stated by Professor Fick, formerly cited, namely, that the Aryan *Akman*, Greek *Akmon*, signifies a " stone," then " heaven," also " anvil," being finally the title of one of the Cabiri, and a personal name of the *father of Ouranus*, or Heaven. Thus *Elyon* and *Akmon* are both put for the father of heaven, to which might probably be added *Ge*, the earth. As cosmical powers, therefore, the two personages are to be identified, and from the language of Isaiah, assigning *Elyon* to the eighth celestial region, it is obvious that *Akmon* should be assimilated to the same region, in company with the Phœnician *Eshmun*. As two of these names were titles of a Cabirus, it is quite certain that such was originally the case with *Elyon*, as already suggested.

The reader has been made acquainted with the character of the Assyrian Hercules, at one time assimilated to the sun of the lower hemisphere, at another time to the planet Saturn, etc., but whose supposed voluntary death and resurrection to a new life constitute the chief matters of interest in his history. Dr. Movers proves that the name *Akmon*, denoting a Cabirus, was otherwise a title of Hercules. The evidence has been already introduced to the effect that a Cabirus was supposed to have suffered a violent death at the hands of his brother or brothers; and in order to complete the analogies in all such traditions, it is necessary to suppose that he

[1] Vid. Kenrick, *Phœnicia*, p. 333.

was afterwards raised from the dead. Finally, the reader has noticed that the Phœnician *Elyon*, or *Elioun*, was thought to have been killed in conflict with wild beasts; that from this circumstance he was consecrated, and worshiped by libations and sacrifices. The Orientalist detects at once the import of such language. The Tammuz-Adonis was killed by a wild boar while hunting on the mountains, but was brought to life again. M. Lenormant, in a critical and very satisfactory investigation, has shown that the Phœnician *Elyon*, or *Elioun*, presents the more ancient and pure version of the character of Adonis, corresponding better with the Assyrian Hercules, who suffers a violent but voluntary death, is burned upon the funeral pile, but finally raised again to a new and glorious existence.[1] It will be seen that the various facts grouped into one view in the present section go to establish some very important conclusions, which we now proceed to formulate.

SEC. 177. The inferences to be drawn from the data, most of which are already familiar to us, are chiefly the following: —

1st. The Hebrew and Phœnician *Elyon* must be regarded as one and the same personage, whose primitive conception had been inherited alike by the two peoples speaking the same language. The two different phases of character thus transmitted, which appear to be equally ancient, afford material aid in interpreting each other.

2d. The thoroughly mystical character of *Elyon* is abundantly proved by the two orders of conceptions found centring in him. The notion of ascending stages, like the ladder of Jacob's vision, coupled with the idea of the chambers of heaven, is plainly involved in the Hebrew etymology of this name and the terms radically connected with it. To this is to be added the direct assimilation of *Elyon*, by Isaiah, to the eighth celestial region, to which the Phœnician Eshmun and the Shemal of the Haranite mysteries appertained, affording thus the plainest indications of an original connection with the Cabiriac worship. On the Phœnician side of his character, *Elyon* appears as the highest cosmical potency, from whom the genesis of the heaven and earth proceeds; at the same time he represents the primitive idea of a divinity who suffers voluntarily a violent death, but is raised to life again.

3d. But *Akmon* is the father of Ouranus or Heaven, the same as

[1] *Lettres Assyriologiques*, t. ii. pp. 291–301.

Elyon ; is the name even of a Cabirus, is a title of Hercules; and this term etymologically involves the notion of "heaven," "a stone," evidently a dressed stone, and is cognate with *Aktan,* "eight," and *Aktama,* "the eighth," like the Semitic *Eshmun.* It is obviously this very Cabirus, whose symbol was a dressed stone, subsequently replaced by a meteoric stone, used as an anvil, who was reputed to have been slain by his brethren.

4th. Melchizedek was a priest of *Elyon,* identified with *El,* the "Strong One," as seen in the following: "And Melchizedek king of Salem brought forth bread and wine: and he was the priest of the most high God. And he blessed him, and said, Blessed be Abram of the most high God, possessor of heaven and earth" (Gen. xiv. 18, 19). In this passage *El* and *Elyon* are identified as the same personage, in opposition to any polytheistic conception; hence, the original has *El-elyon.* The phrase, "*possessor* of heaven and earth," in the nineteenth and twenty-second verses, is unusual and somewhat peculiar. The idea of *possession* here is not that acquired by purchase, by gift, nor by creation even, except in the sense of a *generation,* as in the language attributed to Eve: "I have *gotten a man* from the Lord" (Gen. iv. 1), where the same Hebrew radical appears in the original. This term, *Qa-nah,* "to get, to obtain," never means "to create," and can be compared with the Hebrew *Bara,* only as the latter means "to beget." The notion of possession, then, here proceeds from that of generation. Thus, we show a direct relation of the Hebrew *Elyon,* as "*possessor* of heaven and earth," to the Phœnician *Elyon,* as "*father* of heaven and earth."

5th. The assimilation of *Elyon* to the eighth celestial region, like *Eshmun,* his character as father of Ouranus, or Heaven, like that of *Akmon,* his reputed death by violence and subsequent resurrection, like the Cabirus murdered by his brethren, all go to establish beyond doubt his Cabiriac character. But Melchizedek was a priest of *Elyon,* identified with *El,* the Strong One, and was at the same time king of Salem, being thus a priest-king. The conclusion cannot be avoided, — Melchizedek belonged to the Cabiriac priesthood. The name *Melchi-zedek* signifies "king of justice," the last element, *zedek,* is radically the same as the Phœnician *Sydyc,* "the just," title of the father of the Cabiri. The ancient order of priest-kings was founded upon the notion of *justice,* that is to say, *righteousnes*, and it was obviously this fact that constituted the motive for selecting the term *zedek* in composition with the name of Melchizedek. Sargon the ancient assumed the Assyrian title of *Sar-Kitti*, "king of justice," and everything indicates that he belonged to the Cabiriac fraternity. Thus, it is evident that this notion of "the just" or "justice," coupled with certain distinguished personages, founders of new epochs, had a technical reference to the Cabiri.

6th. If we would realize now the culminating point of the foregoing inferences, it is necessary to recall the facts connected with the celestial pole marking the very period to which all these traditionary conceptions appertained; facts that were fully verified in the fifteenth chapter. This was, at the period referred to, the rotating centre of the heavens, around which the stars of *El* revolved; it was the eighth celestial region to which *Elyon*, equally with *Eshmun* and *Akmon*, pertained; it was in fact the originally conceived abode of the divine powers, the celestial paradise united to the terrestrial by the *Har-Moad*. It was, finally, the point from which the heavens and earth, and all their principal divisions, were generated, giving rise to the notion of the "father of heaven," associated alike with *Elyon*, God of Melchizedek, and *Akmon*, the Cabirus, whose symbol was the dressed stone. But that which is most important to observe as connected with this celestial region, and in close proximity to the asterism representing the great mother, is the figure of Hercules, clad in the lion's skin, and in the very act of bruising the serpent's head with his heel; a reference to the promised seed of the woman so plain as not to be doubted.

If, now, we would find a character that most completely and perfectly realizes this entire circle of traditionary ideas, it is necessary to go to the Christian Scriptures. (1) Christ is represented as a cosmical agent. All things are made by him, and he is the especial centre of a new creation, a new heaven and earth. (2) He is both king and priest, thus definitely related to Melchizedek, and to the ancient order generally of the priest-kings. (3) He is the "Strong One," the "Lord mighty in battle," the "Lion of the tribe of Juda," the true Hercules. (4) He suffers voluntarily a violent death, and is raised again on the third day. (5) He is the "chief corner-stone," the "stone which the builders rejected." (6) His ascent into heaven was like the high-priest's entrance into the Holy of Holies, which represented the eighth celestial region, assimilated to *Elyon*, and usually to the eighth Cabirus. (7) The fundamental idea of his character and mission was that of *justice* or *righteousness*, on which also the Cabiriac priesthood was founded. Thus, all is fulfilled and realized in this one character. The personage to whom this remarkable character appertains is not an ideal one, but strictly historical, as admitted by the best modern critics of the rationalistic school. Such being the case, it must be admitted, I think, that no such extraordinary coincidences as here established could possibly occur, except by the special intervention of a Divine Providence. There is many an expression attributed to the Saviour in the gospel narrative that proves him to have been perfectly conscious of his relation to the traditions of all past ages, and that it was his mission to realize and fulfill all. But we have not the space here to treat these matters in a manner such as their importance deserves. We have desired merely to indicate to whom these ancient foregleams of prophecy appertained.

SEC. 178. The aim of all true religion is to establish an intelligent relationship between God and man, a conscious union between the infinite and finite mind. But no such intelligent union can subsist between things merely general; no more between the finite mind and something purely universal, as a certain recondite essence filling all space. If the Divine Mind be not apprehended as a concrete, personal existence, and this under the limitations of time and space, it is impossible to conceive of an intelligent communication between God and man, of a conscious relation subsisting between them, of anything, in fact, but vague impressions and emotions arising from religious or æsthetical contemplations. Neither science nor philosophy allows of aught beyond this, but religion aspires to an actual and personal relation with the Deity. If the Divine Mind exists, yet never communicates with his rational creatures, his must be the loneliest and saddest heart in the wide universe. In such case infinite space itself must be to the Deity but a solitary confinement, worse than the convict's cell. The sweetest, highest happiness is to love and be loved; yea, and to tell of it. If God cannot love and reveal it, his must be a hard lot. Thus, while science and philosophy, with their heartless generalizations and merciless logic, continually break down the ladder of communication between heaven and earth, religion, with its infinite longings, ever strives to replace it.

In the primeval doctrines, the effort was to build down from God, as the universal and infinite, to man as the individual and finite. To this end, the first principle laid down was that God dwells, inhabits, the same as man. As well attempt to communicate with empty space as with a Deity who merely dwells *everywhere*, since for all such purposes this would be the same as *nowhere*. Conceived in its broadest sense, the dwelling of God was the cosmos, more especially that chief division of the cosmos, — the starry heaven. These points were long since established by Dr. Bähr, as heretofore cited. But to understand here the world in general, or the heavens in general, is to give to these doctrines a pantheistic interpretation. The allusion is definitely to the traditional heaven and earth as known to the first men, to that particular celestial region penetrated by the summit of the *Har-Moad*, which the *Beth-Moad* was designed to reproduce; in other terms, to that identical region assumed by all the Asiatic nations as the seat of the divine hierarchy.

The next stage in building down from heaven to earth, or from God to man, was the temple, conceived as an image of the cosmos, at the same time as a dwelling of the Deity. But the temple, in order to become such abode, must truly represent the cosmos, and be constructed upon the same principles. Hence, the theory of both was the same. Here, again, it is necessary to avoid the pantheistic interpretation of the world in general, and to assume a direct reference to the traditional heaven and earth, identified with the celestial and terrestrial paradise. The "*house* of the assembly" must replace the "*mount* of the assembly," or, as Dr. Lowth would have it, the "mount of the divine presence."

The final stage in this process was to assume man as the true temple, as the real cosmos, as, in fact, the most befitting habitation of the Divine Mind. This stage is peculiar and fundamental to the Christian system, as appears from numerous passages like the following: "In whom ye also are builded together for a habitation of God through the Spirit" (Eph. ii. 22). But man is likewise the real cosmos; the threefold division of the human powers, so uniformly adopted by the sacred writers, into spirit, soul, and body, obviously proceeded originally from the three chief divisions of the cosmos, — the heaven, the atmosphere, and the earth.[1] It was on the same principle that the regenerated man was deemed the new creation, or the new heaven and earth.

Such, then, in general terms, was the theory according to which it was attempted to bridge over the chasm between the infinite and finite, the divine and human, and to establish an intimate, personal relation between them. Such were the principles upon the basis of which, at the opening of new epochs, when the old order of things had gone to ruins, it was attempted to replace the ladder of communication between heaven and earth; between the ancient seat of the divine powers on the summit of the Har-Moad and the heavenly kingdom which was to be reinaugurated among men. It is a fact very remarkable, and not less important, that among so many widely separated nations in antiquity, and through so many different epochs, the same celestial region was uniformly regarded as the particular abode of the divine hierarchy, from whence the government of the world proceeded, and the heavenly influences descended upon the earth. It is sufficient to recall here the data summarized in the 174th section. That the Hebrew Scriptures indorse generally the same traditional doctrine is already quite apparent, and admits of being even more clearly demonstrated. The allusions to this region by the prophet Isaiah, the candlestick with seven branches in the Hebrew tabernacle, obviously representing the seven stars of *El*, the definite relation of the *Beth-Moad* to the *Har-Moad*, — all tend to support the doctrine common to the entire Asiatic world. The Christian Scriptures also contain many allusions pointing in the same direction. As, however, the celestial pole was constantly changing its position, the more usual reference to this region was by means of the seven stars of the chariot, which never leave the visible heavens, but roll

[1] I find that the Jewish rabbis have some notion of the relation of the microcosm to the macrocosm. Dr. Cox observes : "Another authority declares that 'God created three worlds, — the upper world, the middle world, and the lower world.' Farther, it is said in Zohar that God created three souls, answering to the three worlds, namely, the *Nephesh*, that is, the soul ; and the *Ruach*, that is, the spirit; and the *Neshama*, that is, the precious soul." (*Bib. Antiquities*, p. 423.) It is obvious that the primitive doctrine is here somewhat obscured, that the "lower world" should be considered the earth. to which corresponds the *Baras*, or body; while the *Neshama* should be excluded from the category, as a later arrangement.

forever around that divine summit from whence the holy mountains of all ages have been derived.

Sec. 179. If any doctrine was more fundamental than another in nearly all the systems of antiquity, it was that relating to the active and passive powers of the universe, or the male and female, appearing primarily in the cosmogony as *mind and matter*, whose union in one androgynous divinity constituted the absolute *first principle*. The conception of a pure *spirituality*, prior to and independent of all matter, does not appear to have been primitive, but rather the product of later philosophical abstraction. The cosmogonies, so far as I am able to perceive, do not teach any such doctrine. The most primitive titles of divinity prevailing among men do not involve etymologically the notion of a pure spiritual existence, but quite the contrary. The Chinese *Tien*, the Sanskrit *Dyu* or *Dyaus*, and the Accadian *An*, all denoted the "heaven" or the "sky," conceived as the Deity. Even the substantive verb *to be*, which has given birth to so many divine names in various languages, relates rather to that which *really is* than to the abstract *pure being* of the philosophers. Matter *really is* the same as mind, and this verb involves equally the notion of both. The idea of the poverty of language does not apply here; for the truth is, that the first men had no difficulty in finding words to express exactly their *real notion* of the Deity, and among these were those just noticed. They signified "heaven," and heaven was taken for the Deity. The ancients, however, *did not divorce mind from matter*, as is done in modern thought. They were conceived in eternal marriage. Thus, the primitive man had no difficulty in associating mind, and even personality, with the material heaven, considered as the Divinity. The notion of *heaven-he*, or of heaven regarded as a personal deity, was perfectly consistent with the prevailing habit of thought. Again, the conception of a deity having his abode *beyond* the material heaven, and *outside* the material creation, was obviously the result of later philosophical speculations, and wholly foreign to the primitive doctrines. It is a sufficient proof of this that, when the original androgynous principle separated into the two chief divisions of the cosmos, or heaven and earth, these were universally conceived as the *heaven father* and the *earth mother*, from whose commerce all things had their birth. This separation was, in all the cosmogonies, the first act of creation. Now it is evident in this case that the *heaven father* was assimilated to the material heaven, and the *earth mother* to the material earth; and that nothing was imagined as beyond or outside until metaphysical abstraction began to subvert the ancient doctrines. The notions of the *heaven father* and of the *earth mother* naturally proceeded from the doctrine, common to all the cosmogonies, that *creation* was in some sense a *generation;* a doctrine distinctly recognized in the Mosaic system, as in the phrase, "The *generations* (*Toledoth*, from *Yalad*, "to beget") of the heavens and of the earth" (Gen. ii. 4). The same fundamental ideas as those here set forth were absolutely primitive in China, in Egypt, in the Euphrates valley, among the Aryans of India and Persia, among the Phœnicians, and probably all the Semitic races.

It is hardly necessary to insist upon the point, in view of the facts familiar to us, that the *heaven father* and *earth mother*, as conceptions appertaining to the ancient cosmogonies, are not to be interpreted of the heaven and earth in general, and thus in the purely philosophical and pantheistical sense, but of that particular and traditional heaven and earth known to the first men, and supposed to be united by the paradisiacal mountain. It was here, indeed, as heretofore shown, that all the cosmogonies centred. It was here, too, that the birth of man, usually conceived as the result of the commerce of Heaven and Earth, actually did take place. The *heaven father*, then, is to be associated with that celestial region which, in the traditions of all Asia, was regarded as the especial abode of the divine powers, and of which the Holy of Holies in the Hebrew tabernacle was a symbolical representation. The *earth mother*, on the other hand, must be assimilated to the terrestrial paradise; and of this the outer court of the tabernacle was an image, as previously established. The primitive relation of the earth-goddess to the "celestial earth," and even to the church itself, is sufficiently established in the ninth chapter; and we may interpret the man born of the *heaven father* and *earth mother* in a special sense of the regenerated man. Thus, we see how very important it is to a proper understanding of these primitive conceptions that we restore to them their traditional elements. We perceive, then, at once their thoroughly religious import; while otherwise they appear as purely philosophical dogmas, whose pantheistical character is very apparent. It is the same of the temple, regarded as an image of the cosmos. The origin of the temple in the primitive *Har-Moad* is all important to be considered, a point upon which we have heretofore insisted. In general, it is necessary to supply to the systems of antiquity their traditional element, if we would correctly interpret them. It is a striking characteristic of the two religions of the Bible that they preserve intact the traditional science inherited from the very beginning, and that this constitutes the solid foundation of their respective epochs, the indestructible basis upon which the communication between heaven and earth is to be maintained. It is uniformly from the same celestial region that the holy influences descend, and to which everything is adjusted. The assumption of the entire expanse of the sky as the abode of the Heavenly Father was just that pantheistical conception expressly discountenanced by the sacred writers, and the assumption of a purely ideal region beyond the material heavens as such abode found no justification in the sacred tradition; but this was rather an additional world, superinduced upon the real cosmos, when philosophy and science had consummated the divorce between mind and matter, and it is just this divorce which is to-day undermining the faith of mankind. It has substituted for the *heaven father* the speculative doctrine of the absolute. It has substituted for the *earth mother* a dead matter governed by a blind law. It has given us a Divinity who *dwells everywhere*, and this is as much as to say *nowhere*. But faith cannot subsist in such a vacuity as this. It is on the *Har-Moad* that the Father dwells, and it is there that our beloved have gone up.

LIST OF AUTHORS AND THEIR WORKS.

Asmus (P.). Die indogermanische Religion in den Hauptpunkten ihrer Entwickelung. Halle, 1875. 8vo.

Bahr (K. C.). Symbolik des Mosaischen Cultus. Heidelberg, 1837. 2 vols.
8vo.

Benfey (Theo.) and Stern (M. A.). Ueber die Monatsnamen einiger alter
Völker. Berlin, 1836. 8vo.

Bernard (Th.). Dictionnaire Mythologique Universel. On the basis of Jacobi.
Paris, 1846. 7mo.

Birch (S.). Ancient History from the Monuments : Egypt. New York, 1875.
12mo.

Bresslau (M. H.). Hebrew and English Dictionary. On the basis of Fürst.
London, 1855. 12mo.

Broca (Paul de). In Smithsonian Reports.

Brocklesby (J.). Elements of Astronomy. New York, 1874. 7mo.

Brugsch (H.). Matériaux pour servir à la Reconstruction du Calendrier des
Anciens Egyptiens. Leipzig, 1864. 4to.

——— Grammaire Hiéroglyphique. Leipzig, 1872. 4to.

Bunsen (Ernst v.). Biblische Gleichzeitigkeiten, oder Uebereinstimmende
Zeitrechnung bei Babyloniern, Assyrern, Aegyptern, und Hebräern. Berlin,
1875. 8vo.

Burritt (E. H.). The Geography of the Heavens. Mattison's edition. New
York, 1856. 7mo.

Bush (George). Notes Critical and Practical on the Book of Genesis. New
York, 1838. 2 vols. 7mo.

Calmet (Aug.). Dictionary of the Holy Bible. Robinson's edition. Boston,
1832. Large 8vo.

Campbell (G.). The Four Gospels ; translated from the Greek, with Preliminary Dissertations. Boston, 1824. 4 vols. 8vo.

Carré (L.). L'Ancien Orient ; Etudes Historiques, Religieuses, et Philoso-
phiques. Paris, 1875. 4 vols. 8vo.

Chabas (F.). Etudes sur l'Antiquité Historique d'après les Sources Egyptiennes
et les Monuments Préhistoriques. Paris, 1873. Large 8vo.

——— Les Papyrus Hiératiques de Berlin. Paris, 1863. 8vo.

——— Observations sur le chapitre vi. du Rituel Funéraire Egyptien, etc. Paris,
1863. 4to.

CHWOLSOHN (D.). Die Ssabier und der Ssabismus. St. Petersburg, 1856. 2 vols. 8vo.

—— Ueber die Ueberreste der Altbabylonischen Literatur. St. Petersburg, 1859. 4to.

COX (F. A.). Biblical Antiquities, with some collateral subjects, etc. London, 1852. 7mo.

CUDWORTH (R.). The True Intellectual System of the Universe. Andover, 1837. 2 vols. 8vo.

CURTIUS (G.). Grundzüge der Griechischen Etymologie. 4e Aufl. Leipzig, 1873. Large 8vo.

DELITZSCH (FRIED.). Assyrische Studien. II. I. Leipzig, 1874. 8vo.

—— Assyrische Lesestücke. Leipzig, 1876. 4to.

DILLMANN (Dr.). In Schenkel's Bibel-Lexikon.

DUPUIS (CH. F.). L'Origine de tous les Cultes, ou Religion Universelle. Paris, 1795. 3 vols. 4to.

FABER (G. S.). The Origin of Pagan Idolatry. London, 1816. 3 vols. 4to.

FICK (A.). Vergleichendes Wörterbuch der indogermanischen Sprachen. 3e Aufl. Göttingen, 1874–1876. 4 vols. 8vo.

FURST (J.). Hebräisches und Chaldäisches Schulwörterbuch, über das Alte Testament. Leipzig, 1872. 16mo.

GESENIUS (WILLIAM). Hebrew and English Lexicon. Translated by Robinson. Boston, 1850. 8vo.

—— Der Prophet Jesaia. Neu übersetzt. Leipzig, 1820. 2 vols. 7mo.

—— Scripturae Linguaeque Phoeniciae, Monumenta quotquot supersunt. Leipsiæ, 1837. 4to.

GOLDZIHER (I.). Der Mythos bei den Hebräern, und seine geschichtliche Entwickelung, etc. Leipzig, 1876. 8vo.

GRÉBAUT (E.). In Révue Archéologique, Paris.

GRILL (J.). Die Erzväter der Menschheit ; ein Beitrag zur Grundlegung einer hebräischen Alterthumswissenschaft. Leipzig, 1875. 8vo.

GRIVEL (Jos.). In Trans. Bib. Arch. Society, London.

GROTE (GEORGE). The History of Greece. New York, 1856. 12 vols. 7mo.

HALÉVY (Jos.). In Journal Asiatique, Paris.

HAMMER (J. DE). Mémoire sur le Culte de Mithra. Paris, 1833. 8vo. Plates, 4to.

HERZOG (Dr.). In Protestant Theological and Ecclesiastical Encyclopedia, Philadelphia.

HOPKINS (S.). In Bibliotheca Sacra, Andover.

JOSEPHUS (FLAVIUS). The Works of Josephus. Translated by Whiston. Cincinnati, 1837. 8vo.

KENRICK (J.). Phoenicia. London, 1855. 8vo.

KURTZ (Dr.). In Protestant Theological and Ecclesiastical Encyclopedia, Philadelphia.

LAJARD (M. F.). Recherches sur le Culte, les Symboles, les Attributs, et les Monuments figurés de Vénus. Paris, 1848. 4to. Plates, folio.

LE BAS (PH.). L'Univers : Suède et Norwège. Paris, 1841. 8vo.

LENORMANT (FRANÇOIS). Essai de Commentaire des Fragments Cosmogoniques de Bérose. Paris, 1872. 8vo.

—— Les Premières Civilisations ; Etudes d'Histoire et d'Archéologie. Paris, 1874. 2 vols. 8vo.

—— Lettres Assyriologiques et Epigraphiques, sur l'Histoire et les Antiquités de l'Asie antérieure. Paris, 1872. 2 vols. 4to, autographie.

—— Lettres Assyriologiques. Seconde Série. Etudes Accadiennes. Paris, 1873. Vol. I. 4to, autographie.

—— Essai sur un Document Mathématique Chaldéen. Paris, 1868. 8vo, autographie.

—— La Magie chez les Chaldéens et les Origines Accadiennes. Paris, 1874. 8vo.

—— La Divination et la Science des Présages, chez les Chaldéens. Paris, 1875. 8vo.

—— Manuel d'Histoire Ancienne de l'Orient. Paris, 1869. 6me édition. 3 vols. 12mo. Atlas.

LUBBOCK (J.). The Origin of Civilization, and the Primitive Condition of Man. New York, 1870. 7mo.

MAINE (H. S.). Ancient Law : Its Connection with the Early History of Society, and its Relation to Modern Ideas. New York, 1864. 8vo.

MARIETTE-BEY (AUG.). Musée d'Antiquités Egyptiennes à Boulaq. Paris, 1869. 8vo.

—— Aperçu de l'Histoire Ancienne d'Egypte. Paris, 1867. 8vo.

—— Mémoire sur Mère d'Apis. Paris, 1856. 4to.

MASPERO (G.). Histoire Ancienne des Peuples de l'Orient. Paris, 1875. 7mo.

—— In Bibliothèque de l'Ecole des Hautes Etudes.

MÉNANT (J.). Babylone et la Chaldée. Paris, 1875. Large 8vo.

MEIER (E.). Hebräisches Wurzelwörterbuch, nebst drei Anhängen. Mannheim. 1845. 8vo.

MOVERS (F. C.). Untersuchungen ueber die Religion und die Gottheiten der Phönizier. Bonn, 1841. 8vo.

MÜLLER (M.). Lectures on the Science of Languages. 1st and 2d Series. New York, 1865. 2 vols. 7mo.

—— Chips from a German Workshop. New York, 1869. 2 vols. 7mo. Other volumes since issued.

—— Lectures on the Science of Religion. In Littell's Living Age. Aug. 20, 1870.

NAVILLE (E.). La Litanie du Soleil. Inscriptions recueillies dans les Tombeaux des Rois à Thèbes. Leipzig, 1875. 4to. Plates, 4to.

NISSEN (H.). Das Templum. Antiquarische Untersuchungen. Berlin, 1869. 8vo.

NORRIS (E.) Assyrian Dictionary. London, 1868–1872. 3 vols. 4to.

OBRY (J. B. F.). Du Berceau de l'Espèce Humaine, selon les Indiens, les Perses, et les Hébreux. Paris, 1858. 8vo.

OPPERT (J.). L'Immortalité de l'Ame chez les Chaldéens. Paris, 1875. 8vo.

444   LIST OF AUTHORS AND THEIR WORKS.

Oppert (J.). Etudes Assyriennes. Textes de Babylone et de Ninive, déchiffrés et interprétés. Inscription de Borsippa. Paris, 1857. 8vo.
——— Histoire des Empires de Chaldée et d'Assyrie d'après les Monuments. Versailles, 1865. 8vo.
——— In Journal Asiatique, Paris.
Rawlinson (Geo.). The History of Herodotus. A new English version. New York, 1861-1866. 4 vols. 8vo.
——— The Five Great Monarchies of the Ancient Eastern World. London, 1871. 3 vols. 8vo. 2d edition.
Rawlinson (H. C.). The Persian Cuneiform Inscription at Behistun. London. 1846. 8vo. [Extract from Journal of Royal Asiatic Society, London.]
——— A Commentary on the Cuneiform Inscriptions of Babylonia and Assyria. London, 1850. 8vo. [Extract Jour. Roy. As. Society, London.]
——— In Rawlinson's Herodotus.
Renan (Ernest). Histoire Générale et Système Comparé des Langues Sémitiques. Paris, 1863. 8vo. 4th edition.
——— De l'Origine du Langage. Paris, 1864. 8vo. 4th edition.
——— In Journal Asiatique, Paris.
Reinisch (L.). Der Einheitliche Ursprung der Sprachen der Alten Welt. Wien, 1873. Vol. I. Large 8vo.
Rich (E.). In Occult Sciences. London, 1855. 8vo.
Roediger (E.) and Pott (A. F.). In Zeitschrift für die Kunde des Morgenlandes. Bonn, 1846.
Romieu (A.). Lettres à Monsieur Lepsius, sur un Décan du Ciel Egyptien. Leipzig, 1870. 4to.
Rougé (E. de). Chrestomathie Egyptienne. Paris, 1867. 4to.
——— Same. 1868. 1st part, autographie. 2d part, 8vo.
Rougé (J. de). Monnaies des Nomes de l'Egypte. Paris, 1873. 8vo.
Sayce (A. H.). In Trans. Bib. Arch. Society, London.
Schlegel (G.). Uranographie Chinoise. Ouvrage accompagné d'un Atlas Céleste Chinois et Grec. La Haye, 1875. Large 8vo. Atlas, folio.
Schlottmann (K.). Die Inschrift Eschmunazars, Königs der Sidonier, etc. Halle, 1868. 8vo.
Schrader (E.). Die Keilinschriften und das Alte Testament. Giessen, 1872. 8vo.
——— Die Assyrische-Babylonischen Keilinschriften. Leipzig, 1872. 8vo.
Smith (George). Assyrian Discoveries. New York, 1875. 8vo.
——— The Chaldean Account of Genesis. New York, 1876. 8vo.
——— Same. Revised Edition. 1880. 8vo.
Smith (William). Classical Dictionary of Greek and Roman Biography, Mythology, and Geography. Author's edition. New York, 1859. 8vo.
Stallo (J. B.). General Principles of the Philosophy of Nature. Boston, 1848. 7mo.
Stewart (M.). A Commentary on the Apocalypse. Andover, 1845. 2 vols. 8vo.
Talbot (F.). In Trans. Bib. Arch. Society, London.

TAYLOR (B.). Central Africa. New York, 1859. 7mo.
TUCH (F.). Commentar über die Genesis. Halle, 1871. 8vo. 2d edition.
UHLEMANN (M.). Handbuch der gesammten ägyptischen Alterthumskunde. Leipzig, 1857, 1858. 4 vols. 8vo.
VOGUÉ (Cte DE). Mélanges d'Archéologie Orientale. Paris, 1868. 8vo.
———— Syrie Centrale. Inscriptions Sémitiques, etc. Paris, 1869. 4to.
WARBURTON (Wm.). The Divine Legation of Moses Demonstrated. London, 1846. 3 vols. 8vo.
WHITNEY (W. D.). Oriental and Linguistic Studies. New York, 1873. 7mo.
———— The Same. 2d Series. New York, 1874. 7mo.
WILFORD (F.). In Asiatic Researches.
WILKINSON (G.). In Rawlinson's Herodotus.

OTHER PUBLICATIONS.

Asiatic Researches. London. Vols. ii., vi. 4to. 7 vols. 1799-1803.
Bibliothèque de l'Ecole des Hautes Etudes. 12e Fascicule. Du Genre Epistolaire chez les Egyptiens, etc. Paris, 1872. 8vo.
Bulletin du Congrès International des Orientalistes. Session de 1876, à St. Petersbourg. 8vo.
Cuneiform Inscriptions of Western Asia. British Museum Series. London. 1861-1875. 4 vols. Folio.
Das Todtenbuch der Aegypter, etc. Herausg. von Dr. R. Lepsius. Leipzig, 1842. 4to.
Encyclopedia Americana. Boston. Vol. ix.
Journal Asiatique. Paris, 1874-1876. 8vo.
Protestant Theological and Ecclesiastical Encyclopedia. Edit. by Bomberger. Vol. I. Philadelphia, 1858. 8vo.
Records of the Past ; being English translations of the Assyrian and Egyptian Monuments. Vol. I. London, 1873. 16mo.
Révue Archéologique. Paris, 1873, 1874. 8vo.
Rituel Funéraire des Anciens Egyptiens. Texte complet, etc. Par M. E. de Rougé. Paris, 1864, etc. Folio.
Schenkel's Bibel-Lexikon. Leipzig, 1869-1875. 5 vols. 8vo.
Smithsonian Reports. Washington, 1868. 8vo.
The Vishnu Purana. Translated by H. H. Wilson. London, 1840. 4to.
Translations of the Society of Biblical Archæology. London, 1872-1874. 8vo.
Zeitschrift der Deutschen Morgenländischen Gesellschaft. Leipzig, 1873. 8vo.

www.ingramcontent.com/pod-product-compliance
Lightning Source LLC
Chambersburg PA
CBHW031814270326
41932CB00008B/424